METHODS IN MOLECULAR BIOLOGY

Series Editor
John M. Walker
School of Life and Medical Sciences
University of Hertfordshire
Hatfield, Hertfordshire, UK

For further volumes:
http://www.springer.com/series/7651

For over 35 years, biological scientists have come to rely on the research protocols and methodologies in the critically acclaimed *Methods in Molecular Biology* series. The series was the first to introduce the step-by-step protocols approach that has become the standard in all biomedical protocol publishing. Each protocol is provided in readily-reproducible step-by-step fashion, opening with an introductory overview, a list of the materials and reagents needed to complete the experiment, and followed by a detailed procedure that is supported with a helpful notes section offering tips and tricks of the trade as well as troubleshooting advice. These hallmark features were introduced by series editor Dr. John Walker and constitute the key ingredient in each and every volume of the *Methods in Molecular Biology* series. Tested and trusted, comprehensive and reliable, all protocols from the series are indexed in PubMed.

Iron Metabolism

Methods and Protocols

Edited by

Oleh Khalimonchuk

Department of Biochemistry, University of Nebraska-Lincoln, Lincoln, NE, USA

Humana Press

Editor
Oleh Khalimonchuk
Department of Biochemistry
University of Nebraska-Lincoln
Lincoln, NE, USA

ISSN 1064-3745 ISSN 1940-6029 (electronic)
Methods in Molecular Biology
ISBN 978-1-0716-4042-5 ISBN 978-1-0716-4043-2 (eBook)
https://doi.org/10.1007/978-1-0716-4043-2

This Humana imprint is published by the registered company Springer Science+Business Media, LLC, part of Springer Nature.
The registered company address is: 1 New York Plaza, New York, NY 10004, U.S.A.

Preface

Iron is a vital micronutrient utilized in a plethora of cellular functions under normal and/or pathological conditions. The ability of iron to easily switch between its ferrous (reduced, Fe^{2+}) and ferric (oxidized, Fe^{3+}) states is an important feature utilized by various processes within a cell. Being arguably one of the oldest and most common and versatile inorganic cofactors, iron ions can be utilized in a form of chelated iron or as more complex metal-organic molecules such as iron-sulfur clusters, wherein iron is coordinated by additional sulfur atoms, and heme—a versatile metalolipid-like organic compound. Whereas iron ions can serve as an inorganic cofactor that typically binds to proteins' mononuclear binding sites, the more sophisticated molecules mentioned above offer many advantages including higher reactivity and redox tuning to their cognizant enzymes. These features, however, come with a price—unbuffered free iron or its derived cofactors easily participate in uncontrolled redox reactions such as the Fenton reaction that produce highly reactive oxygen radicals. Because of such inherent reactivity, iron levels in a cell must be tightly controlled and kept within a narrow range. Therefore, iron is stored, trafficked, and utilized in cells through a variety of intricate and carefully orchestrated mechanisms.

Indeed, even though iron metabolism has been extensively studied since the 1950s, new and exciting discoveries are still being made, oftentimes due to new technological or methodological breakthroughs being introduced to the field. However, to address the sheer complexity and significance of iron metabolism in a cell, it is important to use the entire arsenal of methods and approaches that build on decades of previous research and method development. This volume of Methods in Molecular Biology provides a compendium of both classical—"near-lost-art"—and cutting-edge methods, optimized and validated to analyze various aspects of iron metabolism from in vitro to multi-organ level. The collection covers methods applicable to a variety of systems, ranging from bacteria to cultured mammalian cells and tissues. A number of chapters focus on mitochondrial iron and its derivatives. This is not surprising considering the fact that mitochondria are the primary site for the synthesis of iron-sulfur and heme cofactors and harbor one of the largest pools of intracellular iron, accounting for over 50% of total cellular iron content. Unsurprisingly, cellular and mitochondrial iron homeostasis are interdependent and heavily intertwined. For these reasons, significant attention is given to mitochondria-oriented methods. Much remains to be discovered in this rapidly growing research area, and many lessons and experimental approaches from bacteria, yeast, and roundworms as well as experimental approaches established and optimized in these models have proven to be instrumental in unraveling the mysteries of iron metabolism in humans. Therefore, it is expected that this book will be of interest to general scientific readership and will help in demystifying and boosting iron metabolism-related research.

In closing, I would like to thank all the authors who have graciously agreed to invest their time and contribute to this exciting volume and share their knowledge and expertise on the subject matter. Finally, I would also like to express my gratitude to the series editor, Dr. John Walker, for his helpful guidance, support, and patience during the process of putting this volume together.

Lincoln, NE, USA ***Oleh Khalimonchuk***

Contents

Contributors

NISANSALA S. ABEYRATHNA • *Department of Chemistry and Biochemistry, The University of Texas at Dallas, Richardson, TX, USA*
SAMEERA S. ABEYRATHNA • *Department of Chemistry and Biochemistry, The University of Texas at Dallas, Richardson, TX, USA*
VICTORIA ADEDOYIN • *Department of Chemistry and Biochemistry, Montana State University, Bozeman, MT, USA*
DANIEL W. BAK • *Boston College, Department of Chemistry, Chestnut Hill, MA, USA*
ANTONI BARRIENTOS • *Department of Biochemistry and Molecular Biology, University of Miami Miller School of Medicine, Miami, FL, USA; Department of Neurology, University of Miami Miller School of Medicine, Miami, FL, USA*
HECTOR A. BERGONIA • *Division of Hematology and Hematologic Malignancies, Department of Internal Medicine, University of Utah School of Medicine, Salt Lake City, UT, USA*
BREANN L. BROWN • *Department of Biochemistry, Vanderbilt University School of Medicine, Nashville, TN, USA; Center for Structural Biology, Vanderbilt University School of Medicine, Nashville, TN, USA*
PAUL A. COBINE • *Department of Biological Sciences, Auburn University, Auburn, AL, USA*
ROSE M. CURTIS • *Department of Chemistry and Biochemistry, The University of Texas at Dallas, Richardson, TX, USA*
RONIVALDO RODRIGUES DA SILVA • *Department of Chemistry and Biochemistry, Montana State University, Bozeman, MT, USA*
HARRY A. DAILEY • *Department of Biochemistry and Molecular Biology, University of Georgia, Athens, GA, USA; Department of Microbiology, University of Georgia, Athens, GA, USA*
MINH ANH DANG • *Department of Biochemistry & Molecular Biology, Michigan State University, East Lansing, MI, USA*
IRAMOFU M. DOMINIC • *School of Chemistry and Biochemistry, Georgia Institute of Technology, Atlanta, GA, USA; Parker Petit Institute for Bioengineering and Biosciences, Atlanta, GA, USA*
REBECCA K. DONEGAN • *Department of Chemistry, Barnard College, New York, NY, USA*
JENNIFER L. DUBOIS • *Department of Chemistry and Biochemistry, Montana State University, Bozeman, MT, USA*
CHRISTOPH J. FAHRNI • *Petit Institute for Bioengineering and Bioscience, School of Chemistry and Biochemistry, Georgia Institute of Technology, Atlanta, GA, USA*
ABIGAIL K. FREER • *School of Chemistry and Biochemistry, Georgia Institute of Technology, Atlanta, GA, USA; Parker Petit Institute for Bioengineering and Biosciences, Atlanta, GA, USA*
SAMANTHA GILLIS • *Department of Biological Sciences, University of Delaware, Newark, DE, USA; Division of Hematology/Oncology, Department of Medicine and Vascular Medicine Institute, University of Pittsburgh, Pittsburgh, PA, USA*
HUMERA GULL • *Department of Chemistry and Biochemistry, The University of Texas at Dallas, Richardson, TX, USA*

DAVID A. HANNA • *School of Chemistry and Biochemistry, Georgia Institute of Technology, Atlanta, GA, USA; Parker Petit Institute for Bioengineering and Biosciences, Atlanta, GA, USA*
ERIC L. HEGG • *Department of Biochemistry & Molecular Biology, Michigan State University, East Lansing, MI, USA*
MADUJIKA A. HORADIGALA GAMAGE • *Department of Chemistry and Biochemistry, The University of Texas at Dallas, Richardson, TX, USA*
ALICIA N. KREIMAN • *Department of Biological Sciences, University of Delaware, Newark, DE, USA*
PAUL A. LINDAHL • *Departments of Chemistry and of Biochemistry and Biophysics, Texas A&M University, College Station, TX, USA*
OSIRIS MARTINEZ-GUZMAN • *School of Chemistry and Biochemistry, Georgia Institute of Technology, Atlanta, GA, USA; Parker Petit Institute for Bioengineering and Biosciences, Atlanta, GA, USA*
AMY E. MEDLOCK • *Department of Biochemistry and Molecular Biology, University of Georgia, Athens, GA, USA; Augusta University/University of Georgia Medical Partnership, Athens, GA, USA*
GABRIELE MELONI • *Department of Chemistry and Biochemistry, The University of Texas at Dallas, Richardson, TX, USA*
SOPHIA MILLER • *Department of Molecular and Medical Genetics, Oregon Health & Sciences University, Portland, OR, USA*
COURTNEY M. MOORE • *School of Chemistry and Biochemistry, Georgia Institute of Technology, Atlanta, GA, USA; Parker Petit Institute for Bioengineering and Biosciences, Atlanta, GA, USA*
LORENA NOVOA-APONTE • *Genetics and Metabolism Section, Liver Diseases Branch, National Institutes of Diabetes and Digestive and Kidney Diseases, National Institutes of Health, Bethesda, MD, USA*
EVA NYVLTOVA • *Department of Neurology, University of Miami Miller School of Medicine, Miami, FL, USA*
LAURA E. OLDFATHER • *Department of Biological Sciences, Auburn University, Auburn, AL, USA*
JOHN D. PHILLIPS • *Division of Hematology and Hematologic Malignancies, Department of Internal Medicine, University of Utah School of Medicine, Salt Lake City, UT, USA*
CAROLINE C. PHILPOTT • *Genetics and Metabolism Section, Liver Diseases Branch, National Institutes of Diabetes and Digestive and Kidney Diseases, National Institutes of Health, Bethesda, MD, USA*
MITCHELL A. POPE • *Department of Chemistry and Biochemistry, The University of Texas at Dallas, Richardson, TX, USA*
MARTINA RALLE • *Department of Molecular and Medical Genetics, Oregon Health & Sciences University, Portland, OR, USA*
AMIT R. REDDI • *School of Chemistry and Biochemistry, Georgia Institute of Technology, Atlanta, GA, USA; Parker Petit Institute for Bioengineering and Biosciences, Atlanta, GA, USA*
ELISE D. RIVETT • *Department of Biochemistry & Molecular Biology, Michigan State University, East Lansing, MI, USA*
JAVIER SERAVALLI • *Nebraska Redox Biology Center, University of Nebraska-Lincoln, Lincoln, NE, USA*

MOLLY C. SUTHERLAND • *Department of Biological Sciences, University of Delaware, Newark, DE, USA*
DIANE M. WARD • *Department of Pathology, University of Utah School of Medicine, Salt Lake City, UT, USA*
ERANTHIE WEERAPANA • *Boston College, Department of Chemistry, Chestnut Hill, MA, USA*
MATHILDA M. WILLOUGHBY • *School of Chemistry and Biochemistry, Georgia Institute of Technology, Atlanta, GA, USA; Parker Petit Institute for Bioengineering and Biosciences, Atlanta, GA, USA*
TANIA YEASMIN • *Department of Chemistry and Biochemistry, University of Delaware, Newark, DE, USA*
YVETTE Y. YIEN • *Division of Hematology/Oncology, Department of Medicine and Vascular Medicine Institute, University of Pittsburgh, Pittsburgh, PA, USA*
HUI ZHONG • *Department of Biochemistry and Molecular Biology, University of Miami Miller School of Medicine, Miami, FL, USA*
XINYU ZHU • *Department of Biological Sciences, Auburn University, Auburn, AL, USA*

Part I

General Methods to Study Iron Metabolism

Chapter 1

Iron Homeostatic Regulation in *Saccharomyces cerevisiae*: Introduction to a Computational Modeling Method

Paul A. Lindahl

Abstract

Over the past 30 years, much has been learned regarding iron homeostatic regulation in budding yeast, *S. cerevisiae*, including the identity of many of the proteins and molecular-level regulatory mechanisms involved. Most advances have involved inferring such mechanisms based on the analysis of iron-dysregulation phenotypes arising in various genetic mutant strains. Still lacking is a cellular- or system-level understanding of iron homeostasis. These experimental advances are summarized in this review, and a method for developing cellular-level regulatory mechanisms in yeast is presented. The method employs the results of Mössbauer spectroscopy of whole cells and organelles, iron quantification of the same, and ordinary differential equation-based mathematical models. Current models are simplistic when compared to the complexity of iron homeostasis in real cells, yet they hold promise as a useful, perhaps even required, complement to the popular genetics-based approach. The fundamental problem in comprehending cellular regulatory mechanisms is that, given the complexities involved, different molecular-level mechanisms can often give rise to virtually indistinguishable cellular phenotypes. Mathematical models cannot eliminate this problem, but they can minimize it.

Key words Biochemical reaction networks, Iron–sulfur clusters, Hemes, Mitochondria, Vacuoles, Cytosol, Ordinary differential equations, Cellular regulatory mechanisms, Mössbauer spectroscopy, Soft or continuous Heaviside logistical functions

1 Introduction

Cellular homeostatic regulatory mechanisms are intriguing because they involve a network of biochemical reactions that work together to anoint the cell with special powers, namely, the flexibility to adjust to a changing environment or metabolic state, a fundamental characteristic of life itself. Such mechanisms are often difficult to decipher because they inherently involve groups of reactions that work together to create the observed effect or phenotype. Due to this complexity, as well as the large uncertainties in quantifying concentrations and other kinetic parameters, multiple molecular-level regulatory mechanisms can give rise to indistinguishable, or

Oleh Khalimonchuk (ed.), *Iron Metabolism: Methods and Protocols*, Methods in Molecular Biology, vol. 2839,
https://doi.org/10.1007/978-1-0716-4043-2_1,

nearly indistinguishable, phenotypes. This difficulty in inferring a molecular-level mechanism from a cellular phenotype is called the *inverse problem* in mathematics.

Homeostatic regulation involves *feedback*. Within the context of cellular physiology, this includes sensing some chemical species *Sen* to be regulated, comparing its concentration [Sen] to an optimal "setpoint" concentration [SP], and shifting [Sen] back to [SP], or near to it, if or when it deviates. Homeostasis is a stimulus: response phenomenon in which the cell is stimulated by a perturbing event, and it responds by returning to, or near, the unperturbed state. For us, the optimal setpoint concentration refers to the unperturbed value; for an evolved cell, it refers to the concentration of the sensor that optimally allowed it to survive and reproduce in a changing and difficult environment.

In terms of event progression, regulatory mechanisms can be described by an initial perturbation of [Sen] followed by a causal chain of biochemical events that ultimately returns [Sen] to the vicinity of [SP]—an *autoregulatory cycle*. *Feedforward* mechanisms might also play a role in homeostatic regulation. In such cases, deviations between [Sen] and [SP] are amplified. Causal chains are routinely complicated and incompletely understood. The long-term goal here is to describe those event chains, on the molecular or biochemical level, used by budding yeast *Saccharomyces cerevisiae* to homeostatically regulate iron trafficking and metabolism.

Why iron and why yeast? Iron is essential for all eukaryotic cells due to its "superpowers" in terms of catalytic abilities and chemical reactivity. This d-block transition metal can participate in a wide range of electron transfer reactions, including multielectron events. It can bind and activate small molecules such as O_2, N_2, or CO (and others) and provide neighboring coordination sites for substrates to bind, thereby dramatically increasing their probability of reaction. Iron is a critical component of mitochondrial respiration; thus, it plays a huge role in cellular energetics. Iron is at the active site of numerous metalloenzymes, where it drives their catalytic abilities. Perhaps surprisingly, iron, often in the form of $[Fe_4S_4]$ clusters, is involved in nucleic-acid–related processes, including transcription, translation, DNA replication, and repair.

Ironically, those same superpowers also make iron dangerous to the cell. Certain Fe^{II} species promote Fenton chemistry {Fe^{II} + $H_2O_2 \rightarrow Fe^{III}(OH) + OH\bullet$}, generating reactive oxygen species (ROS) such as hydroxyl radicals that create havoc for the cell by damaging DNA, lipids, proteins, etc. Regulating iron traffic must have been a high priority for evolving cells, especially once environmental O_2 became plentiful.

Although not as disease-relevant as human cells, budding yeast is arguably the best "workhorse" eukaryote for scientific investigations. It is convenient and inexpensive to grow, virtually all genetic strains are available, and the biochemical mechanisms operating within these cells are better understood as compared to most other cells. Moreover, the mechanisms in yeast are often similar to those operating in mammalian cells, and so mechanistic insights applicable to both types of cells can be gained. However, biochemical processes in yeast are far simpler than the corresponding processes in mammalian cells, rendering such insights more easily comprehended.

"Simpler" and "better understood" are relative terms, and indeed, regulation of iron in this organism is actually incredibly complicated and incompletely understood. *S. cerevisiae* contains about 100 iron-containing proteins, which have been cataloged and individually described [1]. As shown below, there are *layers* of regulation operating simultaneously in the cell, with each layer often interacting with others. Understanding/comprehending this complexity as a unified cellular system remains a challenge.

In this chapter, iron regulation in yeast is first summarized from an experimental perspective in which molecular-level manipulations (deleting or overexpressing particular genes) have been used to generate various cellular behaviors and phenotypes related to iron homeostasis. This summary is incomplete; additional perspectives can be found in other excellent reviews [2–7], not to mention the primary literature. I'll then describe efforts from my group to use mathematical modeling to explain the observed cellular system at the level of interacting molecules. Much of the data needed for this have been extracted from Mössbauer spectroscopy in conjunction with iron determinations of various samples, so these methods are also briefly described.

1.1 The Iron Regulon

The Iron regulon (*IR*) is the major regulator of iron homeostasis in yeast. Discovered more than 25 years ago [8], the *IR* is composed of 20–30 genes that encode proteins that globally regulate iron import, trafficking, storage, and mobilization. These include Fet3, Ftr1, Fet5, Fth1, Fre1-Fre4, Fet4, Smf3, Hmx1, Mrs4, Cth1, Cth2, Mmt1, Mmt2, and Rnr1, among others [9, 10]. The expression level of the *IR* is controlled by the concentration (and form) of iron in the growth medium, with activation occurring under Fe-deficient conditions.

For cells growing in lab cultures (i.e., lacking siderophores that might be excreted by other organisms in the wild), iron is imported via two low-affinity transporters on the plasma membrane, namely, Fet4 and Smf1 [2, 11, 12], and via a high-affinity protein complex referred to as the FET3 system (*see* Fig. 1). The latter system is composed of Fet3 (a multicopper oxidase), Ftr1 (an iron permease), and Fre1 (ferric reductase) [13]. Both low- and high-affinity systems import nutrient iron into the cytosol, which then contributes to the cytosolic labile Fe pool (called Fe^{II}_{cyt} in Fig. 1 or FC in the model described below). All FET3 proteins are part of the *IR* and, thus, are activated under Fe-deficient conditions.

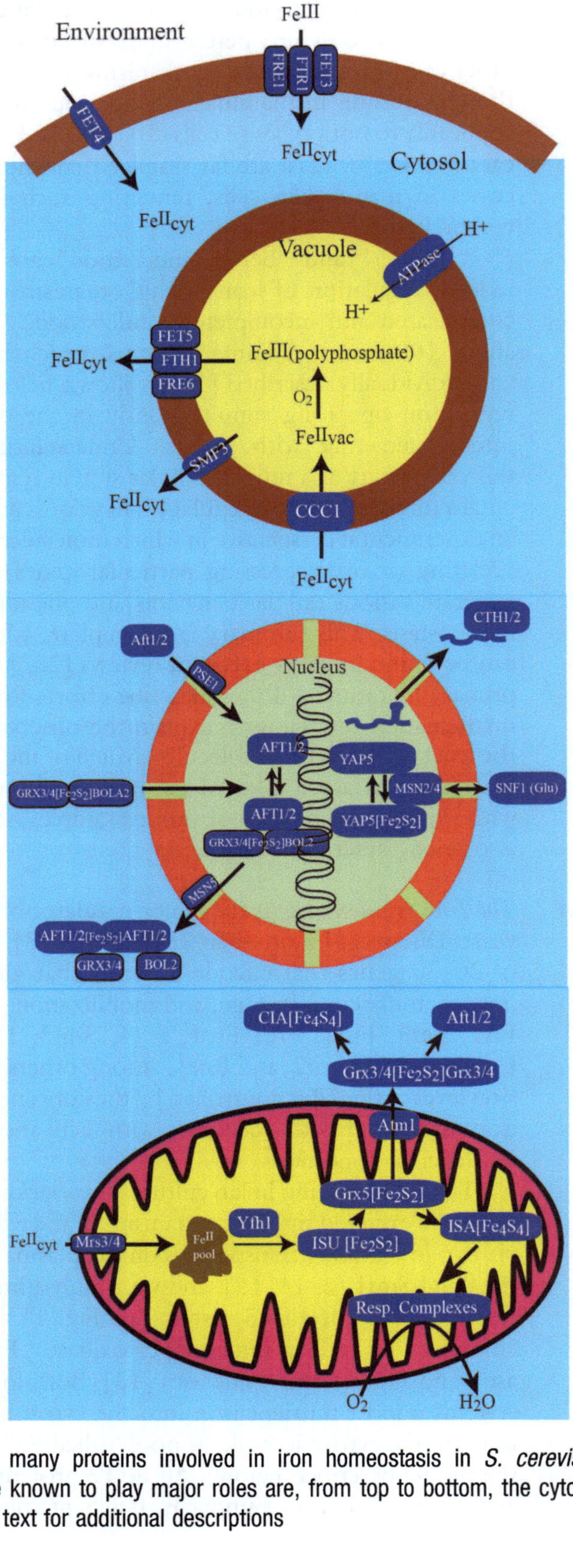

Fig. 1 Subset of the many proteins involved in iron homeostasis in *S. cerevisiae*. The four cellular compartments that are known to play major roles are, from top to bottom, the cytosol, vacuoles, nucleus, and mitochondria. See text for additional descriptions

Vacuoles are spherically shaped organelles that sequester and store iron for the cell [14]. Their lumen is acidic due to a V-ATPase on the vacuolar membrane that imports protons from the cytosol; this acidity is essential for its function. Under Fe-replete conditions, vacuoles import $Fe^{II}{}_{cyt}$ via Ccc1, an iron transporter also on the vacuolar membrane [15]. Much of that iron is oxidized to Fe^{III} and coordinated to polyphosphate [16], which is present at millimolar (mM) concentrations in the vacuolar lumen. The vacuolar membrane contains a low-affinity iron transporter Smf3 and a high-affinity system called FET5, homologous to FET3. The FET5 system includes multicopper oxidase Fet5, permease Fth1, and ferric reductase Fre6. Like FET3, the FET5 system is part of the *IR* and is thus activated under Fe-deficient conditions. Under those conditions, stored vacuolar iron passes through FET5 into the cytosol. Thus, the role of both FET3 and FET5 systems is to move iron into the cytosol, which contributes to the cytosolic Fe^{II} pool.

Activation of the *IR* is controlled by two paralogous transcription factors: Aft1 and Aft2 [17]. These two proteins recognize slight differences in the DNA sequence of the promoters of the *IR* genes [10, 18]; thus, they are not fully redundant. Aft1 is more involved in iron import (e.g., the FET3 system), while Aft2 is more involved in intracellular iron trafficking. For example, Aft2 strongly regulates Smf3 and Mrs4, a high-affinity iron importer on the mitochondrial inner membrane.

Whether or not the *IR* is activated depends on the *location* of Aft1 and Aft2 in the cell rather than on their concentrations. Under Fe-replete conditions, these transcription factors are located in the cytosol, whereas under Fe-deficient conditions, they localize to the nucleus. Only when in the nucleus do they function as transcription factors, binding and activating the promoter regions of the *IR* genes.

Cellular localization involves two nuclear-associated proteins: Pse1 and Msn5 [19–21]. Import into the nucleus is not affected by cellular iron status, but nuclear export is. Under Fe-deficient conditions, Aft1/2 is bound to the promoter regions in the nucleus so as to activate *IR* genes. Under Fe-replete conditions, a complex likely forms with a heterodimer Grx3/4[Fe_2S_2]Bol2 composed of monothiol glutaredoxin Grx3 or Grx4, the Bol2 protein, and a bridging [Fe_2S_2] cluster. This heterodimer promotes the dissociation of Aft1/2 from its DNA binding sites by transferring its cluster to Aft1/2 [22, 23]. This dissociation deactivates the *IR*. Homodimer Aft1/2:[Fe_2S_2]:Aft1/2 dissociates from the DNA and exits the nucleus in a process involving Msn5, perhaps escorted by Grx3/4 and/or Bol2 [2]. A phosphorylation event might also be involved.

Before explaining where Grx3/4[Fe_2S_2]Bol2 comes from, some background iron metabolism of the mitochondria must be introduced. This organelle is the primary site for the assembly of iron–sulfur clusters (ISCs) and the only site in the cell for heme biosynthesis; thus, much cellular iron flows into it. Cytosolic Fe^{II} enters the mitochondria via two high-affinity importers on the inner membrane, Mrs3 and Mrs4. The imported iron contributes to a pool of high-spin Fe^{II} in the matrix [24]. This pool is thought to be the feedstock for the assembly of ISCs and hemes.

ISCs are assembled in the matrix by a protein complex composed of cysteine desulfurase (Nsf1/Isd11), scaffold proteins (Isu1/2), and Yfh1 (yeast frataxin homolog 1) [25]. [Fe_2S_2] clusters are assembled on the scaffold proteins and then transferred to monothiol glutaredoxin Grx5. Some clusters are subsequently transferred to the Isa1/2 proteins, where they are reductively coupled to form [Fe_4S_4] clusters. Others are installed, along with hemes, into respiratory complexes that collectively reduce O_2 to water as they pump protons across the mitochondrial inner membrane.

Still, other [Fe_2S_2] clusters formed on the scaffold proteins are exported from the mitochondria into the cytosol, probably via the inner membrane ABC transporter Atm1. The exact species exported, historically called *X-S*, has not been identified but is generally assumed to be a [Fe_2S_2] cluster coordinated by 4 glutathione ligands. Once in the cytosol, these clusters are received by two monothiol glutaredoxins Grx3 and Grx4. Monomers of these proteins homodimerize, with a [Fe_2S_2] cluster bridging. Most such clusters are delivered to the cytosolic iron–sulfur assembly complex (called the CIA), where they are reductively coupled to form [Fe_4S_4] clusters. Such clusters are ultimately installed into various client proteins in the cytosol and nucleus. Other [Fe_2S_2] clusters bridging the cytosolic glutaredoxin homodimers are indirectly transferred to Aft1 and Aft2 in the nucleus as described above [22, 26].

1.2 Regulation of Ccc1

CCC1 encodes the only known importer of iron into vacuoles. The Ccc1 protein transfers cytosolic Fe^{II} from the labile iron pool into the vacuolar lumen for storage and sequestration. Transcription of *CCC1* depends on the iron and glucose status of the cell. Expression is regulated by transcription factor Yap5 (which also regulates the expression of glutaredoxin *GRX4* and a few other genes). Yap5 is permanently located in the nucleus, where it binds the promoter regions of its target genes. Under Fe-replete conditions, two [Fe_2S_2] clusters bind Yap5, generating a conformational change that activates transcription [27]. Curiously, the clusters are not derived from Grs3/4[Fe_2S_2]Grs3/4 homodimers. The regulation of *CCC1* under high iron conditions protects cells from the toxic effects of high iron.

The low-glucose sensor kinase Snf1 also controls the expression of *CCC1*; it does so indirectly by activating the stress transcription factors Msn2 and Msn4 [28, 29]. *CCC1* is expressed to a greater extent under low-glucose (i.e., respiring) conditions. This process is independent of Yap5. *CCC1* expression is also regulated by Cth1/2, mRNA binding proteins that promote the degradation of *CCC1* mRNA [30–32] under Fe-starvation conditions. This serves to shut down dispensable Fe-requiring processes (e.g., storing Fe) under Fe-starvation conditions. This multilayered form of regulation creates a more graded and nuanced response for *CCC1* expression.

1.3 Translational Regulation by Cth2 and Cth1

Cth2 and Cth1 are paralogous mRNA-binding proteins that are part of the *IR*. Cth2 is expressed under extreme Fe-deficient conditions [33]. These proteins shut down Fe-requiring but unessential processes by degrading the mRNA transcripts associated with them. These processes are generally replaced by Fe-independent processes. Thus, the overall effect of Cth1/2 is to reconfigure the cell to grow in an iron-starved environment.

Cth1/2 shuttle between the cytosol and the nucleus. Once in the nucleus, these proteins bind mRNA strands containing adenosine- and uridine-rich elements (AREs) in their 3′ untranslated regions. Such mRNA strands encode proteins involved in nonessential iron-dependent processes such as the electron transport chain, heme biosynthesis, the TCA cycle, the biosynthesis of unsaturated fatty acids and amino acids, and the biosynthesis of biotin/lipoic acid [5, 32]. As mentioned above, Cth2 also degrades *CCC1* mRNA to prevent Fe from being stored in vacuoles [34]. Perhaps surprisingly, Cth2 degrades its own mRNA, as too much Cth2 is toxic to the cell [30–32]. Cth2 expression is controlled by dual phosphorylation by casein kinase I protein Hrr25; once phosphorylated, it is degraded by the proteasome [35]. Another surprise is that Cth2 indirectly *stimulates* the activity of ribonucleotide reductase, an essential *Fe-requiring* enzyme [36, 37]. Cth2 expression protects against ROS [38] and may play a role in aging [39]. The nonsense-mediated mRNA decay (NMD) pathway regulates mRNA expression involved in iron (and copper) homeostasis [40]. The NMD pathway regulates the ISC protein Rli1, which is involved in ribosome biogenesis [41].

1.4 Metabolic Regulation

Metabolic factors besides cellular iron concentration are also involved in regulating the *IR*. Montella-Manuel et al. [42] found that the TORC2 branch of the TOR signaling pathway, which regulates the cellular metabolic status, and the biosynthesis of sphingolipids affect Aft1 localization. Autophagy (the breakdown of cellular organelles due to nutrient starvation) also plays a role in iron homeostasis [43]; for example, iron released into the cell when the mitochondria are degraded helps alleviate cellular Fe deficiency.

Metabolic growth mode (fermenting vs respiring) appears to override Fe deficiency in controlling *IR* activation. As glucose is exhausted during the diauxic shift, the mitochondria increase in number and volume, and the expression of iron-rich respiratory complexes increases. The *IR* is activated under this condition, increasing iron import and mobilizing vacuolar iron to meet the additional iron needs of the cell [44]. Remarkably, activation occurs *even in Fe-replete conditions* due to control by two glucose-signaling kinases, namely, protein kinase A (PKA) and the AMP-activated protein kinase Snf1 (from the AMPK family). The presence of glucose configures the cell to ferment, which is largely iron-independent. In contrast, iron is essential for respiration, so this metabolic state is linked to the cell having sufficient stores of the metal. PKA appears to prevent excess iron import during fermentation. Snf1 is activated when cells are glucose-depleted (e.g., during the diauxic shift), and this in turn activates the *IR*. Aft1 is phosphorylated when cells undergo the diauxic shift (but not when the cell is Fe-deficient), and this might be part of the override strategy. Simpson-Lavy et al. [45] recently found that the low-glucose sensor Snf1 has a poly-histidine region that is used to control carbon metabolism. Under Fe-deficient conditions, this region interacts with Aft1 (but curiously not Aft2) at the nuclear membrane, and this inhibits Snf1, preventing the cell from switching to respiration.

The acidity of the vacuoles also plays a role in iron homeostasis; diminished acidity activates the IR *even in Fe-replete conditions.* Loss of acidity is associated with defects in ISCs [46]. The mitochondria and vacuoles are functionally linked and are connected to aging and disease [47, 48].

There is also a link between phosphorylation and sphingolipids. HOG1 is a mitogen-activated protein (MAP) kinase that regulates the cell's response to oxidative stress including high osmotic pressure and mitochondrial dysfunction. This kinase phosphorylates Aft1, which promotes its export from the nucleus [2, 49].

Many proteins involved in DNA replication and repair contain ISCs. The role of the clusters in these proteins remains controversial, but they likely connect iron homeostasis with other forms of regulation involving cell cycle, genome stability, and DNA replication and repair [50]. DNA polymerases bind ISCs, and defects in ISC biogenesis cause DNA damage and genome instability. Yeast cells regulate ribonucleotide reductase, an enzyme used to synthesize deoxyribonucleotides, in response to fluctuations in iron bioavailability [51].

1.5 Heme Regulation

In Fe deficiency, the expression of heme oxygenase Hmx1 increases. This enzyme catalyzes heme degradation. The iron released in the process contributes to the cytosolic labile iron pool [2]. Heme degradation causes a decline in the activity of the Hap1 and Hap2/3/4/5 transcriptional activators, which are stimulated by binding heme [52]. The Hap1 regulon includes genes that encode numerous cytochromes. HAP1 activity increases as heme levels increase, suggesting feedforward regulation. Labile heme levels help regulate mitochondrial biogenesis via the HAP2/3/4/5 complex [53]. This complex regulates respiration, ROS production, and iron homeostasis [54].

1.6 Regulation of Ergosterol Biosynthesis, Fatty Acid Biosynthesis, and ERMES

Ergosterol biosynthesis (ERG) genes are regulated by iron availability, depending on sterol and heme levels [55]. ERG genes are activated in Fe-deficient cells; the HAP1 system is also involved. A combination of factors fine-tunes the expression of ERG genes to iron status so as to avoid accumulating toxic sterol intermediates and to enable cells to quickly adapt to changing environmental conditions. In Fe-deficient cells, Mga2, a transcription factor anchored to endoplasmic reticula, activates the transcription of the Fe-O-Fe containing Δ9-fatty acid desaturase *OLE1* [56]. This allows cells to grow with nondefective membranes under Fe-limited conditions.

The protein complex ERMES tethers the endoplasmic reticulum and mitochondria, and this allows communication between the organelles, including processes involving iron homeostasis. Loss of ERMES (Endoplasmic Reticulum Mitochondria Encounter Structure) activates the *IR even in Fe-replete conditions* [57], which leads to the accumulation of excess cellular iron.

1.7 Mitochondria and Iron Regulation

Many advances in understanding iron regulation in yeast have been derived from efforts to understand Friedreich's ataxia, a mitochondrial disease characterized by a deficiency of frataxin. This protein, like Yfh1, is part of the ISC assembly protein complex in the mitochondrial matrix [25]. Thus, a deficiency of Yfh1 results in an inability (or a diminished ability) to generate ISCs in the cell. Part of the Yfh1-deficiency phenotype is a massive accumulation of Fe^{III} oxyhydroxide phosphate-associated nanoparticles in the matrix of the mitochondria. This is due to a surge of iron entering the cell and the mobilization of iron stored in vacuoles. Reactive oxygen species (ROS) also increase and iron homeostasis becomes dysregulated such that the *IR* is activated *even under Fe-replete conditions.* Yfh1 deficiency alters the expression of ca. 70 genes [58]. The presence or absence of O_2 majorly affects this phenotype [59, 60]. There is general consensus that these phenotypic properties result from the dysregulation of iron homeostasis, but by what mechanism exactly?

Among the most influential articles in this regard was Chen et al. [61] who studied Yfh1-deficient yeast cells. Prior to that, the sensor for activating the *IR* was assumed to be cytosolic labile Fe^{II}, leading to the conclusion that Yfh1-deficient cells accumulated iron because the cytosol Fe^{II} concentration was lower than normal. In another article, Kaplan and coworkers provided evidence that activation of the *IR* in *Yfh1Δ* mutants is sensed by the activity of the mitochondrial ISC assembly, not the concentration of cytosolic iron. They created a strain of yeast in which an iron-dependent enzyme (Erg25-2) was genetically installed to monitor cytosolic

Fe^{II} concentrations. They first demonstrated that the enzyme accurately senses low cytosolic Fe^{II} concentrations in Fe-deficient cells and confirm that this stimulates the *IR*. They then showed that disruption of the mitochondrial ISC assembly by Yfh1 deficiency activated the *IR independent of cytosolic iron concentration*. Their suggestion that Aft1 "senses a mitochondrial product" dovetailed nicely with the earlier proposal of Lill's for the Atm1-dependent export of X-S [62]. Accordingly, X-S was assigned a second function (besides being a substrate for the CIA), namely, as a sensor of mitochondrial ISC assembly activity. The following year, Rutherford [63] found that the *IR* was induced in cells depleted of the mitochondrial inner membrane transporter Atm1, but not by repression of proteins associated with the CIA proteins in the cytosol. They concluded that iron sensing by Aft1/Aft2 is linked to the mitochondria via the export X-S from Atm1. The implication was that X-S provides the cluster found in Grx3/4[Fe_2S_2]Grx3/4 homodimers and Grx3/4[Fe_2S_2]Bol2 heterodimers.

This complex and the associated regulatory mechanism are supported by much evidence, but perhaps not enough that it should be regarded as settled knowledge. Tamarit and coworkers have presented evidence that the *IR* can be activated in the *absence* of ISC deficiency [64]. They attribute the Yfh1-defiency phenotype as involving "anomalous iron species (in the mitochondria) that compromise iron bioavailability and activate a signaling cascade that results in Aft1 activation" and metabolic remodeling. By showing that the *IR* is activated in otherwise healthy cells with low concentrations of cytosolic Fe^{II}, the results of Chen et al. [61] would actually seem to be consistent with cytosolic Fe^{II} concentrations serving as a sensor for the *IR* in some conditions. We have also had difficulty rationalizing some of our Mössbauer results and in designing mathematical models that adhere precisely to all aspects of this popular scenario. Although more systematic studies are needed, mitochondria isolated from Fe-limited cells exhibit Mössbauer spectra that suggest similar concentrations of ISCs relative to those isolated from Fe-replete cells [65], yet the *IR* is activated exclusively in the former. Mitochondria from healthy fermenting cells contain ~3× lower concentrations of ISCs, yet this does not seem to affect the activation state of the IR. Vacuoles seem to be filled or emptied according to the concentration of iron in the growth medium, suggesting control by cytosolic Fe^{II}. We raise the possibility that *IR* activation may be controlled by more than one sensor, for example, both cytosolic Fe^{II} and X-S. There might be layers of regulation, with some overriding others, such that the *IR* can become activated independent of the concentration of cytosolic Fe^{II}. Admittedly, testing this hypothesis has been (and continues to be) challenging. Before explaining our approach to addressing this, some of the techniques used should be introduced.

2 Materials

2.1 Cell Growth and Metal Determinations

The methods that we use to grow cells, isolate mitochondria and vacuoles, and determine metal concentrations have already been published [66–68]. We typically grew strain W303 in minimal medium with glucose or glycerol/ethanol as carbon sources, for fermenting and respiring metabolic modes, respectively. The medium is typically supplemented with 1 or 40 μM $^{57}Fe^{III}$ citrate to create Fe-limiting or Fe-replete conditions, respectively. For severe Fe-deficient conditions, the chelator bathophenanthroline disulfonate must be added. For aerobic conditions, cultures are rapidly shaken in air; for hypoxic growths, an O_2/N_2 mixture yielding ~5% O_2 is bubbled through cultures in sealed flasks [59]. The exponential growth rate, termed α, is taken as the slope of the ln(OD600) vs. time for the growing culture.

Measuring absolute iron concentrations in whole cells and isolated organelles is essential for our analysis. Our method for doing this requires calculating the fractional volume of packed wet cells due to the cells themselves (excluding the interstitial buffer). Typical packing efficiencies are 0.6–0.7. Packed cells are quantitatively transferred to screw-top plastic tubes, and trace metal-grade HNO_3 is added. The tubes are sealed, wrapped with electrical tape, and heated overnight. Digested samples are diluted with water, filtered, and metal concentrations are determined by ICP-MS. Packing efficiency and dilution factors are used to back-calculate the iron concentrations within whole cells or organelles.

2.2 Mössbauer Spectra

We use Mössbauer spectroscopy to investigate the iron content in various strains of yeast, grown under a variety of media conditions [69]. Like NMR spectroscopy, Mössbauer involves nuclear spin-state transitions. For ^{57}Fe, there are 6 allowed transitions from the $I = ½$ ground state to the $I = 3/2$ excited state. The energies of these transitions depend on the chemical and electronic environment around the ^{57}Fe nucleus. This includes the oxidation state and electronic state of the iron (e.g., Fe^{II} vs. Fe^{III}), the spin sate (high spin vs. low spin), coordination number (e.g., 4, 5, or 6 ligands), nature of the donor atoms (S, N, or O), and whether the iron is part of an ISC or heme. Different types of centers exhibit different spectra, but resolution is limited such that the overlap of spectral features can be significant and complicate the analysis. Another strength is that the intensity associated with a particular spectral feature is approximately proportional to the fraction of iron in the sample associated with that feature. Thus, it is possible to discern (within limits) the quantitative distribution of iron centers within a complex sample such as a cell or an organelle.

The Mössbauer spectrum of whole yeast cells grown under respiring conditions in minimal media supplemented with 40 μM $^{57}Fe^{III}$ citrate is shown as the black hashmarks in Fig. 2. The

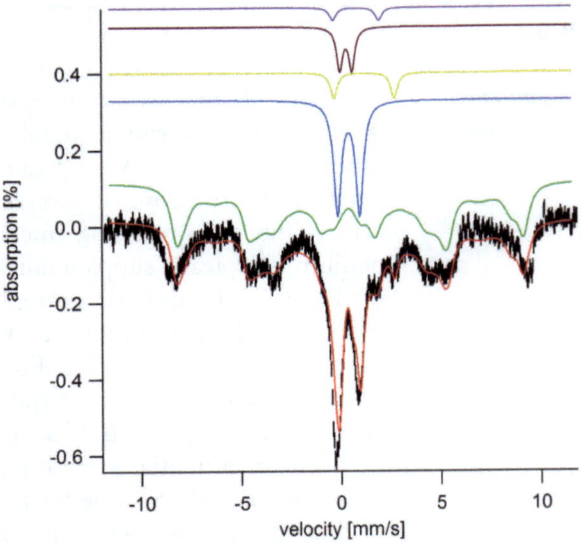

Fig. 2 Mössbauer spectrum (5 K, 0.05 T magnetic applied parallel to the gamma rays) of whole respiring yeast cells grown in minimal media supplemented with 40 μM $^{57}Fe^{III}$ citrate. Simulation lines indicate the following species: purple, high-spin Fe^{II} hemes; maroon, $S = 0$ $[Fe_2S_2]^{2+}$ clusters; gold, nonheme high-spin Fe^{II}; blue, $S = 0$ $[Fe_4S_4]^{2+}$ clusters plus low-spin Fe^{II} hemes; green, high-spin $S = 5/2$ Fe^{III}. The red line is the composite simulation

spectral simulation, shown as the red line, is the sum of 5 components. The dominant component, shown in green, is due mainly or entirely to the high-spin $S = 5/2$ Fe^{III} polyphosphate found in vacuoles under Fe-replete conditions such as obtained here. Isolated vacuoles show essentially the same spectrum [16]. The blue line is a simulation of the second-most intense spectral feature, called the central quadrupole doublet. The fitting parameters of this doublet are typical of $S = 0$ $[Fe_4S_4]^{2+}$ clusters and low-spin Fe^{II} hemes (the two cannot be distinguished). Most of the iron centers giving rise to this doublet arise from the mitochondria, followed by the cytosol, and then the nuclei. Spectra of isolated mitochondria are dominated by the same doublet. We simulated the spectrum of whole WT cells by summing the contributions of all 100 iron-containing proteins found in yeast [1]. The minor-intensity doublet shown as the gold line has parameters typical of nonheme high-spin Fe^{II} ligated by 5/6 ligands with O/N donors. The doublet shown in maroon is typical of $S = [Fe_2S_2]^{2+}$ clusters; that shown in purple is due to nonheme high-spin (NHHS) Fe^{II} heme centers. The percent intensity of each component relates to the fraction of each species present in the cell. In this case, vacuolar Fe^{III} and $[Fe_4S_4]^{2+}$ clusters represent ca. 65% and 20%, respectively. The other 3 features compose the remaining ~15% of spectral intensity. The doublet due to NHHS Fe^{II} likely arises from many species, including (a) the cytosolic labile iron pool (FC), (b) all cellular proteins bound to a mononuclear Fe^{II} ion, and (c) the mitochondrial labile iron pool (FM). This component represents ~5% of cellular iron, corresponding to ~25 μM in a cell containing 500 μM of iron, a typical concentration. The cytosolic labile Fe^{II} pool in yeast has been estimated at 10–50 μM [1]. The mitochondrial labile Fe^{II} pool is obvious in isolated mitochondria [24] but would not contribute much intensity in whole cells because the fractional volume of the mitochondria is only 0.05–0.1.

We collected spectra of whole cells, including WT and various strains, in media containing different levels of iron supplementation (40 μM vs. 1 μM ^{57}Fe) and different carbon sources (glucose for fermenting; glycerol/ethanol or lactate for respiring). We also determined the iron concentration of similarly grown whole cells using ICP-MS. Isolated mitochondria were also investigated. Overall, our results indicate the following general patterns [1]. First, the spectra of fermenting vs. respiring cells when grown under Fe-replete conditions are similar; both are dominated by vacuolar $S = 5/2$ Fe^{III} and include a strong central doublet. The major difference is that in respiring cells, the central doublet is more intense, reflecting an increase in the fractional volume of the mitochondria in the cell and an increased expression of respiratory complexes in the mitochondria. The spectra of fermenting vs. respiring cells when grown under Fe-limited conditions are similar as well. Both are dominated by the central doublet, and both lack features due to vacuolar iron. This undoubtedly is due to the emptying of vacuolar iron under Fe-deficient conditions. The concentration of cellular iron is also consistent with this. To obtain Mössbauer spectra, ~1 μM ^{57}Fe must be added to the growth media, and so these growth conditions are not rigorously Fe-deficient as would be obtained when an Fe chelator is included in unsupplemented minimal media; rather, *iron-limited* is a more accurate description. Iron-limited cells grow at the same exponential growth rate as Fe-replete cells, so there is no evidence that they are stressed, only that they don't store iron in vacuoles [65]. We also examined Mössbauer spectra of isolated mitochondria from such cells. Overall, mitochondria isolated from respiring cells exhibit spectra dominated by the central doublet, regardless of whether the cells are Fe-limited or Fe-replete. The central doublet is more intense in respiring mitochondria than in the mitochondria from fermenting cells, consistent with a greater level of ISC assembly. We also analyzed spectra of various mutants in which ISC assembly in the mitochondria has been compromised; this includes Yfh1-deficient [59] and Yah1-deficient [70] cells. We also examined Atm1-deficient [71] and Aft1-1up cells and organelles [72]. In each study, we attempted to understand how changes in Fe regulation could give rise to the observed phenotypes. We had limited success in explaining our results according to the current understanding of iron homeostasis. Our difficulties prompted us to employ mathematical modeling to explore various possible regulatory mechanisms that might explain our data.

3 Methods

3.1 Mathematical Modeling of Iron Homeostasis in Yeast

Given the complexity of the regulatory processes described above, it is unrealistic to model these processes according to the actual molecular-level regulatory mechanisms used by the cell. Such mechanisms are often not understood on the molecular level, and/or there are simply too many unknown fitting parameters. Rather, we use *surrogate* regulatory mechanisms; these can exhibit the same overall system-level effect but at a fraction of the complexity. Tyson et al. [73] suggested the use of continuous or "soft" Heaviside functions for this purpose. We recently adopted these functions [74], described in Eq. (1),

$$\frac{1}{1 + e^{n([\mathrm{SP}] - [\mathrm{Sen}])}}, \tag{1}$$

to control the rate constants of modeled reactions. View these functions like valves controlling the flow of a reaction. In Eq. (1), [Sen], [SP], and n are, respectively, the concentration of the sensed species, the set-point concentration of that species, and the sensitivity of the response—that is, how dramatically the valve opens or closes when [Sen] ≠ [SP]. Parameter n is negative for feedback regulation and positive for feedforward regulation.

To understand their utility, consider a simple model of iron trafficking in a growing yeast cell in which nutrient iron (called IRON) enters the cell through an importer on the plasma membrane, becoming component FC ($\underline{\mathrm{F}}\mathrm{e}^{\mathrm{II}}$ in the $\underline{\mathrm{c}}$ytosol) (*see* Fig. 3). The cell is presumed to be growing, so its volume is expanding in accordance with an apparent rate constant α defined as the slope obtained by plotting the natural logarithm of the absorbance at 600 nm for cells growing exponentially in batch culture vs. time. Accordingly, the cellular concentration of [FC] is determined by two opposing processes, namely, the rate of import R_{in} and the rate of dilution D_{FC} due to cell growth. The time-dependent change in the concentration of [FC] is given by

$$\frac{d[\mathrm{FC}]}{dt} = R_{\mathrm{in}} - D_{\mathrm{FC}}. \tag{2}$$

To illustrate, we assign the simple rate-law expression $R_{\mathrm{in}} = k_{\mathrm{in}}[\mathrm{IRON}]$ and $D_{\mathrm{FC}} = \alpha\,[\mathrm{FC}]$, where [FC] is the concentration of iron in the cytosol. [IRON] and α are fixed for a particular batch of cells. At steady state, the two opposing rates balance, and time-dependent changes in [FC] become 0; thus,

$$\frac{k_{\mathrm{in}}[\mathrm{IRON}]}{\alpha} = [\mathrm{FC}]. \tag{3}$$

If $k_{\mathrm{in}} = 20\ \mathrm{h}^{-1}$, [IRON] = 1 μM, and $\alpha = 0.2\ \mathrm{h}^{-1}$, [FC] would be 100 μM. Accordingly, if [IRON] would double or halve, steady-state [FC] would also double or halve. This unregulated condition is shown in Fig. 4.

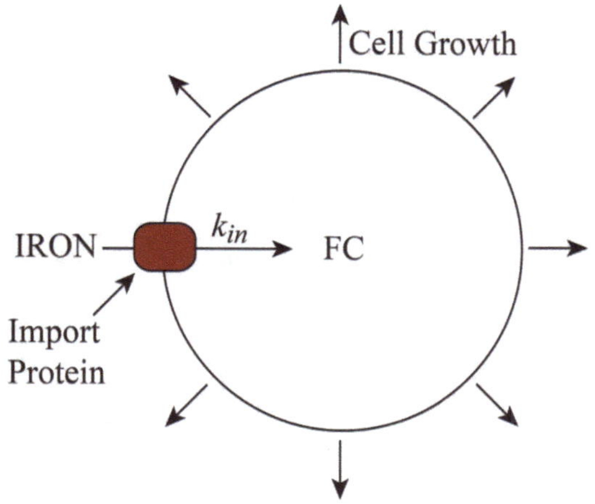

Fig. 3 Illustration of a simple growing cell model. The cell contains a single species called FC (in later models, this is the cytosolic Fe^{II} pool). Two opposing reactions determine the steady-state concentration of [FC], namely, the rate of import (R_{in}) and the rate of dilution due to cell growth (D_{FC}). The effect of regulating the import rate is illustrated in Fig. 4. The growth rate is determined experimentally by measuring the absorbance at 600 nm of a culture of growing cells and plotting the natural logarithm of the absorbance vs. time. The slope of the line during the exponential growth phase is α; the dilution rate $D_{FC} = \alpha \cdot [FC]$

To regulate the system, we recently developed a method in which rate-constants are augmented by the logistical function in Eq. (1) [74]. In this case, k_{in} would be multiplied by this function, with [FC] serving as the sensed species. The rationale for regulating k_{in} is as follows. In real cells, regulation often involves a transcription factor binding the promotor site of the gene, in this case encoding the iron importer on the plasma membrane. This serves to adjust the expression level of the importer protein. The rate of the import reaction would be equal to a rate-constant (k_{cat}) multiplied by the concentration of the protein/enzyme (e.g., $V_{max} = k_{cat}[E]$). Since the importer is not explicitly included in the model, regulating k_{in} is tantamount to transcriptionally regulating the expression level of the gene encoding the protein. The system becomes

$$\frac{d[\mathrm{FC}]}{dt} = k_{\mathrm{in}}\left(\frac{1}{1 + e^{-1([100] - [\mathrm{FC}])}}\right)[\mathrm{IRON}] - \alpha[\mathrm{FC}] \qquad (4)$$

For best regulation, the logistical function should have the value ½ in the unperturbed steady state. For the set-point concentration of FC to equal 100 μM, k_{in} must be assigned 40 h^{-1} to match the unregulated steady-state conditions. The difference is that with the logistical function included, when [IRON] is doubled or halved, [FC] transiently changes but quickly adjusts near to its initial steady state (*see* Fig. 4). If regulation were perfect, the resulting steady-state concentration of [FC] would equal [SP].

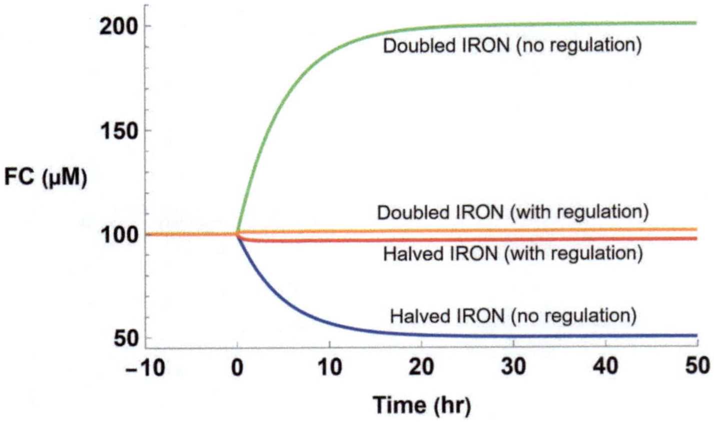

Fig. 4 These plots were obtained by numerically integrating Eq. (4), using *Mathematica*, with and without the logistical function included. Parameters used are given in the text

This is an example of *autoregulation* in which some component within the system, in this case, FC, is the sensed species. As [FC] changes, the regulatory valve(s) that FC controls adjust automatically. It would of course be simpler for the human modeler to manually adjust rate-constants in response to different environmental conditions, but that would be distinctly nonbiological. Autoregulation may also be imperfect, in that some discrepancy in the steady-state concentration of the regulated species may remain relative to the set-point concentration.

Admittedly, logistical functions are simple mathematical functions with only a symbolic connection to biochemical reality. However, for many investigations, the most pressing issues are the identity of the sensed species, the type of regulation (feedback or feedforward), and the sensitivity of the response, all of which are given by these functions.

3.2 Modeling the Transition from Healthy to the Fe-Deficient and Yfh1-Deficient Diseased States

My lab has been interested in modeling how healthy cells transition into the diseased phenotype characteristic of Friedreich's ataxia, when the expression of the *Yfh1* gene declines as the primary causal event [75, 76]. Our most recent model and approach [74] are especially relevant to this chapter because they involve a novel strategy for identifying viable autoregulatory mechanisms.

The core model, excluding any regulatory mechanisms, is shown in Fig. 5 (*see* top panel). Similar to the model described above, nutrient IRON enters the cell and becomes FC (now considered to be the labile iron pool in the cytosol). FC has a number of fates. It can (a) traffic into the mitochondria, becoming FM (labile <u>F</u>e^{II} in the <u>m</u>itochondria); (b) traffic into the vacuoles, becoming component F2 (Fe^{II}); or (c) be used for all other processes in the cytosol, nucleus, ER, etc. Collectively we refer to this iron as component "CIA" since much of that iron ends up being

ISCs generated by the CIA (as described above). Most iron in the vacuoles is stored as Fe^{III} (component F3); this requires the oxidation of F2 by O_2.

A fundamental assumption of the model, known as the "respiratory shield" effect [77], limits the concentration of O_2 in the mitochondrial matrix. Accordingly, the matrix in healthy cells is hypoxic, which protects the FM and FS (the component that symbolizes ISCs and hemes combined) from attack by O_2. The iron centers reflected by FS are installed primarily in the respiratory complexes on the inner membrane, and they (viewed collectively) catalyze the reduction of O_2 to water as protons are pumped and ATP is ultimately generated. Yfh1 is a member of the complex that catalyzes the assembly of ISCs using FM as feedstock. When the expression level of Yfh1 declines, as occurs in Friedreich's ataxia, the rate by which FS is synthesized declines, which results in a decline in respiration. This in turn allows O_2 to penetrate the matrix, where it reacts with FM to generate Fe^{III} oxyhydroxide nanoparticles. O_2 in the matrix also inhibits the FM → FS reaction. The net result is a vicious cycle in which less and less FS leads to greater and greater penetration by O_2, which leads to more and more nanoparticles, ultimately ending in the diseased state.

Transforming Chemistry into Mathematics The next step in our recent modeling study [74] was to translate the 9 chemical reactions given in Fig. 5 into a set of 8 ordinary differential equations (ODEs), one for each component of the model. Then these equations can be rewritten in the matrix form [Stochiometric matrix] [vector of rates] = [vector of component derivatives]. The result, for this model, is given in Eq. (5).

$$\begin{bmatrix} 1 & -1 & -1 & -1 & 0 & 0 & 0 & 0 & 0 & -1 & 0 & 0 & 0 & 0 & 0 & 0 & 0 \\ 0 & 0 & 0 & 1 & 0 & 0 & 0 & 0 & 0 & 0 & -1 & 0 & 0 & 0 & 0 & 0 & 0 \\ 0 & 0 & 8 & 0 & 0 & 0 & 0 & -1 & 0 & 0 & 0 & -1 & 0 & 0 & 0 & 0 & 0 \\ 0 & 0 & 0 & 0 & 0 & 0 & 0 & 1 & 0 & 0 & 0 & 0 & -1 & 0 & 0 & 0 & 0 \\ 0 & 8 & 0 & 0 & -1 & -1 & 0 & 0 & 0 & 0 & 0 & 0 & 0 & -1 & 0 & 0 & 0 \\ 0 & 0 & 0 & 0 & 1 & 0 & 0 & 0 & 0 & 0 & 0 & 0 & 0 & 0 & -1 & 0 & 0 \\ 0 & 0 & 0 & 0 & 0 & 1 & 0 & 0 & 0 & 0 & 0 & 0 & 0 & 0 & 0 & -1 & 0 \\ 0 & 0 & 0 & 0 & 0 & -1 & 1 & 0 & -1 & 0 & 0 & 0 & 0 & 0 & 0 & 0 & -1 \end{bmatrix} \begin{bmatrix} R_{cyt} \\ R_{mit} \\ R_{vac} \\ R_{cia} \\ R_{isu} \\ R_{mp} \\ R_{O2} \\ R_{23} \\ R_{res} \\ D_{FC} \\ D_{CIA} \\ D_{F2} \\ D_{F3} \\ D_{FM} \\ D_{FS} \\ D_{MP} \\ D_{O2} \end{bmatrix} = \begin{bmatrix} FC' \\ CIA' \\ F2' \\ F3' \\ FM' \\ FS' \\ MP' \\ O2' \end{bmatrix} \tag{5}$$

Then we found the steady-state solutions (technically called the null space) by setting the vector on the right side of the equation to

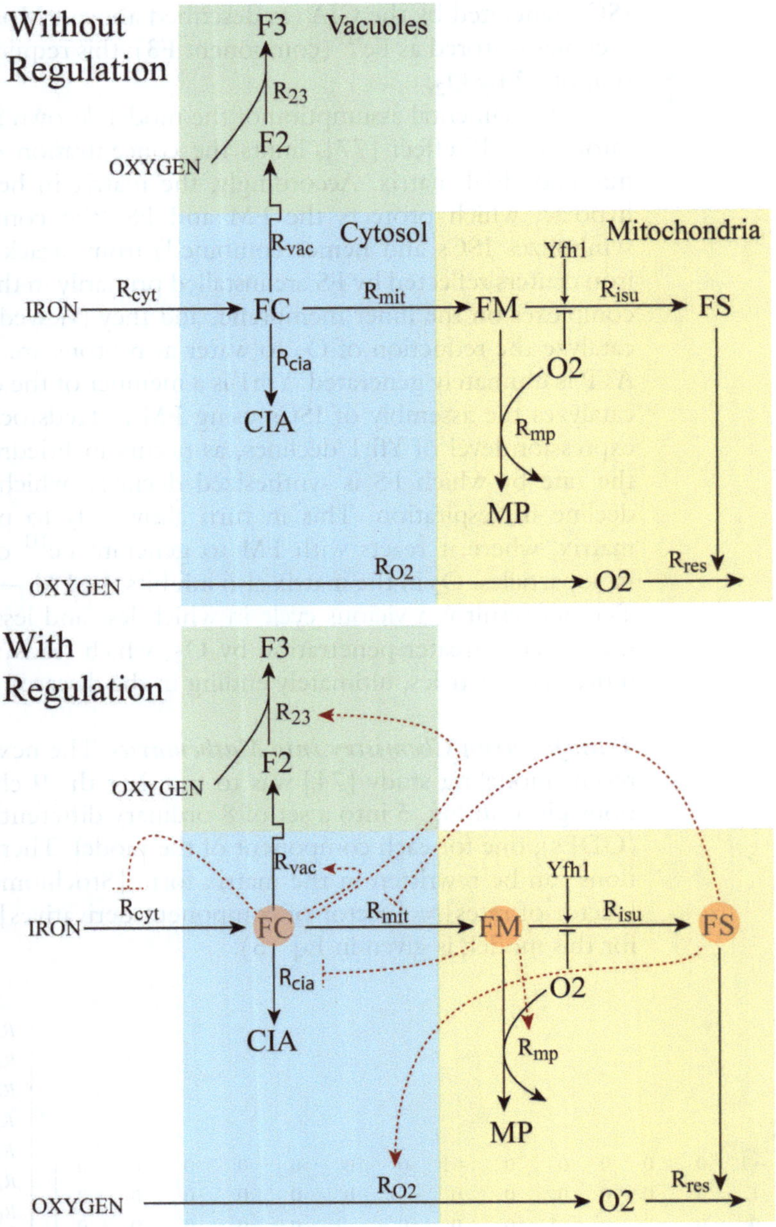

Fig. 5 Iron trafficking biochemical network assumed by Thorat et al. [74]. **Top Panel:** Nutrient IRON enters the cell to form FC, the cytosolic Fe^{II} pool. Iron from this pool can enter the vacuoles, becoming F2 (vacuolar Fe^{II}), entering the mitochondria and becoming FM (mitochondrial Fe^{II}), or being converted into a component called CIA. In this model, the CIA reflects all of the iron in the cell besides the specified components (and thus is defined differently than is used in the main text or in the primary literature). Once in the vacuole, F2 is oxidized to F3 (Fe^{III}). In the mitochondria, FM is converted into ISC and heme centers, collectively called FS. This reaction is promoted by Yfh1 (and inhibited by O_2). FS is installed in respiratory complexes (implied), which then catalyze the reduction of O_2 to water during respiration. The rate of respiration is presumed to be so fast that it partially prevents O_2 from diffusing into the matrix. Thus, in healthy cells, the matrix is hypoxic. Under Yfh1 diseased conditions, O_2 penetrates the matrix, due to insufficient respiration, where it reacts with FM to generate nanoparticles (MP). This creates a vicious cycle that leads to the diseased state. **Bottom Panel:** The

zero. Steady-state rates were divided into those that are *independent* (they could be assigned any value) and those that are *dependent* (immediately calculatable once the independent rates are assigned). This was done by obtaining the *reduced row echelon form* of the stoichiometric matrix and then distinguishing pivot from nonpivot columns.

Most of the independent rates were for dilution reactions, and these could be calculated as the product of α, an easily measured experimental quantity, and the steady-state concentrations for all the components. We were able to determine or estimate most of these from the Mössbauer spectra of various samples in conjunction with iron-concentration determinations. This was done for three cellular states, including healthy Fe-replete WT cells (called the *W* state), healthy but Fe-deficient WT cells (the *D* state), and the diseased Yfh1-deficient state (the *Υ* state). We were required to estimate the rate of respiration and the concentration of O_2 for each cellular state.

We next assigned simple rate-law expressions for each nondilution rate. Most of these were Michaelis–Menten-like expressions that included rate-constant k_{obs} and K_m parameters. For a few reactions, in which simple nonenzymatic chemistry was likely involved, we assumed mass-action kinetics. At that point, the system became nonlinear, and many types of behavior became possible and difficult to predict. We wanted the system (in each of the three cellular states *W*, *D*, and *Υ*) to be stable toward perturbations. To test this, we generated the corresponding Jacobian matrices and found that all eigenvalues in all three states had negative real parts. This guaranteed that the system in any of the three states would recover from a (modest, limited) perturbation of any component concentration from their steady-state value. At that point, both the steady-state and dynamical systems had been solved such that we could numerically integrate the set of ODEs for all three cellular states. However, our objective had not yet been met.

Transitioning from One State to Another Our last objective was to show how a healthy cell, upon becoming deficient in Yfh1 as the primary event, would develop the phenotype that characterizes Friedreich's ataxia as the cell's response. The same ODEs would be operational in both *W* and *Υ* states, but the set of rates would be different. These differences would generate the observed phenotype in terms of altered component concentrations. Thus, the issue

Fig. 5 (continued) best-selected cellular regulatory mechanism is shown in the red dashed lines. FC, FM, and FS are all sensors for one reaction or another. FC controls the import rate of iron into the cell and the rate of vacuolar iron import (and subsequent oxidation). FS regulates the rate of iron entering the mitochondria and the rate of O_2 entry into the cell. FM regulates the rate of nanoparticle formation and the rate of CIA formation. The regulatory mechanism shares some but not all features with what is thought to occur in real cells. (Reproduced with permission from (74))

became how to shift the rates smoothly from the $W \rightarrow$ Y state. This could of course be done manually by the user, but we wanted to employ an autoregulatory mechanism.

As rationalized above, we augmented rate-constants with logistical functions. We hoped that a single pair of parameters ([SP] and n) for each function associated with the 3 rate-constants for a given reaction and using the same sensed species *Sen* could be found such that observed rate-constants ($k_{obs(W)}$, $k_{obs(Y)}$, and $k_{obs(D)}$) could be described as the product of a common rate-constant k_{reg} multiplied by a common logistical function. Mathematically, this meant

$$k_{obs(W)} = \left(\frac{k_{reg}}{1 + e^{n([SP]-[Sen]_w)}}\right); k_{obs(Y)} = \left(\frac{k_{reg}}{1 + e^{n([SP]-[Sen]_Y)}}\right); k_{obs(D)} = \left(\frac{k_{reg}}{1 + e^{n([SP]-[Sen]_D)}}\right) \tag{6}$$

where k_{reg}, n, and [SP] are common for all three states but $k_{obs(W)} \neq k_{obs(Y)} \neq k_{obs(D)}$. This situation could be realized if an appropriate sensed species *Sen* could be identified.

For autoregulation, *Sen* needed to be one of the 8 components in the model, but which *should* be used? We decided to ask a simpler question, namely, which component *could* be used as the sensor. Once these were identified, we could then screen all viable possibilities to find the best as defined by various justifiable criteria. For example, if $k_{obs(W)} > k_{obs(Y)} > k_{obs(D)}$, the logistical function "valve" would need to be shut down to different extents but following the same trend. Specifically, we would need $[Sen]_W < [Sen]_Y < [Sen]_D$ for feedback regulation or the reverse for feedforward regulation.

We first considered all $8^9 = 134{,}217{,}728$ possible *cellular regulatory mechanisms* (CRMs); this included each of the 9 rate-constants being regulated by any of the 8 components. Then we applied various screens or filters to exclude those CRMs that were either mathematically impossible or that generated undesirable behavior. First, for a given reaction, the three rate-constants for the three cellular states needed to have distinct or unique values to be regulatable. We called this the *Uniqueness* filter. After applying it, 2,097,152 (i.e., 8^7) CRMs remained viable.

The second filter, alluded to above, was called *Trending*. Only sensed species that trended appropriately were viable; only 576 of 2,097,152 CRMs survived this filter. The third filter was called *Targeting*. Here, we applied nonlinear regression to optimize the logistical functions—finding the best values for k_{reg}, n, and [SP] that would start at the *W* state and come closest to "hitting the target" in terms of simulating the set of component concentrations that defined the other two states. Informally, imagine a harpoon shot from a boat with the intent of hitting a fish; the final position of the spear relative to the fish indicates its accuracy. In our filter, the 146 CRMs hit within 1% of the Y and D targets, and they were selected. But this wasn't the only important property of the transition; the *pathway* of the harpoon line used in that trajectory was also important. The fourth filter, called *Wandering*, selected CRMs that transitioned from W to Y or D without wandering unnecessarily during the transition. For this filter, 26 CRMs were selected for wandering least. Related to that was the *Smoothness* filter. Due to the nonlinearity of our system, 24 of those CRMs exhibited abrupt nonphysical spikes during transitions, so they were excluded. The remaining 2 CRMs were filtered according to whether n and [SP] were chemically reasonable. Values for n deviating from 1 were considered to be unreasonable, as were values of [SP] that were outside of the range of steady-state concentrations used for the component. However, both top cases passed the *Reasonableness* filter.

One of the two top CRMs is illustrated in Fig. 5 (*see* bottom panel). Regulatory steps are shown as red dashed lines, and sensed species are indicated by a red dot. Feedback regulation is indicated by a perpendicular line at the distal end of the lines, whereas feedforward regulation is indicated by an arrow. Component FC is used to feedback-regulate the import of nutrient IRON and to feedforward-regulate the import of FC into the vacuoles. Thus, when [FC] becomes excessive relative to its set-point concentration, it will simultaneously shut down iron import and stimulate the transfer of cytosolic iron into the vacuoles for storage. Component FM feedback-regulates the rate of CIA formation while feedforward-regulating the formation of nanoparticles (perhaps reflecting the vicious cycle associated with the respiratory shield). Component FS feedback-regulates the import of cytosolic iron into the mitochondria while stimulating the rate of O_2 entering the cell.

Figure 6 (top panel) shows the simulation plots of the model assuming either of the two top autoregulatory cases (one solid line, one dashed line) as the cell transitions from the W state (on the left) to the Y state (on the right). There are virtually no differences in the behavior. The simulated phenotypic changes associated with this transition mirror those observed in real cells. The concentration of [FS], reflecting the ISC assembly, declines as expected since Yfh1 is involved in the reaction forming this component (FM $\rightarrow$ FS). Mitochondrial nanoparticles increase dramatically, and vacuolar iron decreases majorly as the diseased state forms, also as observed.

The bottom panel of Fig. 6 shows the transition of Fe-replete to Fe-deficient healthy cells, and again the simulated behavior mirrors the observed behavior well. Most dramatically, vacuolar iron (F2 and F3) decline majorly as the cell becomes Fe-deficient. Other Fe component concentrations show a resistance to change during this period but eventually decline as the cell becomes Fe-starved. The effects of these transitions on the O_2 levels in the cells have not been experimentally measured, and so these simulations serve as predictions.

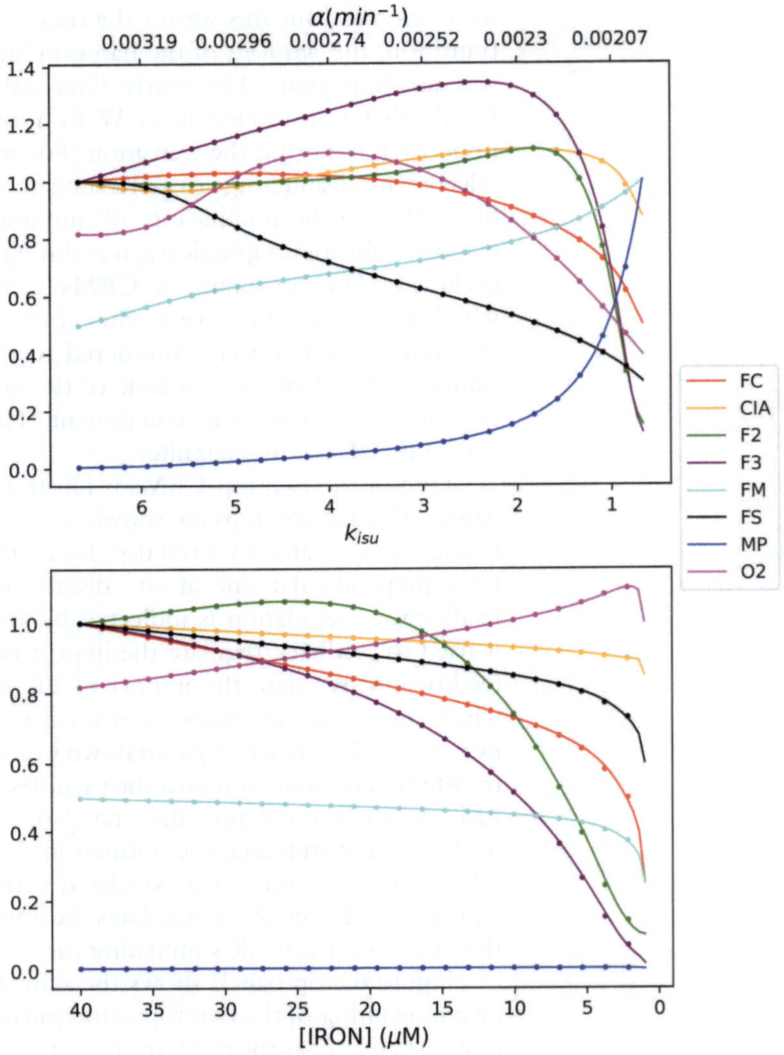

Fig. 6 Plot of all component concentrations in the model of Fig. 5. Both top cellular regulatory mechanisms are shown (one in solid line; one in dotted line). The behavior was nearly identical. **Top Panel:** W → Y transition (from left to right), assuming the same cellular regulation mechanism shown in Fig. 5. Refer to color-coding for each component. Faithful to the observed Yfh1-deficient phenotype, FS declines as k_{isu} declines. This is the primary event driving the transition. There is a dramatic increase in mitochondrial nanoparticle (MP) as the rate of ISC/hemes declines. Also observed is the emptying of vacuolar Fe^{II} and Fe^{III}. ROS levels follow the same trend as the nanoparticles. **Bottom Panel:** W → D transition (left to right) using the same model. As observed in real cells, the vacuoles empty as the cells become iron-deficient. The steady-state concentrations of the other components are fairly constant, as one might expect in a well-regulated cell, until it becomes "Fe-starved" near the right-hand side of the plot. The regulatory mechanism assumed does not include any of the more sophisticated mechanisms that real cells use to survive in an Fe-starved state. (Reproduced with permission from [74])

4 Conclusions and Future Directions

There are some obvious differences/inconsistencies between the regulation assumed by the model vs. experimental results regarding iron homeostatic regulation in yeast. The experimental approach to studying regulation is far better established than the computational one, and the presented model is far simpler than the current state of knowledge. However, the method of modeling regulation at the cellular level is more important as it illustrates the potential for expansion and improvement. We suggest that the development of mathematical models will increasingly become a valuable compliment to the experimental approach.

Clearly, more advanced and realistic mathematical models are required, as are further experimental advances. The two go hand-in-hand. Solving ODE-based models generally requires reliable values for many kinetic parameters (rates, rate-constants, rate-law expressions, Michaelis–Menten parameters, component concentrations, etc.), but these are often unavailable. To minimize the difficulty arising from this shortage, models must be simplified by lumping components and reactions such that they have sufficient complexity to simulate the sought-after behavior and thus provide mechanistic insight. As more is discovered experimentally and more kinetic parameters are established, more accurate and realistic models will be forthcoming.

In our view, the next most intriguing aspect to simulate will be *layers* of regulation, as there is evidence for this in real cells. Doing so will incorporate levels of iron supplementation in the media, metabolic modes (carbon source, glucose vs. glycerol, and perhaps nitrogen source), and O_2 levels (aerobic vs. hypoxic). Some of these effects appear to *override* others. For example, the cytosolic Fe^{II} pool might serve as a regulatory sensor in healthy cells, but its effects might be overridden by metabolic remodeling, such as the initiation of mitophagy when the mitochondria are in distress, or a shift between respiring and fermenting metabolic modes.

Finally, perhaps an intellectual barrier of sorts must be surmounted for greater and more rapid advancement. Experimentalists as a group may want to consider mathematical models more seriously than they have because complex problems involving regulation *may not be solvable without them*. Given the astronomical biochemical complexity of cells, inferring molecular-level mechanisms from observed cellular phenotypes or behaviors is far more difficult and fraught with greater ambiguity than is commonly appreciated. Conversely, computationalists as a group may need to recognize the difficulties, complexities, and limitations of experimental cellular biochemistry and genetics. They need to develop practical models that are directly relevant to experimentalists and explain such models on a level that does not presume expertise in advanced mathematics. By working together, the challenges presented by the inherent complexities of regulatory mechanisms operating within biological cells might be overcome.

Acknowledgments

I thank Drs. Jay R. Walton, Joshua D. Wofford, Shaik Waseem Vali, and Mr. Shantanu Thorat for their critical and essential efforts in developing the computational approaches described here and the many lessons they have taught me. I also thank Mr. Thorat for preparing the plot of Fig. 4 and Dr. Sergi Puig (csic, Valencia Spain) for critically reading the manuscript. These studies have been supported by the National Science Foundation (MCB-1817389), the National Institutes of Health (GM127021), and the Robert A. Welch Foundation (A-2149-20230405).

References

1. Lindahl PA, Vali SW (2022) Mössbauer-based molecular-level decomposition of the *Saccharomyces cerevisiae* ironome, and preliminary characterization of isolated nuclei. Metallomics 14: mfac080
2. Martins TS, Costa V, Pereria C (2018) Signaling pathways governing iron homeostasis in budding yeast. Mol Microbiol 109:422–432
3. Gupta M, Outten CE (2020) Iron–sulfur cluster signaling: the common thread in fungal iron regulation. Curr Opin Chem Biol 55:189–201
4. Talib EA, Outten CE (2021) Iron–sulfur cluster biogenesis, trafficking, and signaling: roles for CGFS glutaredoxins and BolA proteins. Biochim Biophys Acta Mol Cell Res 1868: 118847. https://doi.org/10.1016/j.bbamcr.2020.118847
5. Ramos-Alonso L, Romero AM, Martinez-Pastor MT, Puig S (2020) Iron Regulatory Mechanisms in *Saccharomyces cerevisiae*. Front Microbiol 11:Article Number 582830. https://doi.org/10.3389/fmicb.2020.582830
6. Martinez-Pastor MT, Perea-Garcia A, Puig S (2017) Mechanisms of iron sensing and regulation in the yeast *Saccharomyces cerevisiae*. World J Microbiol Biotech 33:75. https://doi.org/10.1007/s11274-017-2215-8
7. Bird AJ (2015) Cellular sensing and transport of metal ions: implications in micronutrient homeostasis. J Nutr Biochem 26:1103–1115
8. Yamaguchi-Iwai Y, Stearman R, Dancis A, Klausner RD (1996) Iron-regulated DNA binding by the AFT1 protein controls the iron regulon in yeast. EMBO J 15:3377–3384. https://doi.org/10.1002/j.1460-2075.1996.tb00703.x
9. Philpott CC, Protchenko O (2008) Response to iron deprivation in *Saccharomyces cerevisiae*. Eukaryot Cell 7:20–27
10. Rutherford JC, Jaron S, Winge DR (2003) Aft1p and Aft2p mediate iron-responsive gene expression in yeast through related promoter elements. J Biol Chem 278:27636–27643
11. Dix DR, Bridgham JT, Broderius MA, Byersdorfer CA, Eide DJ (1994) The Fet4 gene encores the low-affinity Fe(II) transport protein of *Saccharomyces cerevisiae*. J Biol Chem 269:26092–26099
12. Chen XZ, Peng JB, Cohen A, Nelson H, Nelson N, Hediger MA (2000) Yeast Smf1 mediates H+-coupled iron uptake with concomitant uncoupled cation currents. J Biol Chem 274:35089–35094
13. Kosman DJ (2003) Molecular mechanisms of iron uptake in fungi. Mol Microbiol 47:1185–1197
14. Kane PA (2006) The where, when, and how of organelle acidification by the yeast vacuolar H+-ATPase. Microbiol Mol Biol Rev 70:177–191
15. Li LT, Chen OS, Ward DM, Kaplan J (2001) Ccc1 is a transporter that mediates vacuolar iron storage in yeast. J Biol Chem 276: 29515–29519

16. Cockrell AL, Holmes-Hampton GP, McCormick SP, Chakrabarti M, Lindahl PA (2011) Mössbauer and EPR study of iron in vacuoles from fermenting *Saccharomyces cerevisiae*. Biochemistry 50:10275–10283
17. Rutherford JC, Jaron S, Ray E, Winge DR (2001) A second iron-regulatory system in yeast independent of Aft1p. Proc Natl Acad Sci USA 98:14322–14327
18. Courel M, Lallet S, Camadro JM, Blaiseau PL (2005) Direct activation of genes involved in intracellular iron use by the yeast iron-responsive transcription factor Aft2 without its paralog Aft1. Mol Cell Biol 15:6760–6771
19. Ueta R, Fukanaka A, Yamaguchi-Iwai Y (2003) Pse1p mediates the nuclear import of the iron responsive transcription factor Aft1p in *Saccharomyces cerevisiae*. J Biol Chem 278:50120–50127
20. Ueta R, Fujiwara N, Iwai K, Yamaguchi-Iwai Y (2007) Mechanism underlying the iron-dependent nuclear export of the iron-responsive transcription factor Aft1p in *Saccharomyces cerevisiae*. Mol Biol Cell 18:2980–2990
21. Yamaguchi-Iwai Y, Ueta R, Fukanaka A, Sasaki R (2002) Subcellular localization of Aft1 transcription factor responses to iron status in *Saccharomyces cerevisiae*. J Biol Chem 277:18914–18918
22. Li HR, Outten CE (2019) The conserved CDC motif in the yeast iron regulator Aft2 mediates iron–sulfur cluster exchange and protein–protein interactions with Grx3 and Bol2. J Biol Inorg Chem 24:809–815
23. Chi CB, Tang YJ, Zhang JH, Dai YN, Abdalla M, Chen Y, Zhou CZ (2018) Structural and biochemical insights into the multiple functions of yeast Grx3. J Mol Biol 430:1235–1248
24. Holmes-Hampton GP, Miao R, Morales JR, Guo YS, Münck E, Lindahl PA (2010) A nonheme high-spin ferrous pool in mitochondria isolated from fermenting *Saccharomyces cerevisiae*. Biochemistry 49:4227–4234
25. Tsai CL, Barondeau DP (2010) Human frataxin is an allosteric switch that activates the Fe-S cluster biosynthetic complex. Biochemistry 49:9132–9139
26. Poor CB, Wegner SV, Li H, Dlouhy AC, Schuermann JP, Sanishvili R, Hinshaw JR, Riggs-Gelasco PJ, Outten CE, He C (2014) Molecular mechanism and structure of the *Saccharomyces cerevisiae* iron regulator Aft2. Proc Natl Acad Sci USA 111:4043–4048
27. Rietzschel N, Pierik AJ, Bill E, Lill R, Mühlenhoff U (2015) The basic leucine zipper stress response regulator Yap5 senses high-iron conditions by coordination of [2Fe-2S] clusters. Mol Cell Biol 35:370–378
28. Li L, Kaplan J, Ward DM (2017) The glucose sensor Snf1 and the transcription factors Msn2 and Msn4 regulate transcription of the vacuolar iron importer gene CCC1 and iron resistance in yeast. J Biol Chem 292:15577–15586
29. Li LT, Ward DM (2018) Iron toxicity in yeast: transcriptional regulation of the vacuolar iron importer Ccc1. Curr Genet 64:413–416
30. Bird AJ (2008) Metallosensors, the ups and downs of gene regulation. (Editor, Poole, RK). Adv Microb Physiol 53:231–267
31. Puig S, Askeland E, Thiele DJ (2005) Coordinated remodeling of cellular metabolism during iron deficiency through targeted mRNA degradation. Cell 120:99–110
32. Puig S, Vergara SV, Thiele DJ (2008) Cooperation of two mRNA-binding proteins drives metabolic adaptation to iron deficiency. Cell Metab 7:555–564
33. Ramos-Alsono L, Romero AM, Polaina J, Puig S, Martinez-Pastor MT (2019) Dissecting mRNA decay and translation inhibition during iron deficiency. Curr Genet 65:139–145
34. Martinez-Pastor M, Vergara SV, Puig S, Thiele DJ (2013) Negative feedback regulation of the yeast Cth1 and Cth2 mRNA-binding proteins is required for adaptation to iron deficiency and iron supplementation. Mol Cell Biol 33:2178–2187
35. Romero AM, Martinez-Pastor M, Du G, Sole C, Carlos M, Vergara SV, Sanvisens N, Wohlschlegel JA, Toczyski DP, Posas F, de Nadal E, Martinez-Pastor MT, Thiele DJ, Puig S (2018) Phosphorylation and proteasome recognition of the mRNA-binding protein Cth2 facilitates yeast adaptation to iron deficiency. MBio 9:e01694–e01618
36. Sanvisens N, Bano MC, Huang MX, Puig S (2011) Regulation of ribonucleotide reductase in response to iron deficiency. Mol Cell 44: 759–769
37. Ros-Carrero C, Ramos-Alonso L, Romero AM, Bano MC, Martinez-Pasto MT, Puig S (2020) The yeast Aft1 transcription factor activates ribonucleotide reductase catalytic subunit RNR1 in response to iron deficiency. Biochim Biophys Acta Gene Reg Mech 1863:194522
38. Matsuo R, Mizobuchi S, Fujii M (2017) Central roles of iron in the regulation of oxidative stress in the yeast *Saccharomyces cerevisiae*. Curr Genet 63:895–907
39. Patnaik PK, Beaupere C, Barlit H, Romero AM, Tsuchiya M, Muir M, Martinez-Pastor MT, Puig S, Kaeberlein M, Labunskyy VM

(2022) Deficiency of the RNA-binding protein Cth2 extends yeast replicative lifespan by alleviating its repressive effects on mitochondrial function. Cell Rep 40:Article Number 111113

40. Zhang X, Kebaara BW (2022) Nonsense-mediated mRNA decay and metal ion homeostasis and detoxification in *Saccharomyces cerevisiae*. Biometals 35:1145–1156
41. Peccarelli M, Scott TD, Kebaara BW (2019) Nonsense-mediated mRNA decay of the ferric and cupric reductase mRNAs FRE1 and FRE2 in *Saccharomyces cerevisiae*. FEBS Lett 593: 3228–3238
42. Montella-Manuel S, Pujol-Carrion N, Angeles de la Torre-Ruiz M (2023) Aft1 nuclear localization and transcriptional response to iron starvation rely upon TORC2/Ypk1 signaling and sphingolipid biosynthesis. Int J Mol Sci 24: Article Number 2438
43. Kakhlon O (2021) Casting iron into the cell fate mold. Biochem J 478:1879–1883
44. Raguzzi F, Lesuisse E, Crichton RR (1988) Iron storage in *Saccharomyces cerevisiae*. FEBS Lett 231:253–258
45. Simpson-Lavy KJ, Kupiec M (2023) The poly-HIS tract of yeast AMPK coordinates carbon metabolism with iron availability. Int J Mol Sci 24:Article Number 1368
46. Chen KL, Ven TN et al (2020) Loss of vacuolar acidity results in iron–sulfur cluster defects and divergent homeostatic responses during aging in *Saccharomyces cerevisiae*. Geroscience 42: 749–764
47. Hughes CE, Coody TK, Jeong MY, Berg JA, Winge DR, Hughes AL (2020) Cysteine toxicity drives age-related mitochondrial decline by altering iron homeostasis. Cell 180:296–310
48. Kim JE, Vali SW, Nguyen TQ, Dancis A, Lindahl PA (2021) Mössbauer and LC-ICP-MS investigation of iron trafficking between vacuoles and mitochondria in vma2Δ *Saccharomyces cerevisiae*. J Biol Chem 296:Article 100141, 1–14
49. Martins TS, Pereira C, Canadell D, Vilaca R, Teixeira V, Moradas-Ferreira P, de Nadal E, Posas F, Costa V (2018b) The Hog1p kinase regulates Aft1p transcription factor to control iron accumulation. Biochim Biophys ACTA Mol Cell Biol Lipids 1863:61–70
50. Liu LL, Huang MX (2015) Essential role of the iron–sulfur cluster binding domain of the primase regulatory subunit Pri2 in DNA replication initiation. Protein Cell 6:194–210
51. Puig S, Ramos-Alonso L et al (2017) The elemental role of iron in DNA synthesis and repair. Metallomics 9:1483–1500
52. Krueger A, Keppel M, Sharma V, Frunzke J (2022) The diversity of heme sensor systems - heme-responsive transcriptional regulation mediated by transient heme protein interactions. FEMS Microbiol Rev 46:Article Number fuac002
53. Bouchez CL, Youboue ED et al (2020) "Labile" heme critically regulates mitochondrial biogenesis through the transcriptional co-activator Hap4p in *Saccharomyces cerevisiae*. J Biol Chem 295:5095–5109
54. Mao YH, Chen CB (2019) The Hap complex in yeasts: Structure, assembly mode, and gene regulation. Front Microbiol 10:1645. https://doi.org/10.3389/fmicb.2019.01645
55. Jorda T, Barba-Aliaga M, Rozes N, Alepuz P, Martinez-Pastor MT, Puig S (2022) Transcriptional regulation of ergosterol biosynthesis genes in response to iron deficiency. Environ Microbiol 24:5248–5260
56. Romero AM, Jorda T, Rozes N, Martinez-Pastor MT, Puig S (2018b) Regulation of yeast fatty acid desaturase in response to iron deficiency. Biochim Biophys Acta Mol Cell Biol Lipids 1863:657–668
57. Xue Y, Schmollinger S, Attar N, Campos OA, Vagelauer M, Carey MF, Merchant SS, Kurdistani SK (2017) Endoplasmic reticulum–mitochondria junction is required for iron homeostasis. J Biol Chem 292:13197–13204
58. Foury F, Talibi D (2001) Mitochondrial control of iron homeostasis—a genome wide analysis of gene expression in a yeast frataxin-deficient strain. J Biol Chem 276:7762–7768
59. Fernandez S, Wofford JD, Shepherd RE, Vali SW, Dancis A, Lindahl PA (2022) Yeast cells depleted of the frataxin homolog Yfh1 redistribute cellular iron: studies using Mössbauer spectroscopy and mathematical modeling. J Biol Chem 298:101921
60. Ast T, Meisel JD, Patra S, Wang H, Grange RMH, Kim SH, Calvo SE, Orefice LL, Nagashima F, Ichinose F, Zapol WM, Ruvkun G, Barondeau DP, Mootha VK (2019) Hypoxia rescues frataxin loss by restoring iron–sulfur cluster biogenesis. Cell 177: 1507–1521
61. Chen OS, Crisp RJ, Valachovic M, Bard M, Winge DR, Kaplan J (2004) Transcription of the yeast iron regulon does not respond directly to iron but rather to iron–sulfur cluster biosynthesis. J Biol Chem 279:29513–29518
62. Kispal G, Csere P, Prohl C, Lill R (1999) The mitochondrial proteins Atm1p and Nfs1p are essential for biogenesis of cytosolic Fe/S proteins. EMBO J 18:3981–3989

63. Rutherford JC, Ojeda L, Balk J, Muhlenhoff U, Lill R, Winge DR (2005) Activation of the iron regulon by the yeast Aft1/Aft2 transcription factors depends on mitochondrial but not cytosolic iron–sulfur protein biogenesis. J Biol Chem 280:10135–10140
64. Alsina D, Ros J, Tamarit J (2018) Nitric oxide prevents Aft1 activation and metabolic remodeling in frataxin-deficient yeast. Redox Biol 14:131–141
65. Holmes-Hampton GP, Jhurry NG, McCormick AP, Lindahl PA (2013) Iron content of *Saccharomyces cerevisiae* cells grown under iron-deficient and iron-overload conditions. Biochemistry 52:105–114
66. Lindahl PA, Morales JG, Miao R, Holmes-Hampton GP (2009) Isolation of *Saccharomyces cerevisiae* mitochondria for Mössbauer, Epr, and electronic absorption spectroscopic analyses. In: Allison WS, Scheffler IE (eds) Methods in enzymology, vol 456. Academic, Burlington, pp 267–285. isbn:978-0-08-087776-1
67. Nguyen TQ, Dziuba N, Lindahl PA (2019) Isolated *Saccharomyces cerevisiae* vacuoles contain low-molecular-mass transition-metal polyphosphate complexes. Metallomics 11:1298. https://doi.org/10.1039/c9mt00104b
68. Moore MJ, Wofford JD, Dancis A, Lindahl PA (2018) Recovery of mrs3Δmrs4Δ *Saccharomyces cerevisiae* cells under iron-sufficient conditions and the role of Fe580. Biochemistry 57:672–683
69. Chakrabarti M, Lindahl PA (2014) The utility of Mössbauer spectroscopy in eukaryotic cell biology and animal physiology (Chapter 4). In: Rouault TA (ed) Iron-sulfur clusters. Walter de Gruyter Publisher, Berlin, pp 49–75
70. Miao R, Martinho M, Morales JG, Kim HS, Ellis EA, Lill R, Hendrich MP, Münck E, Lindahl PA (2008) EPR and Mössbauer spectroscopy of intact mitochondria isolated from Yah1p-depleted *Saccharomyces cerevisiae*. Biochemistry 47:9888–9899
71. Miao R, Kim HS, Koppolu UMK, Ellis EA, Scot RA, Lindahl PA (2008) Biophysical characterization of the iron in mitochondria from Atm1p-depleted *Saccharomyces cerevisiae*. Biochemistry 48:9556–9568
72. Miao R, Holmes-Hampton GP, Lindahl PA (2011) Biophysical investigation of the iron in aft1-1up and gal-yah1 *Saccharomyces cerevisiae*. Biochemistry 50:2660–2671
73. Tyson JJ, Laomettachit T, Kraikivski P (2019) Modeling the dynamic behavior of biochemical regulatory networks. J Theor Biol 462:514–527
74. Thorat S, Walton JR, Lindahl PA (2023) A kinetic model of iron trafficking in growing *Saccharomyces cerevisiae* cells; applying mathematical methods to minimize the problem of sparse data and generate viable autoregulatory mechanisms. PLOS Comput Biol 19: e1011701
75. Wofford JD, Lindahl PA (2015) Mitochondrial iron–sulfur cluster activity and cytosolic iron regulate iron traffic in *Saccharomyces cerevisiae*. J Biol Chem 290:26968–26977
76. Wofford JD, Lindahl PA (2019) A mathematical model of iron import and trafficking in wild-type and Mrs3/4ΔΔ yeast Cells *BMC*. Syst Biol 13:Article number 23
77. Wofford JD, Bolaji N, Dziuba N, Outten FW, Lindahl PA (2019) Evidence that a respiratory shield in *Escherichia coli* protects a low molecular mass Fe^{II} pool from O_2-dependent oxidation. J Biol Chem 294:50–62

Chapter 2

ICP-MS Analysis of Iron from Biological Samples

Javier Seravalli

Abstract

Elemental analysis can provide trace concentrations of iron and other transition elements at nanomolar (μg/L) concentrations in whole bacterial and mammalian cells, subcellular compartments, biological fluids, and tissues. The best method of analysis is by far Inductively Coupled Plasma Mass Spectrometry (ICP-MS). I describe here a very general method for the sample preparation, instrument settings, method development, and analysis. The method can be extended to up to 20 common elements in biological samples.

Key words ICP-MS, Iron, Elemental analysis, Inductively coupled plasma, Nitric acid, Hydrogen peroxide, Mass spectrometry

1 Introduction

1.1 Description of the Instrument

The analysis of iron in biology has been essential for the study of iron-containing species in solution such as metal oxide crystals, mineralized Fe (ferritin), mono, and polynuclear complexes (mainly Fe-S clusters, heme groups, and other Fe oxo clusters) in proteins and other complexes [1–3]. Although there are methods for the destructive analysis of tissue and purified protein samples using Fe-chelators such as ferrozine and 1,10-phenanthroline (which require reduction of Fe^{3+} to Fe^{2+}), which are sensitive enough for low micromolar concentrations of Fe, they require external calibrations with sample matching. With the advent in the last 50 years of ICP-type instruments (ICP-AAs, ICP-OES, and ICP-MS), a more direct method of analysis of Fe in solution is feasible, with the added advantage of high-throughput, wider linear range (4–6 orders of magnitude), suitability for all redox states of the element and improved sensitivity. The ICP utilizes an argon-based plasma, generated at a device known as a torch, which is surrounded by a working coil that carries the electromagnetic radiation that generates and maintains the gas of ions (plasma) so long as the Ar is flowing through the instrument and ions are

Oleh Khalimonchuk (ed.), *Iron Metabolism: Methods and Protocols*, Methods in Molecular Biology, vol. 2839, https://doi.org/10.1007/978-1-0716-4043-2_2,

generated at a constant steady-state rate. The Ar plasma feeds the sample coming from a peristaltic pump through the torch, although the effluents of a GC column or HPLC instrument can also be used for the purpose of speciation analysis (hyphenated ICP-MS). The inductively coupled plasma (ICP) performs four sequential functions at the typical temperatures of operation (7500 K): desolvation of the species, destruction of complexes, atomization of compounds, and ionization to the singly ionized state (e.g., Fe^{1+}). At this point, the ions are excited by a UV/visible monochromatic light, with the plasma plume acting as the equivalent of a standard absorbance cuvet. The wavelengths are selected to match the excitation of the element of choice. Likewise, the emission of the ions in the plasma can be monitored at various spectroscopic lines (238.204 nm, 239.562 nm, and 259.940 nm for Fe). Nevertheless, both of these techniques can suffer from spectroscopic interferences (Ru, Co, W, Cr, Hf, and Nb) at some of the excitation lines. ICP-MS, on the other hand, depends on the detection of the specific ion mass (m/z), which invariably is the singly charged isotope of the element (m/z = 56 and m/z = 57 for Fe). The corresponding peaks have significant polyatomic interferences ($^{40}Ar^{15}N^{1}H$, $^{40}Ar^{16}O$, $^{38}Ar^{18}O$, $^{37}Cl^{18}O^{1}H$, $^{40}Ca^{16}O$ for ^{56}Fe and $^{40}Ar^{16}O^{1}H$ and $^{40}Ca^{16}O^{1}H$, for ^{57}Fe) [4], which made it difficult to quantitate trace levels (<0.1 μg/mL) for earlier versions of the instrument. The polyatomic species are produced post-plasma generation by the reaction of contaminating elements (Ca, Cl) with oxygen from either O_2 or water prior to admission to the lens assembly of the instrument. A previous method by the author, which was published in *Methods in Molecular Biology* [5], describes the instrument in more detail.

1.2 Isotopes of Fe

Iron has an average atomic weight in the natural distribution of 55.85 g/mol. The distribution is composed of 5.845% ^{54}Fe, which is mildly radioactive, 91.754% ^{56}Fe, 2.119% ^{57}Fe, and 0.286% ^{58}Fe. ^{54}Fe is stable enough for quantification, but it is interfered with by too many polyatomic interferences in ICP-MS ($^{40}Ar^{14}N$, $^{38}Ar^{15}NH$, $^{38}Ar^{16}O$), while ^{58}Fe is too low in abundance to show significant sensitivity. Most modern ICP-MS instruments are equipped with a collision cell in front of the mass filter. The collision cell can be operated with UHP He gas, which can reduce the interferences by polyatomic species cited above based on the fact that they have a higher ionic radius, which increases their collision profile. The discriminations against polyatomic species and in favor of ^{56}Fe and ^{57}Fe ions occur by a mechanism known as kinetic energy discrimination (KED). UHP H_2 gas can also be used to more effectively destroy the Ar-derived polyatomic species. The mechanism is known as reactive discrimination, which is also very useful in the analysis of trace levels of Ca and Se. The reactions are illustrated below:

$$^{40}Ar^{+} + H_2 \rightleftharpoons {}^{40}Ar + H_2{}^{+}$$
$$^{40}Ar^{16}O^{+} + H_2 \rightleftharpoons {}^{40}Ar^{16}OH + H^{+}$$

Iron is the fourth most common in the Earth's crust. The metal readily oxidizes in damp air and dissolves in dilute acids (1–5%); therefore, the reagent purities must be considered when carrying out trace analysis. Fe solutions are stable in dilute nitric acid and can be stored in plastic containers. Fe is over 95% ionized in the Ar plasma, thus the typical limit of quantification (LOQ) is around 1 ppb. For lower levels of quantification, iron can be analyzed using a lower power (600–900 W) plasma for ICP-MS since that reduces the contributions of argon and calcium derived polyatomic interferences. The addition of UHP H_2 in the reaction mode allows for the use of high-power plasma, dramatic reduction of contaminant species, and lower limits of detection and quantification, as a result.

2 Materials

All materials, calibrations, and sample preparations can be performed on a standard laboratory bench. However, a metal-free hood is highly recommended to prevent the inhalation of acid vapors as well as sample contamination. Pipetting should be done with metal-free pipettors and tips. My personal preference is to use graduated pipet tips that are color free. In general, plasticware should be used throughout for sample collection, storage, digestion, and analysis. Glass and quartz tend to leach elements like Li, B, Al, and even Fe if the containers have been exposed to solutions containing Fe. Acid washing is feasible with 5% nitric acid if glass is required. The plastic should be polyethylene (PE), polyethylene terephthalate (PET), and high-density polyethylene (HDPE) are recommended. For true trace metal analysis, the sample containers and calibrators should be acid-washed overnight, followed by distilled deionized water (18 MΩ-cm at 25 °C).

1. Trace metal nitric acid (70% w/v). Of note the density of 70% w/v HNO_3 is 1.42 at 25 °C, which means it is equivalent to 100% v/v. Thus, a 5% HNO_3 is an aqueous solution where the acid has been diluted by 20-fold.
2. Hydrogen peroxide (30% w/v), also of trace metal grade. Store the H_2O_2 at 4 °C when not in use.
3. Spex Certiprep trace-metal grade water or equivalent. This water should be used in sample preparation, calibrator solutions, ICP carrier solution, and diluents. For ALS rinsing, doubly distilled de-ionized water is sufficient.

4. Metal-free pipettors of volume sizes of 2, 20, 200, and 1000 μL. Also graduated plastic cylinders and metal-free funnels can be useful.
5. High-quality protein work grade Eppendorf tubes (1.5 or 2.0 mL) without any additional color or Falcon tubes (15 mL) for larger samples (>0.5 mL digestions), though any plastic tube that can seal tightly can work most of the time.
6. Hot plate or hot oven, capable of reaching 100 °C. Ideally, a sample digester works better for most tissue samples, like a microwave digester for micro-sized samples (CEM SP-D) that can digest amounts as low as 5 mg, but I don't normally use it.
7. 96-well plate autosampler (ALS): An autosampler with a PTFE probe with one or more placeholders for 96-well plates. I employ one from Elemental Scientific Inc. (Omaha, NE) model SC-DX4 with 4–6 plate positions and a rinse station. The ALS runs using its own software, receiving prompts from the ICP software on when to start an injection sequence and move to the next sample position (*see* Fig. 1).
8. ICP-MS: Agilent 7500cx or equivalent. The instrument should be equipped with either He (99.999%) collision gas accessory, H_2 (99.999%) reaction gas accessory, or a Micromist nebulizer but no autosampler is required on the instrument. Conversely, an ICP-QQQ-MS or ICP-TOF is also suitable, again with

Fig. 1 Elemental Scientific SC-DX4 6 × 96 well plate with dual rinse station (shown on the right) and PTFE 30-cm long probe. The autosampler, injection syringe, and injection valve are controlled by the ESI software

UHP H_2 in the second quadrupole. The method I use for Fe is also suitable for the simultaneous determination of 17 other elements, including Mg, Ca, Mn, Co, Ni, Cu, Zn, As, Se, Mo, and Cd. The instrument should be running with the mix gas mode with 3.5 mL/min of H_2 and 1.5 mL/min of He.

9. 96-well plates from Corning Model 3797. The conical bottoms work better than flat bottoms because they allow the probe to reach the bottom of each well while, at the same time, allowing for a slightly inaccurate alignment of the sample probe.
10. Analytical balance with a precision of up to 0.1 mg.
11. Commercial ICP-MS single or multiple element standards. A suitable standard for Fe is 10 μg/mL (ppm), but concentrated (1000 ppm) standards can be used instead. Standards can be obtained from Inorganic Ventures or SPEX Certiprep.
12. Use Gallium as an internal standard. A 1000 ppm Ga standard is used to prepare a 50 ppm stock solution in 2% HNO_3. From this stock, a working solution is prepared to serve as a diluent for the samples and calibration standards. Use this stock solution at 50 ppm of Ga to prepare all working solutions for a given method to ensure the matching of the Ga concentration.
13. Microtube-size centrifuge capable of reaching 10,000× *g* speeds.
14. UHP Argon. Liquid argon is strongly recommended over compressed gas. A single T-sized Ar bottle can last about 2 h of continuous operation at ~20 L/min, but a Dewar with 180 L of liquefied Argon can last up to 100 h of continuous operation.

3 Methods

3.1 Sample Preparation

1. Store the samples at either −20 °C or −70 °C. Although the metal composition will not change regardless of the storage conditions, significant degradation of metalloenzymes and other complexes can occur, which can affect the dissolution and digestion process and can result in the loss of Fe and other metals.
2. Weigh as much as 100 mg of tissue sample in a metal-free Eppendorf tube. Liquid samples such as blood, serum, and urine can be thawed and pipetted up to 100 μL (*see also* **Note 1**).
3. Add 4× the amount of sample of 68–70% HNO_3. The concentration of the dissolved samples after digestion should be ~200 mg/mL, assuming the sample density is ~1 g/mL. The samples should be incubated at room temperature, inside a hood, for at least 1 h.

4. For harder-to-digest tissues such as bone or brain tissues, add 2× the amount of sample of 70% HNO_3. Incubate for 1 h at room temperature, and then add 2× the amount of sample of 30% H_2O_2. Incubate at room temperature for a second hour.
5. Move samples into the oven for digestion at 70 °C. One can use a standard tube holder to hold the samples plus another tube holder as a cap. Lay a metal block on top of the holder used as a cap to prevent or minimize sample loss during the digestion. Place the assembly on top of another sample holder to prevent direct overheating of the samples from the metal surface at the bottom. Digest for at least 8 h.
6. Move the samples out of the oven and allow them to cool down for at least 30 min. Centrifuge the digestions to settle evaporation and undissolved solids.
7. Pipet 10–20 μL of the digests onto the 96-well plates. I use a pattern where each row will have one blank injection, three injections of the first sample, another blank injection, three injections of the second sample, and a third blank injection, followed by three injections of the third sample. Thus, every row should have three samples in 12 injections. The blank injection is necessary to diminish carryover between samples.
8. Dilute the digests to 200 μL with 2% nitric acid containing gallium, enough to give a final concentration of 50 ppb Ga. The final nitric acid concentration should be at or below 10% v/v.

3.2 Instrument Preparation and Data Acquisition

1. Start the instrument and let the ICP-MS "warm up" for at least 20 min. The Micromist nebulizer should be operating in self-aspiration mode with 2–3% (v/v) HNO_3 at a flow of ~250 μL/min.
2. Tune the instrument according to the manufacturer's instructions (1 ppb Li, Y, Tl, Ce, and Co is recommended by Agilent for the 7500 and later Series). Allow for most of the ICP-MS parameters to be automatically adjusted by the AutoTune routine.
3. Calibrate the detector for the pulse to analog transitions using the recommended P/A solution, also from the manufacturer (100 ppb solution provided by Agilent or other suppliers).
4. Replace the sample tubing with a 0.38 mm ID with three stops of green/orange color (Elemental Scientific or other suppliers) connecting the rinse solution (2% HNO_3) and to the injection valve (port #3). It is important to keep separate tubing for tuning and detector and sample analysis runs to prevent cross-contamination. The tubing connecting port #2 from the valve to the nebulizer should not be the same as the tubing used for tuning and P/A solutions (*see* Fig. 2).

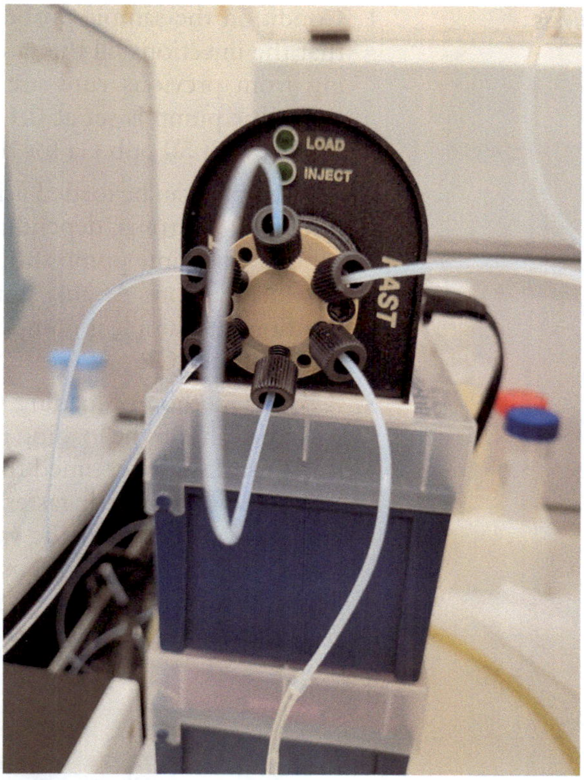

Fig. 2 Elemental Scientific injection valve accessory. The tubing at the bottom right is attached to the carrier solution (2% HNO_3) to port #3. Port #2 goes to the ICP-MS nebulizer inlet. Ports #5 and #6 are attached to the ALS probe and injection syringe, respectively. Ports #1 and #4 are connected to the sample loop (55 uL in this case)

5. The typical parameters for the ICP Tune file are forward power, 1500 W; reflected power, <10 W; sample depth, 8–9 mm; carrier flow, 1.0–1.2 L/min; make-up flow, 0.10.2 L/min; nebulizer pump rate, 0.1 revolutions per second; S/C temperature, 2 °C.
6. The following ICP parameters for the 7500 series are optimal for iron analysis: Extract lens-1, 0–1 V; extract lens-2, –100–130 V; Omega Bias-ce (adjust for sensitivity), −10 to –20 V; Omega Lens-ce, 0 to −4 V (to reduced doubly charged and oxide species below 2%); Cell entrance voltage, −30 V; QP focus, −8 V; cell exit voltage, −38 V; Octopole RF, 170 V; Octopole Bias, −18 V; QP Bias, −15 V. The parameters for the collision cell are He gas flow, 1.5 mL/min; H_2 gas flow, 3.5 mL/min. No optional gas is needed so the flow should be set at zero.
7. Save tuning parameters as a separate file, such as Mix.U. The normal tuning parameters usually save as file Nogas.U.

3.3 Analysis

1. Condition the sample probe by performing 5–6 injections so that the injections fill the sample loop and contaminants coming from previous runs are no longer present. The sample peristaltic pump is set at 0.1 rps. Use the blank diluent of 2% HNO_3 with 50 ppb Ga for these injections.
2. The samples can be loaded using a separate peristaltic pump at a rate of 50–76 rpms, depending on the internal diameter of the tubing. The same pump should be running at ~2 rpms during data acquisition. At this point, the valve is in the injection position; the sample is pumped at 0.1 rps by the main peristaltic pump of the ICP.
3. A better method of loading the samples into the sample loop, with the valve in the loading position, is to use a single mechanical 2.5-mL syringe, timed to aspirate at a rate of 2.0 mL/min. The same flow rate or faster can be used to aspirate the rinse washes (2). Thereafter the valve in front of the syringe can be switched from load to inject so that the syringe can be emptied. The ALS software, separate from the ICP-MS software, controls the loading syringe (*see* Fig. 3).

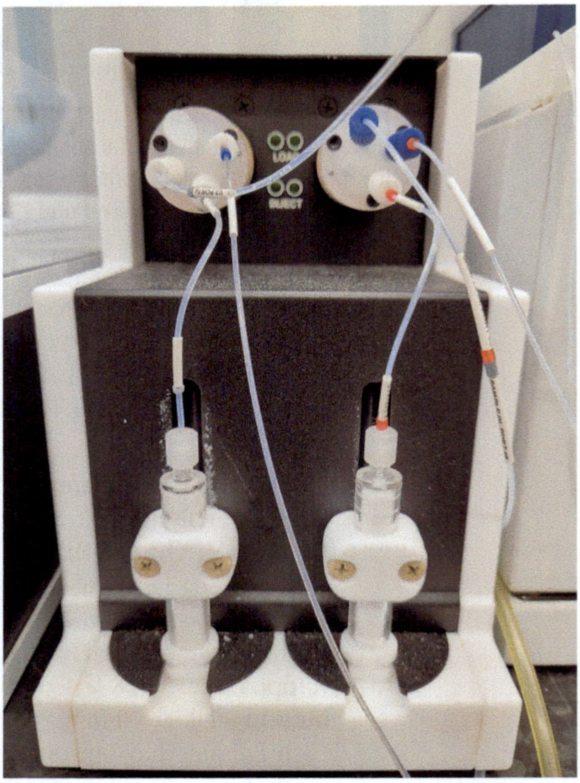

Fig. 3 Elemental Scientific injection syringe with valves. The left-side syringe is used to aspirate the solutions from the well plate and the rinse stations. The 2-position valve is used to direct the outflow to the waste container to empty the syringe

4. The injection valve is a six-port PTFE valve that has two positions (*see* Fig. 2). The sample loop (50–60 μL) is attached to ports #1 and #4. The carrier flow coming from the ICP peristaltic pump is attached to port #3; the outflow of the valve and into the ICP nebulizer is attached to port #2. Connect port #6 to the ALS probe and connect port #5 either to the additional loading peripump or to the 2.5 mL loading syringe.
5. The rinse solution (2% HNO_3) is pumped from a plastic carboy into the bottom of the rinse wells of the ALS. I normally use the ACS grade acid, diluted with distilled deionized water for this rinse.
6. Create a sample sequence in the Agilent ICP-MS software and run the samples after conditioning the probe of the autosampler. For the acquisition method, I use an 18 element protocol that includes the ^{56}Fe and ^{57}Fe isotopes, each monitored for three cycles of 100 ms each, with the same acquisition for the other elements and the internal standard.
7. The method is calibrated separately (either before or after the sample run) with a range of Fe concentrations. I typically use a stock of 10 μg/mL (10 ppm) that is then diluted to 0.333, 1.0, 3.33, 10, 33.3, 100, 333, and 1000 μg/L (ppb). I dilute the stock 10 times and 30 times for the two highest concentrations directly in the plate, followed by tenfold dilutions of those first dilutions for the subsequent dilutions (33.3 and 100 ppb). This cycle is repeated two times more. A couple of diluents (2% HNO_3 with 50 ppb Ga) injections are placed in front of the calibrator solution wells. The injection sequence goes from the blanks to the highest standards. I finish the sequence with a blank, followed by a QCS-50 injection (at 50% of the maximal concentration or 500 ppb), and followed by two blank injections to clear the sample pathway (*see also* **Note 2**).
8. The Offline Data Analysis is used to plot the calibrations, enter the calibrator concentrations, and set up the processing of the data. The calibration regression coefficient should ideally be 0.999, but not lower than 0.99. The limit of detection should be below 0.1 ppb with a background level equivalent or less than 1.0 ppb (1 ug/mL) (*see also* **Note 3**).
9. The sample concentrations are calculated by the ICP-MS software by interpolation. The concentration units are the ones entered during the calibration. The calculation of the concentrations happens on the fly as samples are injected and imported into the Agilent spreadsheet of the ChemStation software. A correction for isotopic distribution can be implemented into the analysis method (*see* **Note 4**).
10. The ICP-MS data acquisition and sample acquisition can be carried out on separate days, but the tuning parameters of the

instrument must not be altered. If running samples for more than 1 day then a sample injection of the calibration curve is included as part of the sample run (quality control sample or QCS).

11. Manually perform 2 blank injections at the end of the experimental run. Disconnect the injection valve from the Nebulizer input; reattach the tubing that you used for tuning. Pass enough 2% HNO_3 (without internal standard) to flush all contaminants that could have come from the samples during the run, followed by trace-metal grade water.
12. Most but not all the steps have to be carried out for Fe analysis in similar ICP-MS instruments. The method development and analysis will however be different for those instruments (*see* **Note 5**).

4 Notes

1. Mammalian cells can be plated onto the 96-well plates and grown to a density of 20 K–50 K per well, using flat bottom well plates. The digestions can be performed directly in the plates, after the careful removal of the supernatant, media using 5% HNO_3 at 65 °C for 6–8 h [6]. This method allows for multiple knockouts and knockdown screens that can reveal novel roles of genes in element trafficking, acquisition, and elimination (ionomics).
2. The method is suitable for up to 20 elements. However, the range of concentrations for the calibration is going to depend on the actual elements that are being monitored and the expected concentrations. For cellular and tissue samples, the following range of calibrations is recommended: 100 ppm Na, P, S, and K; 10 ppm for Mg; 1 ppm for Mn, Fe, Cu, and Zn; and 100 ppb for trace elements like Li, B, Al, Ti, V, Cr, Ni, Co, As, Se, Mo, and Cd [7, 8].
3. The limits of detection (DL) depend on the RSD values for the calibration blank injections and are equal to 3xRSD/slope of the calibration curves (CPS/ppb). The background equivalent concentration (BEC) is the CPS for the blank/slope of the calibration. The limits of quantification (LOQs) for an isotope are estimated from the equation $LOQ = 3.33 \times DL + BEC$.
4. The methods assume that the isotopic distribution of ^{56}Fe and ^{57}Fe is the same for samples and standards, usually the natural abundance. Thus, the calibrator concentrations do not need correction for isotopic distribution. However, the actual concentrations of the two isotopes can be "separated" to get actual

isotope concentrations. This allows the use of this method in Fe studies of uptake and excretion. In such a case, use ^{57}Fe as the isotope tracer.

5. The method described here was developed and used on an Agilent 7500 series. A single quadrupole instrument from another brand should work without many discrepancies from what I presented here, except perhaps the use of different collision/reaction gases (Perkin Elmer instruments use UHP ammonia gas). Newer triple quadrupole (QqQ) instruments improve the method sensitivity as the addition of a second quadrupole ensures that the calibrations are independent of the sample matrix. He gas is used as collision gas in the collision cell [9].

Acknowledgments

I am very grateful for the support from the Redox Biology Center and the Department of Biochemistry at the University of Nebraska-Lincoln for my work.

References

1. Sarpong-Kumankomah S, Gailer J (2021) Application of a novel metallomics tool to probe the fate of metal-based anticancer drugs in blood plasma: potential, challenges and prospects. Curr Top Med Chem 21(1):48–58. https://doi.org/10.2174/1568026620666200628023540
2. Lindahl PA, Moore MJ (2016) Labile low-molecular-mass metal complexes in mitochondria: trials and tribulations of a burgeoning field. Biochemistry 55(30):4140–4153. https://doi.org/10.1021/acs.biochem.6b00216
3. Pluháček T, Lemr K, Ghosh D, Milde D, Novák J, Havlíček V (2016) Characterization of microbial siderophores by mass spectrometry. Mass Spectrom Rev 35(1):35–47. https://doi.org/10.1002/mas.21461
4. Tan SH, Horlick G (1986) Background spectral features in inductively coupled plasma/mass spectrometry. Appl Spectrosc 40(4):445–460
5. Seravalli J (2012) Inductively coupled plasma–mass spectrometry as a tool for high-throughput analysis of plants, Methods in molecular biology, pp 269–288
6. Malinouski M, Hasan NM, Zhang Y, Seravalli J, Lin J, Avanesov A, Lutsenko S, Gladyshev VN (2014) Genome-wide RNAi ionomics screen reveals new genes and regulation of human trace element metabolism. Nat Commun 5(1): 3301. https://doi.org/10.1038/ncomms4301
7. Siming Ma S-GL, Kim EB, Park TJ, Seluanov A, Gorbunova V, Buffenstein R, Seravalli J, Gladyshev VN (2015) Organization of the mammalian ionome according to organ origin, lineage specialization, and longevity. Cell Rep 13(7): 1319–1326
8. Zhang B, Podolskiy DI, Mariotti M, Seravalli J, Gladyshev VN (2020) Systematic age-, organ-, and diet-associated ionome remodeling and the development of ionomic aging clocks. Aging Cell 19(5):e13119. https://doi.org/10.1111/acel.13119
9. Konz T, Migliavacca E, Dayon L, Bowman G, Oikonomidi A, Popp J, Rezzi S (2017) ICP-MS/MS-based ionomics: a validated methodology to investigate the biological variability of the human ionome. J Proteome Res 16(5): 2080–2090. https://doi.org/10.1021/acs.jproteome.7b00055

Chapter 3

Sample Preparation for X-Ray Fluorescence Microscopy of Iron Distribution in Biological Specimen

Sophia Miller and Martina Ralle

Abstract

Characterizing the two- and three-dimensional distribution of trace metals in biological specimens is key to better understand their role in biological processes. Iron (Fe) is of particular interest in these trace metals due to its widespread role in maintaining cellular health and preventing disease. X-ray fluorescence microscopy (XFM) is emerging as the method of choice for investigators to interrogate the cellular and subcellular distribution of Fe. XFM utilizes the intrinsic X-ray fluorescence properties of each element to produce quantitative 2D and 3D distributions of trace metals within a sample. Herein, methods for sample preparation of cells and tissue for the determination of Fe distribution by XFM are described.

Key words X-ray fluorescence microscopy, Trace biological metals, Iron

1 Introduction

Trace metals, in general and iron (Fe), specifically play an essential role in biological processes across all three kingdoms of life. Proper Fe homeostasis is essential for maintaining cellular health and preventing diseases. Much effort has gone into characterizing the structure and function of metalloproteins, determining cellular and tissue concentration of Fe, and elucidating signaling pathways in which Fe is involved. Investigating the two- and three-dimensional distribution of Fe in biological specimens is more complex and "imaging" methods for trace metals are still far from routine. Over the past two decades, synchrotron-based X-ray fluorescence microscopy (XFM) has become the method of choice for investigators to determine the cellular and subcellular distribution of Fe. Examples are Fe accumulations in the reticuloendothelial system in macrophages of mice [1] and the algae *Chlamydomonas reinhardtii* [2]. In this chapter, we will provide detailed protocols on how to prepare biological specimens for XFM measurements.

Oleh Khalimonchuk (ed.), *Iron Metabolism: Methods and Protocols*, Methods in Molecular Biology, vol. 2839,
https://doi.org/10.1007/978-1-0716-4043-2_3,

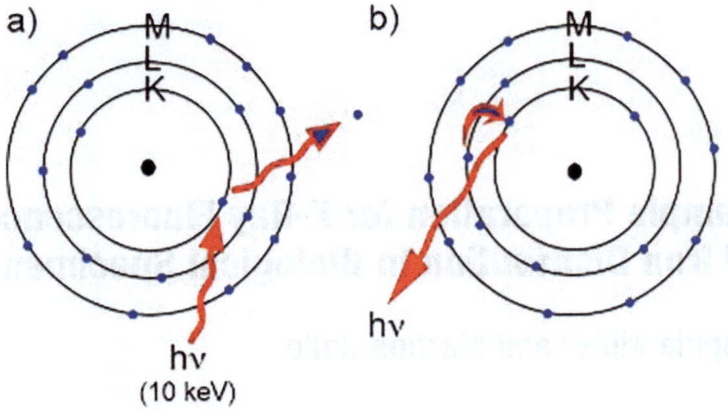

Fig. 1 Principle of X-ray fluorescence. (**a**) An incoming photon produces a photoelectron from a K-shell electron leaving a hole in the K shell. (**b**) An electron from the L shell fills the hole and emits the extra energy as element-specific fluorescent photon

1.1 Principle of X-Ray Fluorescence Microscopy

XFM is ideally suited to obtain elemental distributions in biological specimens [1–7]. XFM takes advantage of the intrinsic X-ray fluorescence properties unique to each element. During an XFM experiment, high-energy X-rays emit an inner shell electron (typically from the K- or L- shell) into the continuum thus creating a vacancy in that shell. X-ray fluorescent photons are generated when this vacancy is repopulated by an electron from a higher energy shell (Fig. 1). The energy of each photon is dependent on the number of protons and the oxidation state of the element. XFM is, therefore, element specific and quantitative, and can be used for element speciation.

1.2 XFM Is a Scanning Technique

Samples are raster scanned in x-y, and the X-ray fluorescence for each pixel is recorded with an energy-dispersive detector. The typical excitation energy of 10 keV for biological experiments will produce 2D distributions ("Maps") for elements from Na to Zn based on their K-absorption line (Fig. 2). Biological XFM experiments are typically carried out at synchrotron sources to take advantage of the high brilliance of the X-rays, but, benchtop instruments with micrometer resolution are increasingly available. Currently, available beamlines at synchrotrons offer spatial resolutions ranging from the tens of μm (microprobes) to 50 nm (nanoprobes) and beyond.

Examples of low- and high-resolution elemental maps are shown in Figs. 3 and 4.

As XFM is a scanning technique, the scanned area and the resolution are often connected. Larger scans at a higher resolution will accumulate more points and require longer scan times. High-resolution scans of 100 nm or less are usually performed on cells, while medium- and low-resolution scans (>500 nm) are most often used for tissue sections.

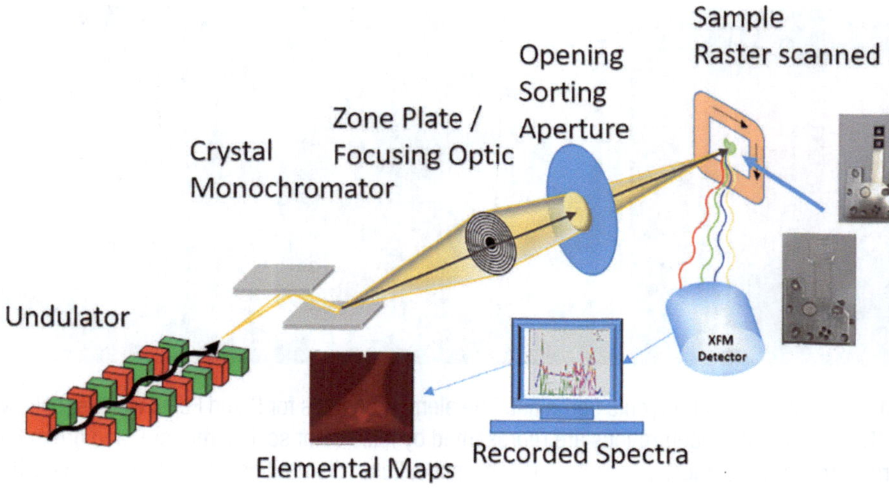

Fig. 2 Schematic of an XFM experiment

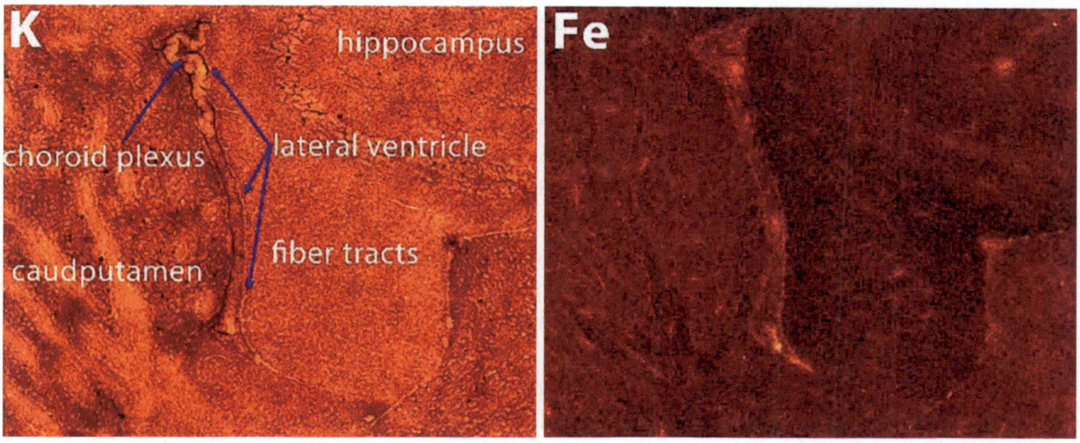

Fig. 3 XFM scan of a saggital wild type mouse brain section. The potassium (K) and iron (Fe) elemental maps are displayed in the red temperature false coloring mode. As expected, Fe levels are elevated in the choroid plexus. Images were recorded at beamline 2-IDE at the Advanced Photon Source. Pixel size 1.5 μm

Ultimately, the choice of beamline depends on the scientific question and the biological specimen to be imaged. For example, to visualize regional differences in Fe distributions in different areas of the brain, experiments would be carried out with microprobes offering 0.5–50 μm resolution (Fig. 3), while visualization of elements in organelles or single cells is best performed with nanoprobes with 15–100 nm resolution (Fig. 4).

1.3 Access to Synchrotrons

XFM imaging of biological specimen will, in most cases, require synchrotron facilities that generate coherent X-rays with high brilliance (high number of photons per pixel). Synchrotrons are usually part of national user facilities and access is provided via a merit-

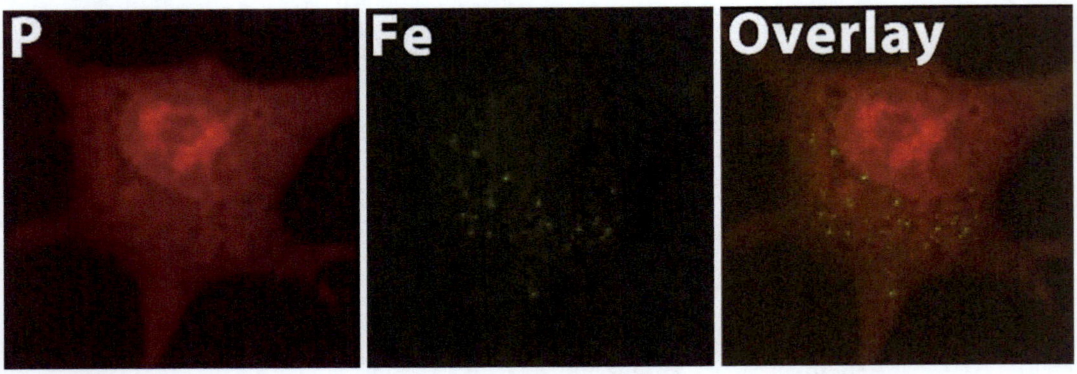

Fig. 4 XFM scan of a mouse embryonic fibroblast. The elemental maps for P and Fe as well as their overlay are displayed. The elemental concentrations are represented by false color scales (red for P and green for Fe) with dark to bright for least to most concentrated. The subcellular distribution of Fe is punctate with a slight preference for perinuclear localization. Data were collected at beamline 2-IDD at the Advanced Photon Source. Pixel size 300 nm

based proposal system. Most review committees at synchrotrons evaluate and score beamtime proposals 2–3 times a year, according to the number of their use cycles. In the United States, the most widely used synchrotrons used for XFM experiments are the Advanced Photon Source at the Argonne National Laboratory, the National Synchrotron Lightsource II at Brookhaven National Laboratory, and the Stanford Synchrotron Radiation Lightsource. Potential users are encouraged to research the specific imaging capabilities of each synchrotron light source and contact beamline staff prior to planning an experiment.

1.4 Sample Preparation for XFM Experiments

Appropriate sample preparation for XFM experiments is crucial to obtain meaningful data and therefore differs for different configurations. Depending on the samples (cells or tissue), how they were harvested (chemical fixation or cryo-fixation), and whether they will be measured at cryogenic or room temperature, the sample preparation procedures have to be adjusted accordingly. The methods section describes the currently two most often used scenarios, preparation of fixed cells and frozen tissue for XFM experiments conducted at room temperature. The described procedures are derived from traditional methods to prepare cells and tissues for visual light microscopy. Because XFM is a sensitive technique (detection limits are typically attograms per voxel), one important aspect of sample preparation is to avoid anything that can introduce Fe (or other trace metal) impurities onto the specimen or the sample mounts.

For cells, sample preparation begins with culturing the cells to harvest density. Adherent cells are seeded directly onto XFM sample supports (silicon nitride, SiN). Once grown to the desired density, cells are chemically fixed, washed, and dried. Nonadherent cells are deposited onto SiN windows after fixation and washes (not described here). For cryo-fixation, cells grown on SiN sample holders are blotted and submersed in liquid ethane using a plunge freezer or a vitrobot. Cells are then kept at liquid nitrogen temperature until transfer into the experimental sample chamber at the beamline. The process of cryo-fixation is described elsewhere [2].

Tissue preparation from frozen blocks starts with harvest and embedding a piece of the tissue into optimum cutting temperature (OCT). The OCT-embedded specimen is then flash-frozen in dry-ice-cooled isopentane for fast heat transfer. Subsequent sectioning using a cryostat and deposition of the sections onto Ultralene®, the preferred mounting media for XFM experiments, are followed by bringing the sample up to room temperature and drying. The sections are then inspected with brightfield and attached to sample holders that will satisfy the requirements of the beamline.

2 Materials

2.1 Materials for Preparation of Tissue

1. Cryomolds (Tissue-Tek Cryomolds).
2. Optimum cutting temperature (OCT) compound.
3. Dry-ice-cooled isopentane in a small dewar.
4. Tweezers.
5. Cryostat for cryo-sectioning.
6. Glass slides.
7. Ultralene® (SPEX™ SamplePrep).
8. Upright microscope for the inspection of the sections and selecting regions of interest to scan.

2.2 Materials for Preparation of Cells on SiN Windows

1. Silicon nitride (SiN) membrane windows (Norcada, frame size: 5 mm x 5 mm, membrane size: 1.5 mm × 1.5 mm, membrane thickness: 500 nm, frame thickness 200 μm).
2. Biosafety cabinet.
3. Thin tip, negative action tweezer.
4. Curved tip tweezers.
5. Three cm cell culture plates or 6-well culture plates.
6. 0.01% poly-L-lysine solution.
7. 200 μL and 1 mL micropipettes.
8. Incubator at 37°C and 5% CO_2 for mammalian cells.
9. Culture medium.
10. Sterile phosphate-buffered saline (PBS) for cell culture.
11. Tryp LE or trypsin to detach cells.

12. Table top centrifuge.
13. Cell counter.
14. 4% paraformaldehyde in PBS.
15. 0.1 M isotonic ammonium acetate solution.
16. Doubly deionized water.
17. Filter paper, cut into circles to fit into a 3 cm culture dish.
18. Upright microscope for selection of cells to scan.

3 Methods

3.1 Preparation of Rodent Tissue for XFM

1. Harvest tissue from saline-perfused animals as usual.
2. Select or cut a block of tissue that does not exceed 10 mm in any dimension. Most sample holders at microprobe beamlines are 6 mm × 6 mm or smaller (*see* **Note 1**).
3. Place the tissue block into an appropriately sized cryo-mold with a small amount of OCT compound added to the bottom. Orient the tissue such that the face to be sectioned is positioned on the bottom of the cryo-mold. Label the cryo-mold with a solvent-resistant marker.
4. Prepare the coolant: Carefully add isopentane to dry ice into a dewar until there is enough liquid for submersion of the cryo-mold (*see* **Note 2**).
5. After positioning the tissue in the mold, add OCT until the entire mold is filled or all the tissue is covered.
6. Pick up one corner of the cryo mold with a pair of tweezers and immerse the cryo-mold in dry-ice-cooled isopentane for 5–10 s (*see* **Note 3**). Hold onto the mold with tweezers while freezing.
7. Place the frozen cryo-mold containing the frozen tissue onto dry ice and, if needed, add more OCT (*see* **Note 4**). Store the samples at −80 °C until sectioning.
8. Before sectioning, remove the OCT block from the cryo-mold and mount it onto an appropriate holder for the cryostat.
9. Prepare Ultralene® wrapped glass slides: Unroll some of the Ultralene® and tightly wrap some around a glass slide. Tape the Ultralene® in the back to keep the film tight on the slide with as few wrinkles as possible (*see* **Note 5**).
10. Cut 10 μm thick sections of the prepared tissues in a cryostat (*see* **Note 6**) and deposit onto slides wrapped with Ultralene® (*see* **Note 7**).
11. Air-dry sections and store them in a covered slide box for 1–2 weeks. For prolonged storage, freeze at −80 °C.

3.2 Preparation of Adherent Cells on Silicon Nitride Windows for XFM

1. Using clean, negative pressure tweezers, carefully remove a silicon nitride (SiN) windows from the gelatin storage capsule and place it in a 3 cm sterile cell culture dish with the membrane facing up (*see* **Note 8**). Use one dish per window. Always hold onto the SiN window by the frame and avoid touching the membrane at all costs. It is extremely fragile.
2. Sterilize the SiN windows by exposing them uncovered to UV for 10 min in the biosafety cabinet.
3. Pipette 20 μL of sterile 0.01% poly-L-lysine solution onto the 1.5 mm × 1.5 mm membrane portion of the SiN window with a long reach or gel loading tip and a 200 μL micropipette. Add only enough poly-L-lysine to cover the membrane. Be careful to not touch the membrane.
4. Incubate the windows at 37 °C for 30 min in closed cell culture dishes.
5. Add additional 20 μL of poly-L-lysine solution to the SiN windows and then remove all of the poly-L-lysine from the membrane.
6. Keep the SiN windows sterile by sealing the culture dish with parafilm until you are ready to add cells (*see* **Note 9**).
7. Grow cells to 70–75% confluency in culture dishes as usual. Wash cells with PBS, and detach cells using TrypLE or your method of choice. Preserve the spent medium.
8. Count and then pellet the cells by centrifugation for 10 min at 200× *g*. Remove the detachment medium.
9. Resuspend the cells in 50% new and 50% spent growth culture medium (*see* **Note 10**) by pipetting them up and down several times to prevent clumping.
10. Ready the poly-L-lysine treated SiN windows in the biosafety cabinet by removing the parafilm and lid from the cell culture dishes.
11. Use a 200 μL micropipette with a long reach or gel loading tip and add ~20 μL of the cell suspension onto the membrane of each SiN window.
12. Cover the cell culture dish and allow the cells to adhere in the incubator at 37 °C for 10 min. Disturb the cells/windows as little as possible after this point. Handle the dishes gently and with care.
13. Dilute the remaining resuspended cells with culture medium 1: 1. Gently add this mixture to cover the entire surface of the cell culture dish (*see* **Note 11**).
14. Allow the cells to fully adhere and recover in the incubator overnight before proceeding with the experiment.

15. If any additional exposures are needed, work carefully to not to disturb the SiN window and surrounding cells. Add any chemicals to one corner of the plates, and mix the media gently by pipetting, rather than rocking (*see* **Note 12**).
16. After the end of the experiment, remove the cell culture media from the cell culture dish using a 1 mL pipette. As before work carefully so the windows move as little as possible around.
17. Fix cells by adding 4% paraformaldehyde in PBS to fully cover the SiN windows (approximately 1 mL) and incubating at 37 ° C for 30 min (*see* **Note 13**).
18. Incubate SiN windows twice with PBS, followed by isotonic ammonium acetate solution (made fresh) and doubly deionized water.
19. After the final rinse, transfer the SiN windows to a new 3 cm culture dish containing filter paper on the bottom.
20. Cover the dish and let the windows dry completely on the filter paper. The filter paper will absorb all residual moisture and prevent the window from sticking to the dish (*see* **Note 14**).
21. After the windows are completely dried, transfer them back into the gel capsules (*see* **Note 15**).
22. To select cells for the XFM experiment, remove the SiN window from the gel capsule, and place it with the membrane side up onto a glass slide. Carefully mark one corner of the window with a sharpie and mount the slide onto an upright microscope.
23. Using a 10× objective, take multiple images with the cells in focus covering the entire area of the membrane. Save the images and return the SiN window to the gelatin capsule.
24. Prior to the XFM experiment, the individual images should be "stitched" back together and suitable cells selected. The marked corner can be used as a fiducial and will be helpful when identifying selected cells during the experiment.

4 Notes

1. The maximum possible scan area in an XFM experiment is usually limited. It is best to check with a beamline scientist at the synchrotron facility and inquire about maximum and minimum sample sizes.
2. This is best done in a small dewar.
3. It is important to flash freeze the samples long enough to preserve tissue morphology but not too long to avoid damage from isopentane.

4. If cracks develop as part of the freezing process more OCT should be added while the sample is on dry ice.
5. This will take a few tries. For this step, it is mandatory to use clean gloves and not touch the front of the slide where the section will be put on. XFM is very sensitive and impurities from fingers or even gloves are easily visible. Ultralene® albeit thin (6 μm) is difficult to cut. It is advisable to use double-edged razor blades (with the appropriate caution) when cutting Ultralene®.
6. The temperature of the cryostat should be set to −20 °C initially and adjusted if the tissue is too soft or too brittle.
7. Depending on the size of the tissue 1–3 sections can be placed on one slide.
8. Look for the "even" surface, this is the side cells should be placed on. The backside looks slightly "carved" out.
9. .At this point the windows might stick to the bottom of the culture dish—however, they will come loose again when media are added. Do not try to move or touch the SiN windows at this point.
10. When resuspending cells consider which experiments you still need to perform and draw off aliquots as needed. If no further experiments have to be done, resuspend the cells in the same amount of culture media as they were harvested from (for 70–75% density). If additional exposures of more than 24 h are planned, consider the growth rate for your resuspension volume accordingly.
11. Do not pipette directly onto the SiN window. Make sure that the window is completely submerged in a growth medium. Continue to minimize any movement of the window within the plates so cells can properly attach and grow on the SIN window. Do not pipette directly onto the SiN window. Make sure that the window is completely submerged in a growth medium. Continue to minimize any movement of the window within the plates so cells can properly attach and grow on the SIN window.
12. Make sure the cells are attached and healthy on the SiN windows before proceeding. Cells do not attach to SiN windows as tightly as to culture plates. Be cautious and minimize disturbances to the windows.
13. It is advised to check windows after each of the fixing and rinsing steps to ensure that the cells are still attached to the SiN membrane.

14. If the SiN windows are dried flat onto the surface in the culture dish despite best efforts, add some water and let them incubate for 2–3 min. Do not use force to loosen the window, as this may destroy the delicate membrane surface.
15. Handling SiN windows when removing them or returning them to the capsules can be done using tweezers (negative pressure tweezers work best). The windows can also be held on their thin edges between thumb and index finger.

References

1. Pek RH, Yuan X, Rietzschel N, Zhang J, Jackson L, Nishibori E, Ribeiro A, Simmons W, Jagadeesh J, Sugimoto H, Alam MZ, Garrett L, Haldar M, Ralle M, Phillips JD, Bodine DM, Hamza I (2019) Hemozoin produced by mammals confers heme tolerance. eLife 8:e49503. https://doi.org/10.7554/eLife.49503
2. Schmollinger S, Chen S, Strenkert D, Hui C, Ralle M, Merchant SS (2021) Single-cell visualization and quantification of trace metals in Chlamydomonas lysosome-related organelles. Proc Natl Acad Sci USA 118:e2026811118. https://doi.org/10.1073/pnas.2026811118
3. Victor TW, O'Toole KH, Easthon LM, Ge M, Smith RJ, Huang X, Yan H, Chu YS, Chen S, Gursoy D, Ralle M, Imperiali B, Allen KN, Miller LM (2020) Lanthanide-binding tags for 3D X-ray imaging of proteins in cells at nanoscale resolution. J Am Chem Soc 142:2145–2149. https://doi.org/10.1021/jacs.9b11571
4. Leary SC, Ralle M (2020) Advances in visualization of copper in mammalian systems using X-ray fluorescence microscopy. Curr Opin Chem Biol 55:19–25. https://doi.org/10.1016/j.cbpa.2019.12.002
5. Domart F, Cloetens P, Roudeau S, Carmona A, Verdier E, Choquet D, Ortega R (2020) Correlating STED and synchrotron XRF nano-imaging unveils cosegregation of metals and cytoskeleton proteins in dendrites. elife 9. https://doi.org/10.7554/eLife.62334
6. Davis CI, Gu X, Kiefer RM, Ralle M, Gade TP, Brady DC (2020) Altered copper homeostasis underlies sensitivity of hepatocellular carcinoma to copper chelation. Metallomics 12:1995. https://doi.org/10.1039/d0mt00156b
7. Schmidt K, Bari B, Ralle M, Washington-Hughes C, Muchenditsi A, Maxey E, Lutsenko S (2019) Localization of the locus coeruleus in the mouse brain. J Vis Exp 145:e58652. https://doi.org/10.3791/58652

Chapter 4

Proximity Ligation Assay for the Analysis of Iron-Mediated Protein–Protein Interactions in the Nucleus

Lorena Novoa-Aponte and Caroline C. Philpott

Abstract

Iron forms essential cofactors used by many nuclear enzymes involved in genome maintenance. However, unchaperoned nuclear iron may represent a threat to the surrounding genetic material as it promotes redox toxicity that may affect DNA integrity. Safely handling intracellular iron implies metal transfer and cofactor assembly processes based on protein–protein interactions. Identifying those interactions commonly occurs via high-throughput approaches using affinity purification or proximity labeling coupled with mass spectrometry analysis. However, these methods do not identify the subcellular location of the interactions. The one-on-one confirmation of proposed nuclear interactions is also challenging. Many approaches used to look at protein interactions are not tailored for looking at the nucleus because the methods used to solubilize nuclear content are harsh enough to disrupt those transient interactions. Here, we describe step-by-step the use of Proximity Ligation Assay (PLA) to analyze iron-mediated protein–protein interactions in the nucleus of cultured human cells. PLA allows the subcellular visualization of the interactions via the in situ detection of the two interacting proteins using fluorescence confocal microscopy. Briefly, cells are fixed, blocked, permeabilized, and incubated with primary antibodies directed to target proteins. Primary antibodies are recognized using PLA probes consisting of one PLUS and one MINUS oligonucleotide-labeled secondary antibody. If the two proteins are close enough (<40 nm), the PLA probes are ligated and used as the template for rolling circle amplification (RCA) with fluorescently labeled oligonucleotides that yield a signal detectable using fluorescence confocal microscopy. A fluorescently labeled membrane-specific stain (WGA) and the DNA-specific probe DAPI are used to identify cellular and nuclear boundaries, respectively. Confocal images are then analyzed using the CellProfiler software to confirm the abundance and localization of the studied protein–protein interactions.

Key words Proximity ligation assay, Protein–protein interactions, Subcellular localization, Iron cofactors, Iron trafficking, Genome maintenance

1 Introduction

Iron is an essential nutrient because it is used to form cofactors that are incorporated into iron-dependent proteins. Cells have hundreds of proteins containing the redox-active iron cofactors heme, iron–sulfur (Fe-S) clusters, and mono- and di-nuclear (non--heme) iron centers. These are extensively used for the catalysis of

Oleh Khalimonchuk (ed.), *Iron Metabolism: Methods and Protocols*, Methods in Molecular Biology, vol. 2839,
https://doi.org/10.1007/978-1-0716-4043-2_4,

intracellular reactions. Yet, the potential for redox activity also accounts for the toxicity of uncoordinated iron in a cellular milieu filled with oxidation-sensitive proteins, nucleic acids, and lipids. Cells have evolved sophisticated mechanisms for the safe assembly and distribution of iron cofactors, thereby supporting metabolism, genome integrity, cell growth, and proliferation [1–3]. Those metabolic processes supported by iron cofactors occur in multiple intracellular organelles, including the cytoplasm, mitochondria, endoplasmic reticulum, and nucleus. Still, the molecular characters distributing and maintaining the different iron pools between those compartments have yet to be studied in detail. Most studies have focused on the mitochondrial and cytoplasmic iron pools [1, 4, 5], and very few have confirmed iron ions' presence in the nucleus [6, 7]. Whether there is a pool of labile/exchangeable nuclear iron is still unknown. This is partly due to the lack of experimental approaches tailored to analyze metal trafficking in the nuclear compartment.

The best evidence for the presence of iron in the nucleus is the existence of many iron-dependent enzymes that reside in the nucleus. Those include Fe(II) center-binding enzymes, such as the DNA demethylases, DNA nucleases, ribonucleotide reductase, and histone demethylases, and Fe-S cluster-binding proteins, such as DNA primase, the replicative DNA polymerases, DNA exonuclease, DNA glycosylases, and several DNA helicases and chromosome-associated kinesins essential for DNA and chromatin metabolism and cell division [8, 9]. These nuclear proteins bind iron cofactors while functioning in the nucleus. However, it is not fully understood how the iron cofactors are handled in the nucleus without compromising the integrity of the DNA. Studies suggest that the metallation of nuclear proteins occurs during the initial protein folding in the cytoplasm before the translocation of the holoprotein from the cytoplasm to the nucleus. For example, biogenesis and maturation of some nuclear Fe-S proteins require the cytosolic iron–sulfur cluster assembly (CIA) machinery as disruption of proteins in the CIA pathway affects the metallation and activity of target proteins (i.e., replicative DNA polymerases, DNA primase, and DNA helicases), resulting in genome instability [10, 11–13]. In fact, the late-acting CIA proteins MMS19 and CIA2B, part of the CIA targeting complex, colocalize in the nucleus with the chromokinesin KIF4A, a component of the mitotic machinery that uses Fe-S clusters for function [14]. Could this metallation occur in the nucleus?

Protein–protein interactions constitute the safest way to transfer highly reactive metal cofactors from metal delivery systems (called metal chaperones) to client proteins, by ensuring the proper delivery of cofactors to cognate apo-enzymes, protecting the cofactor from chemical attack during delivery, and preventing the cofactor from engaging in harmful redox chemistry [14]. Mapping protein–protein interactions occurring during metal transfer reactions usually starts by using iron proteins as baits to capture interactors via coimmunoprecipitation or proximity labeling approaches, followed by mass spectrometry. These high-throughput approaches give an extensive list of candidate-interacting proteins (interactomes) that may constitute a network of proteins working together to mobilize iron inside the cell [5]. However, confirming those protein–protein interactions and analyzing their intracellular localization are challenging due to the very transient nature of the interactions and the limited availability of approaches to track their occurrence.

Proximity Ligation Assays (PLA) are an attractive alternative to evaluate protein–protein interactions once interacting candidates have been identified. PLA allows the detection of pairs of proteins that are less than 40 nm apart (Fig. 1). It requires the use of highly sensitive and specific primary antibodies to detect the two target proteins suspected as interactors. The secondary antibodies used

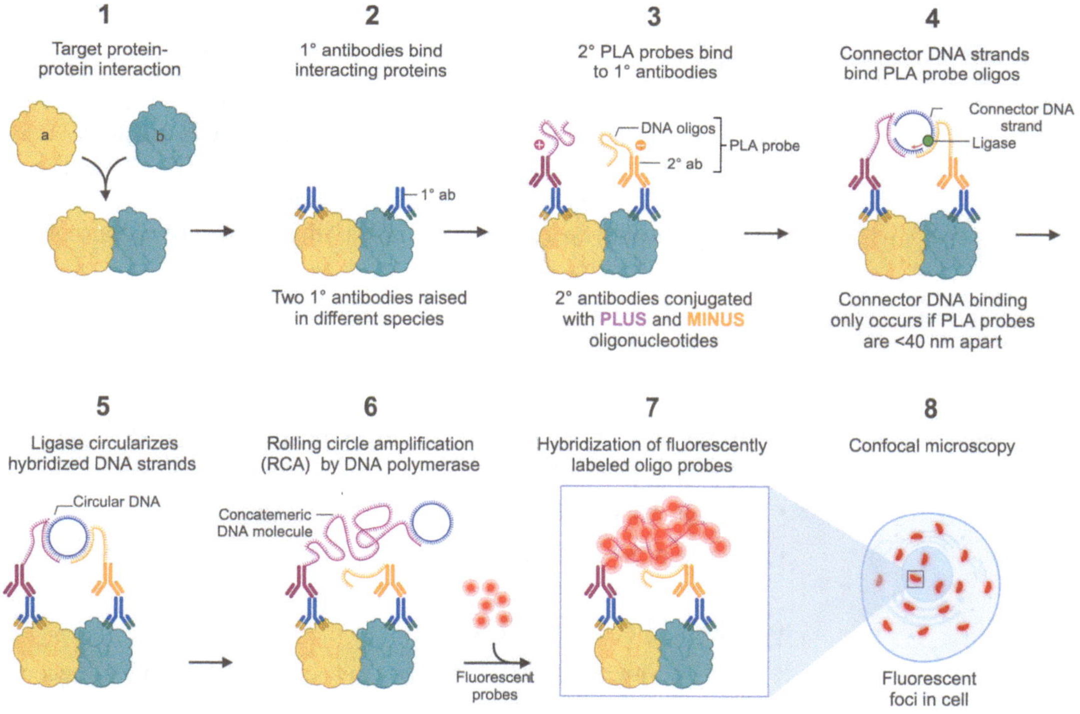

Fig. 1 Rationale behind a PLA experiment. Standard workflow of PLA experiments for analyzing iron-mediated protein–protein interactions in the nucleus. Two interacting proteins (1) are recognized using primary antibodies raised in different species (2). PLA probes consisting of oligonucleotides conjugated with secondary antibodies recognize the primary antibodies (3). One PLUS and one MINUS probe are required for each PLA assay. If the PLA probes are less than 40 nm apart, connector DNA strands will bind the PLA oligonucleotides (4) and serve as substrate for the ligation step (5). The resulting circular DNA is used as the template for RCA amplification (6) and further labeling with fluorescently labeled oligonucleotides (7) that yield a signal detectable using fluorescence confocal microscopy (8). Figure created with BioRender.com

for their recognition have oligonucleotides covalently bound (called PLA probes) that form a duplex DNA structure in the presence of a connector DNA, but only if the primary antibodies are close enough for the DNA to interact. Those are then ligated forming a circular DNA (proximity-dependent DNA ligation) that serves as a template for the amplification via rolling circle amplification (RCA). The amplified DNA is then hybridized with fluorescent probes in situ labeling the protein–protein interaction [15, 16]. Each PLA signal comprises about 1000 fluorescent oligonucleotides bound to the amplified DNA at the place of the original protein–protein interactions. This amplification of the signal allows for detecting PLA via fluorescence confocal microscopy. The main advantage of PLA is the in situ recognition of endogenous proteins (no need for cloning, overexpression, or genetic manipulations). Also convenient is the small amount of material needed, its high specificity and single molecule sensitivity due to the RCA approach, and the possibility of quantification, allowing the quick production of publication-ready results. The timescale for the single technique from cell culture to detection and quantification is approximately 1 week.

Here, we show the use of PLA to evaluate protein–protein interactions between iron proteins in the nucleus of mammalian cells (Fig. 1) and further use of the image analysis software CellProfiler [17] to model and quantify the nuclear PLA signals (Fig. 2). PLA experiments omitting one primary antibody per reaction should be carried out in parallel as negative controls to validate the results.

2 Materials

2.1 Reagents and Recipes

- Sterile 1X Phosphate-Buffered Saline pH 7.4 (PBS). Keep at room temperature.
- μ-Slide 15 Well 3D ibiTreat ibidi plates (ibidi).
- Sterile poly-D-lysine 0.1 mg/mL (ThermoFisher). Store at 4 °C.
- HEK293 cells (*see* **Note 1**).
- DMEM medium: 10% fetal bovine serum (FBS) in Dulbecco's Modified Eagle Media. Mix 50 mL of FBS with 450 mL of DMEM and add 1X of penicillin (100 units/mL) and streptomycin (100 μg/mL). Filter-sterilize using a 0.22 μm filter. Store at 4 °C and warm to 37 °C before use (*see* **Note 1**).
- Gel loading tips.
- Paraformaldehyde (PFA) 20% solution EM grade (Electron Microscopy Sciences).

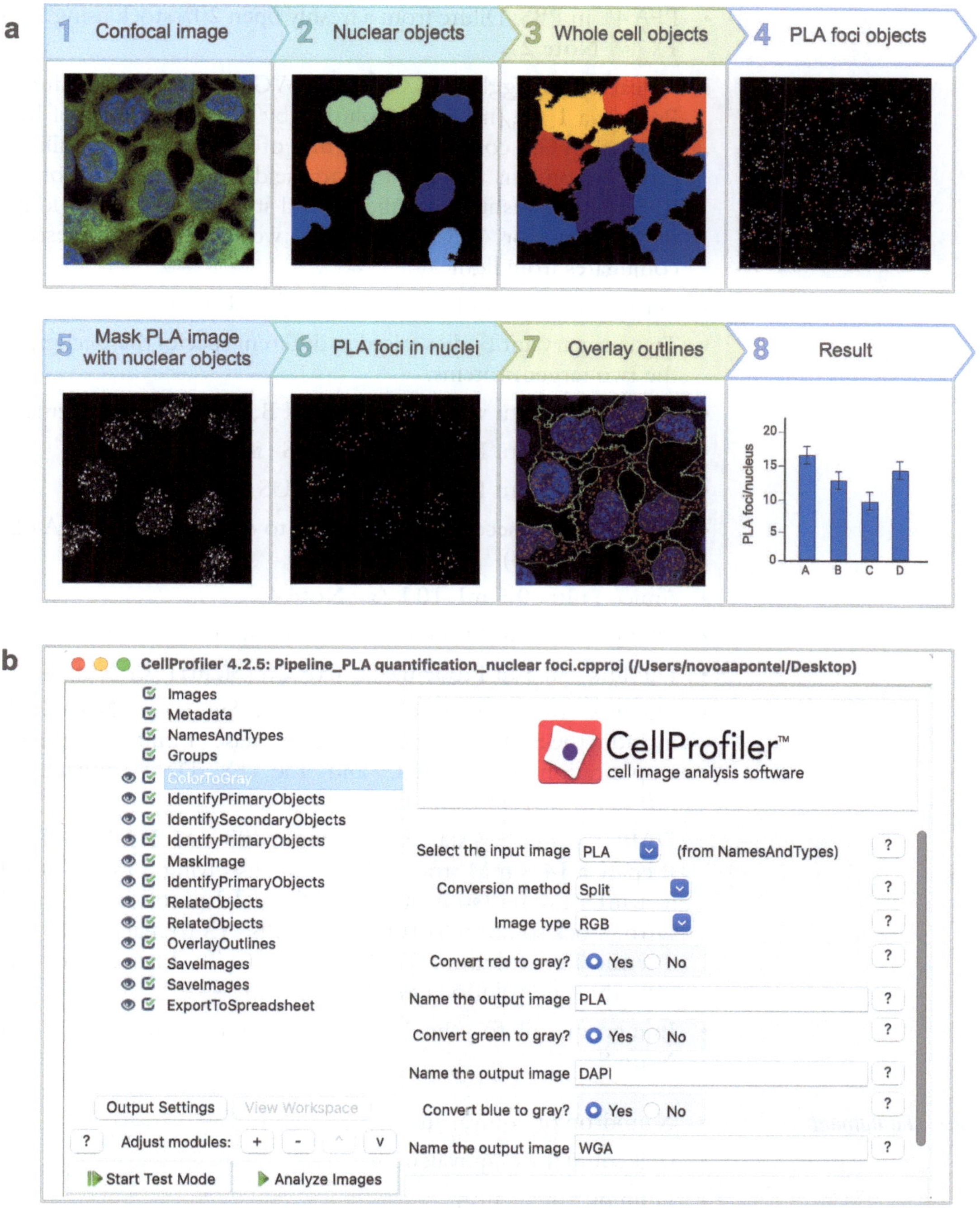

Fig. 2 Pipeline for the quantification of PLA signals using CellProfiler. (**a**) Schematic workflow designed to automatically recognize nuclei labeled with DAPI in the blue channel, whole cells based on fluorescence in the green channel, and PLA foci in the red channel as nuclear, whole cell, and PLA foci objects, respectively (panels 1–4). Cell profiler randomly assigns colors to different objects in an image. Nuclear objects are used to mask the PLA red channel image (panel 5) to selectively identify PLA foci in the nuclear space (panel 6). All the modeled objects are overlaid on the original confocal images to check the modeling quality (panel 6) before quantifying the PLA signals per whole cell and nucleus (panel 8). (**b**) Snapshot of CellProfiler workspace showing the modules for the stepwise identification and quantification of PLA signals. The pipeline was developed using CellProfiler version 4.2.5. Figure created with BioRender.com

- PFA 4% in PBS: Dilute from a freshly open 20% stock using PBS 1X (*see* **Note 2**).
- Wheat Germ Agglutinin CF®488A WGA conjugate (Biotium). Prepare a 1 mg/mL stock solution by dissolving 1 mg of lyophilized WGA conjugate in 1 mL of water. The lyophilized product contains enough PBS to yield a concentration of 1X PBS after reconstitution. Aliquot and store at −20 °C for long-term storage or 4 °C for up to a week. Protect fluorescent conjugates from light.
- Permeabilization buffer: 0.1% Triton® X-100 in PBS.
- Two primary antibodies raised in different species that recognize the two target proteins.
- Duolink® In Situ wash buffers A and B, Fluorescence (Sigma).
- Duolink® In Situ PLA® Probe PLUS (*see* **Note 3**).
- Duolink® In Situ PLA® Probe MINUS (*see* **Note 3**).
- Unconjugated secondary antibodies to conjugate with PLA oligos (*see* **Note 4**).
- Amicon ultra 0.5 mL 10 k (*see* **Note 4**).
- Duolink® In Situ Probemaker PLUS (Sigma) (*see* **Note 4**).
- Duolink® In Situ Probemaker MINUS (Sigma) (*see* **Note 4**).
- Duolink® In Situ detection reagents red (Sigma) (*see* **Note 5**). It includes the 5X ligation buffer, the 1X ligase (1 unit/μL), the 5X amplification buffer red, and the 1X DNA polymerase (10 units/μL) (*see* **Note 6**).
- DAPI for nuclear staining (Sigma): 300 nM DAPI in PBS. Prepare a 14.3 mM stock solution by dissolving 5 mg of DAPI in 1 mL of water (store at 4 °C). Dilute 2.1 μL of the 14.3 mM DAPI stock solution to 100 μL with PBS 1X to make a 300 μM intermediate solution. Dilute 1:1000 with PBS to get the 300 nM stain solution (*see* **Note 7**).
- Immersion oil for the lens of the confocal microscope (*see* **Note 8**).

2.2 Equipment

- Nanodrop or equivalent.
- Cell counter or equivalent.
- Humidity incubator.
- Confocal microscope.

2.3 Software

- Fiji (National Institutes of Health, Bethesda, MD) [18]. Download the software for free from https://fiji.sc.
- CellProfiler (Broad Institute, Cambridge, MA) [17]. Download the software for free from www.cellprofiler.org.

3 Methods

The conditions described here have been optimized for the culture of HEK293 cells in μ-Slide 15 Well 3D ibidi plates and staining using the Duolink® In Situ PLA—Fluorescence Red kit. Proficiency in basic laboratory skills and methods, cell culture, and confocal microscopy is required. This protocol was adapted with modifications from the Duolink® PLA manufacturer instructions (Sigma) and several other references (15, 16, 19).

3.1 Cell Culture

The cell culture conditions will depend on the type of cells used, the treatments to perform, and the slides used to grow the cells, among other factors. The conditions described here are suggested for the culture of HEK293 cells in a μ-Slide 15 Well 3D ibidi plate for 48 h before the fixation step (*see* **Note 9**). Cells must be fixed at 50–70% confluence for efficient reagent penetration. Overconfluent cells give lower signals and tend to detach from the culture surface. In our hands, seeding 6000 cells per well results in a ~70% confluent layer within 2 days. A pretreatment of the slide with poly-D-lysine considerably increases the cell attachment to the surface if working with cell lines that attach poorly to the slides (Subheading 3.1.1). Otherwise, proceed to the cell seeding step (Subheading 3.1.2) (*see* **Note 10**).

3.1.1 Poly-lysine Pretreatment (Optional Step)

1. Cover each culture well of the 15-well plate with 50 μL of sterile poly-D-lysine (*see* **Note 11**).
2. Incubate for 15 min at room temperature.
3. Remove the coating solution by suction (*see* **Note 12**).
4. Rinse the culture surface three times with sterile 1X PBS.
5. Remove the PBS and dry the coated culture plate in the laminar hood for 5 min.

3.1.2 Cell Seeding and Culture

1. Trypsinize and count cells as usual.
2. Dilute the cell suspension to the desired concentration (120 cells/μL if seeding 6000 cells per well) (*see* **Note 13**).
3. Add 50 μL of the cell suspension to the upper well.
4. Cover the slide with the supplied lid.
5. Incubate for 48 h at 37 °C and 5% CO_2 as usual (*see* **Note 14**).
6. Conduct any required treatment for the experiment (*see* **Note 15**).

3.2 Fixation

1. Remove the culture medium using gel loading tips (*see* **Note 16**).
2. Rinse cells once with 50 μL of 1X PBS at room temperature (*see* **Note 17**).

3. Add 50 μL of 4% PFA in 1X PBS for 10 min at room temperature to fix cells (no rocking).
4. Remove PFA and wash the wells with 50 μL of 1X PBS once.
5. Safe stopping point (*see* **Note 18**).

3.3 Cell Membrane Staining (Optional Step)

The quantification of PLA signals is often presented in terms of *PLA dots per cell* (Fig. 3). Defining the cell boundaries during the PLA quantification step is essential for correctly assigning PLA signals to each cell in the confocal microscopy image. This is achieved using a fluorescently labeled cell envelope marker, such as the wheat germ agglutinin (WGA) fluorescent conjugate that labels glycoproteins in the membranes, as suggested here. Cell surface staining must be performed before permeabilization to avoid undesired labeling of membranous intracellular compartments, such as the Golgi structures.

1. Prepare a 1 mg/mL WGA CF®488A conjugate stock solution in water.
2. Prepare a 5 μg/mL WGA CF®488A staining solution in 1X PBS.
3. Remove the PBS buffer from the wells.
4. Stain cell membranes by adding 25 μL of WGA staining solution to each well (*see* **Note 19**).
5. Incubate for 10 min at room temperature (protect from light, no rocking) (*see* **Note 20**).
6. Wash three times with 50 μL of 1X PBS.
7. Safe stopping point.

3.4 Permeabilization and Blocking

1. Permeabilize cells with 50 μL of 0.1% Triton® X-100 for 10 min at room temperature.
2. Block the cells using 50 μL of blocking solution (included in the PLA kit).
3. Incubate for 1 h at 37 °C or overnight at 4 °C.
4. Remove the blocking solution only when ready to add the primary antibodies.
5. Safe stopping point.

3.5 PLA Probes Conjugation (Optional Step) (See Notes 3 and 4)

There are commercially available ready-to-use PLA probes against primary antibodies from mouse, rabbit, and goat (from Sigma). Proceed with the PLA probes conjugation only if commercial PLA probes are unavailable. The antibody to conjugate must be 1 mg/mL and amine-free. If the antibody to conjugate contains more than 0.002% azide, remove it by buffer exchange as follow (Subheading 3.5.1). If no buffer exchange is required, proceed to the probe conjugation step (Subheading 3.5.2).

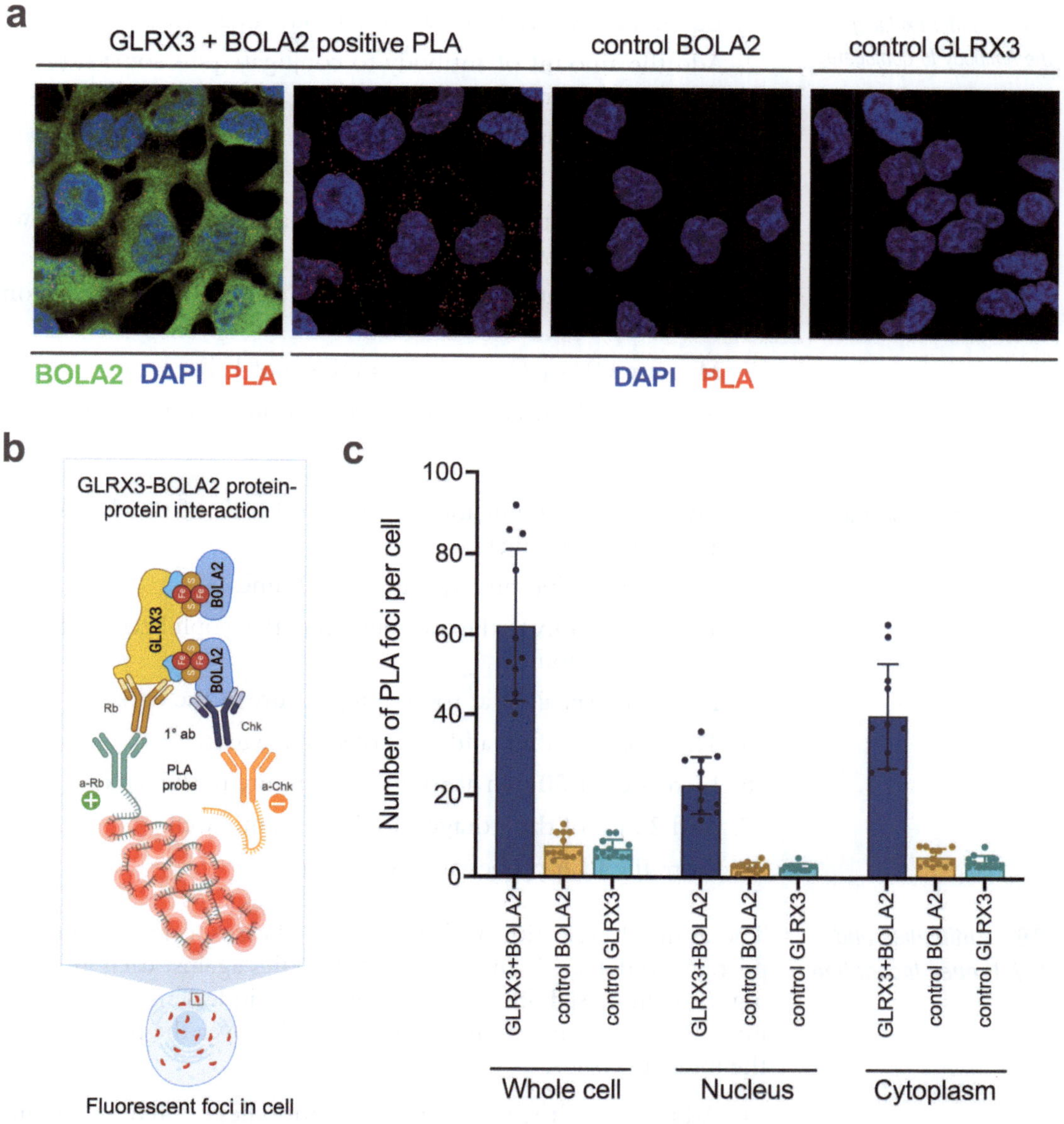

Fig. 3 Expected results for a PLA experiment detecting iron-mediated protein–protein interactions in the nucleus. The interaction between the monothiol glutaredoxin GLRX3 and the BolA-like protein BOLA2 to form the cytosolic [2Fe-2S] chaperone complex in human cells has been demonstrated using coimmunoprecipitation assays [20]. Here, we use PLA to show that this interaction occurs in both the cytosol and the nucleus. (**a**) Representative confocal images for the PLA experiments looking at the interaction between GLRX3 and BOLA2, using the primary antibodies anti-GLRX3 (rabbit) and anti-BOLA2 (chicken) and recognized using PLUS anti-rabbit and MINUS anti-chicken PLA probes, with the Duolink® In Situ detection red kit as schematized in (**b**). Protein complexes, shown as red PLA foci in the positive PLA reactions (top left), are located both in cytoplasm and nucleus. DAPI is used for staining nuclei (blue), and Alexa Fluor™ 488 goat anti-chicken detecting BOLA2 (green) as whole-cell control. Negative controls omitting one primary antibody are also included to assess background signals (top right). Control BOLA2 (anti-GLRX3 is absent) and control GLRX3 (anti-BOLA2 is absent). The pictures were taken with a Zeiss LSM 700 system (Carl Zeiss), using oil-immersion 40X objective lens, 2X zoom in workspace, and the blue, green, and red lasers for 405, 488, and 555 nm, respectively. **c**. Fluorescent PLA foci per cell and per nuclei were quantified using the CellProfiler software, as shown in Fig. 2. Figure created with BioRender.com

3.5.1 Buffer Exchange of the Antibody to Conjugate

1. Equilibrate Amicon by washing it four times with 1X PBS.
2. Add the amount of antibody to conjugate plus some extra to the pre-equilibrated Amicon.
3. Fill up the column with 1X PBS.
4. Spin 5–10 min at 7600 g.
5. Wash twice with 1X PBS, and spin as before. Discard flow-through in between washes.
6. Transfer the antibody to a new tube (aliquot in upper Amicon chamber).
7. Measure OD_{280} (1 mg/mL = OD_{280} of 1.4).
8. Adjust the final antibody concentration to 1 mg/mL using 1X PBS.

3.5.2 Probe Conjugation

1. Add 2 μL of conjugation buffer to 20 μL of the 1 mg/mL antibody to conjugate.
2. Mix well by pipetting up/down ~20 times.
3. Transfer the mix to the vial containing the lyophilized probe for the conjugation.
4. Incubate overnight at room temperature in the dark.
5. The morning after, add 2 μL of stop reagent.
6. Incubate for 30 min at room temperature in the dark.
7. Add 24 μL of the storage solution and mix well.
8. Store the PLA probes at 4 °C until use.

3.6 Antibodies and PLA Probes Incubation

The antibody selection will depend on the suspected interacting proteins to detect. Using primary antibodies against each interacting protein raised in different species is indispensable. Fifteen microlitres is the minimum volume to fully cover the inner well of the ibidi coverslips.

1. Add 15 μL of the mix of primary antibodies diluted in Duolink antibody diluent. The usual dilution is 1:1000 (*see* **Note 21**).
2. Incubate at room temperature for a few hours or overnight at 4 °C (no rocking) (*see* **Note 21**).
3. Wash thrice for 5 min with 50 μL of 1X buffer A at room temperature.
4. Prepare enough PLA probes master mix for all reactions (*see* **Note 22**). The 1:5 dilution suggested here is for the commercially available PLA probes. Standardization for homemade PLA probes conjugated using the probemaker kit may be required (*see* **Notes 23** and **24**):

Reagent	1X reaction
PLA probe PLUS (μL)	3
PLA probe MINUS (μL)	3
PLA ab diluent (μL)	9
Total volume (μL)	15

5. Mix well and add 15 μL of PLA probes, fully covering the inner well.
6. Incubate for 1 h at 37 °C (protect from light, no rocking).
7. Wash thrice with 50 μL of 1X buffer A at room temperature.
8. Safe stopping point.

3.7 Ligation

1. Prepare ligation master mix (*see* **Note 25**):

Reagent	1X reaction
H_2O (μL)	11.7
Ligase buffer 5X (μL)	3.0
Ligase (μL)	0.35
Total volume (μL)	15

2. Mix well and add 15 μL of master mix per well.
3. Incubate for 30 min at 37 °C (protect from light, no rocking).
4. Wash thrice for 5 min with 50 μL of 1X buffer A at room temperature.

3.8 Amplification

1. Prepare amplification master mix (*see* **Note 25**):

Reagent	1X reaction
H_2O (μL)	11.8
Amplification buffer 5X (μL)	3
DNA polymerase (μL)	0.2
Total volume (μL)	15

2. Mix well and add 15 μL of master mix per well.
3. Incubate for 100 min at 37 °C (protect from light, no rocking) (*see* **Note 26**).
4. Wash twice for 10 min with 50 μL of 1X buffer B at room temperature.
5. Rinse once with 50 μL of 0.01X buffer B at room temperature.

3.9 Nuclear Staining

1. Prepare a DAPI stock solution of 300 μM.
2. Prepare DAPI staining solution in 1X PBS, 300 nM final concentration.
3. Remove buffer from wells.
4. Add 25 μL of DAPI staining solution to each well.
5. Incubate for 5 min at room temperature (protect from light, no rocking).
6. Wash twice and replace with 50 μL of 1X PBS.
7. Slide is now ready for confocal microscopy (protect from light).
8. Safe stopping point (*see* **Note 27**).

3.10 Confocal Microscopy

Confocal laser scanning microscopy can be performed with any confocal system using oil-immersion 40X objective lens and a three-channel filter set appropriate for the reagents used in the experiment. The blue, green, and red lasers convenient for the reagents suggested in this protocol are 405, 488, and 555 nm, respectively (*see* **Note 28**).

1. Turn on and set up the pipeline in the confocal microscope to acquire images in the appropriate channels using the 40X objective lens.
2. Apply immersion oil, load the PLA slide in the holders, and adjust the height of the lens for the oil to contact the slide.
3. Use brightfield to locate the area to scan (*see* **Note 29**).
4. Zoom ~2x the scan area in the workspace to get around 10–15 cells per field (*see* **Note 30**).
5. Take about 20 pictures per well once the proper imaging setup and channels have been adjusted (*see* **Note 31**). Also capture Z-stack images to confirm if the detected PLA signals are inside the nucleus of the cell rather than in the nuclear surroundings.
6. Save the images in the *.czi* file format (*see* **Note 32**).

3.11 Quantification of PLA Signal

The ultimate goal of this experiment is to discriminate protein–protein interactions occurring in the nuclear space from those that may happen in other parts of the cell. We optimized a pipeline for detecting and quantifying nuclear and cytoplasmic PLA signals via confocal microscopy using the open-source software CellProfiler (https://cellprofiler.org/) (*see* **Note 33**). This pipeline is a sequential set of image-processing modules organized to model biological entities (i.e., cells, nuclei, foci) using the confocal images as input. Each image processing step requires an initial optimization of the parameters to accurately detect the nucleus, whole cell, cytoplasm, and PLA foci in each cell for each image obtained in the confocal microscope. Signals in each channel are modeled as "objects" (i.e., nucleus, whole cell, cytoplasm, PLA foci). The main parameters used to identify and differentiate objects in images are size and shape. The *nuclear objects* are detected using the DAPI signal in the blue channel. The *whole-cell objects* are detected using the WGA signal in the green channel. The *PLA foci objects* are assigned based on the signals from the red channel. Therefore, the *cytoplasm objects* result from subtracting the *nuclear objects* from the *whole-cell objects* (*see* **Note 34**).

The CellProfiler parameters showed here were optimized for images of HEK293 cells scanned using a Zeiss LSM 700 system (Carl Zeiss) confocal microscope with excitation lasers 405 nm, 491 nm, and 555 nm, and using 40X objective lens with a 2X zoom in the workspace. Because each set of images exhibits variability, these imaging parameters should be optimized. Optimization will allow an accurate analysis of each of the defined objects in the images. One of the main advantages of CellProfiler is that once the optimal settings for identifying each object of interest have been optimized for an image, hundreds or even thousands or millions of images can be automatically processed using the same pipeline. Additional modules are added at the end of the pipeline to export the data to a spreadsheet and to draw outlines of the modeled objects in the original images to evaluate the accuracy of the modeling. Because all pictures are processed as part of a single analysis, the final measurements for all pictures populate a single Excel file, which facilitates the downstream calculations. The protocol is explained module per module as it should be organized in the CellProfiler workspace (Fig. 2).

1. Open the CellProfiler software.
2. Load the images to analyze in the *Images* module (*see* **Note 35**).
3. Define the desired *Input* and *Output* folders in the *Output Settings* tab (left, bottom corner).
4. Create a pipeline by adding one by one the following modules using the *Adjust modules* "+" tool (*see* **Note 36**).
5. Color to gray: Only gray images can be processed. Use *split* as the *conversion method*. This module converts each channel from the original picture to a single-channel gray picture. Give a meaningful name to each channel picture. Use, for example, DAPI for the blue channel, WGA for the green channel, and PLA for the red channel (*see* **Note 34**).
6. Identify primary objects—nuclear objects: use the DAPI staining in the blue channel to identify the nuclei. The default settings must be adjusted to choose the best thresholding method and the expected range size of the nuclei based on the cell type and the zoom used when acquiring the confocal images. Suggested optimal parameters under advanced settings:

(a) Select DAPI as input image (from color to gray step).
(b) Give a name to the object to identify (for example, *nuclear_objects*) (*see* **Note 37**).
(c) Typical diameter of objects: 300–600 pixels (*see* **Notes 38** and **39**).
(d) Discard objects outside the diameter range? Yes.
(e) Discard objects touching the border of the image? No.
(f) Threshold strategy: Global.
(g) Thresholding method: Minimum Cross-Entropy.
(h) Threshold smoothing scale: 1.3.
(i) Threshold correction factor: 1.0.
(j) Lower and upper bounds on threshold: 0.0–1.0.
(k) Log transform before thresholding? No.
(l) Method to distinguish clumped objects: Shape.
(m) Method to draw dividing lines between clumped objects: Shape.
(n) Automatically calculate the size of the smoothing filter for declumping? Yes.
(o) Automatically calculate the minimum allowed distance between local maxima? Yes.
(p) Speed up by using lower-resolution images to find the local maxima? Yes.
(q) Display accepted local maxima? No.
(r) Fill holes in identified objects? After both thresholding and declumping.
(s) Handling of objects if an excessive number of objects are identified. Continue.

7. Identify secondary objects—whole-cell objects: Use the *nuclear objects* identified in the blue channel in **step 6** and the WGA staining to identify the whole-cell boundaries using the green channel image. This identification is performed by propagation, assuming every cell has a single nucleus. Optimal conditions:
   (a) Select WGA as the input image (from color to gray step).
   (b) Select nuclei as input objects (from Identify Primary Objects—*nuclear_objects*).
   (c) Give a name to the object to identify (for example, *whole_cell_objects*).
   (d) Select the method to identify the secondary objects: Propagation.
   (e) Threshold strategy: Global.

(f) Thresholding method: Minimum Cross-Entropy.
(g) Threshold smoothing scale: 0.0.
(h) Threshold correction factor: 1.0.
(i) Lower and upper bounds on the threshold: 0.0–1.0.
(j) Log transform before thresholding? No.
(k) Regularization factor: 0.05.
(l) Fill holes in identified objects? Yes.
(m) Discard secondary objects touching the border of the image? No.

8. Identify primary objects—PLA foci objects: Identify the PLA foci using per-object thresholding to compute a threshold for each individual PLA signal object. Some manual adjustment of the smoothing filter size and maxima suppression distance is required to optimize segmentation. Optimal conditions for the *PLA foci objects* identification under the experimental conditions are described here:
   (a) Select advanced settings.
   (b) Select PLA as the input image (from color to gray step).
   (c) Give a name to the object to identify (for example, *PLA_foci_objects*).
   (d) Typical diameter of objects: 10–40 pixels (*see* **Note 38**).
   (e) Discard objects outside the diameter range? Yes.
   (f) Discard objects touching the border of the image? No.
   (g) Threshold strategy: Global.
   (h) Thresholding method: Otsu. Two classes.
   (i) Threshold smoothing scale: 1.3488.
   (j) Threshold correction factor: 1.0.
   (k) Lower and upper bounds on threshold: 0.1–1.0.
   (l) Log transform before thresholding? No.
   (m) Method to distinguish clumped objects: Intensity.
   (n) Method to draw dividing lines between clumped objects: Intensity.
   (o) Automatically calculate the size of the smoothing filter for declumping? No.
   (p) Size of smoothing filter: 4.
   (q) Automatically calculate the minimum allowed distance between local maxima? No.
   (r) Suppress local maxima that are closer than this minimum allowed distance: 4.

(s) Speed up by using lower-resolution images to find the local maxima? Yes.

(t) Display accepted local maxima? No.

(u) Fill holes in identified objects? After both thresholding and declumping.

(v) Handling of objects if an excessive number of objects are identified. Continue.

9. Mask image: Use the *nuclear objects* identified in the blue channel in step number 6 to mask the PLA foci image:

   (a) Select PLA as the input image (from color to gray step).

   (b) Give a name to the masked image (for example, Masked_nuclear_PLA).

   (c) Use objects or an image as a mask? Objects.

   (d) Select the object for mask: Nuclei (from Identify Primary Objects—*nuclear_objects*).

   (e) Invert the mask? No.

10. Identify primary objects—PLA foci in nuclei: Identify the nuclear PLA foci as in **step 8** but using the Masked_nuclear_-PLA image as the input. Give it a name, for example, PLA_nuclei. Keep all other parameters as in **step 5**.

11. Relate objects—PLA to whole cell: Establish a parent–child relationship between the PLA foci and the cells to determine which foci belong to which cell. Then, calculate the mean foci counts for each cell. Suggested parameters:

   (a) Parent objects: *Whole_cell_objects* (from Identify Secondary Objects - *whole_cell_objects*).

   (b) Child objects: *PLA_foci_objects* (from Identify Primary Objects—*PLA_foci_objects*).

   (c) Calculate per-parent means for all child measurements? Yes.

   (d) Calculate child–parent distances? None.

   (e) Do you want to save the children with the parents as a new object set? Yes.

   (f) Name the output object: PLA_foci_in_whole_cell.

12. Relate objects—PLA to nucleus: Establish a parent–child relationship between the PLA foci and the nuclei to determine which foci belong to which nuclei. Then, calculate mean foci measurements for each nucleus. Suggested parameters:

   (a) Parent objects: Nuclei (from Identify Primary Objects—*nuclear_objects*).

   (b) Child objects: PLA_whole_cell (from Identify Primary Objects—PLA foci in nuclei).

(c) Calculate per-parent means for all child measurements? Yes.

(d) Calculate child-parent distances? None.

(e) Do you want to save the children with the parents as a new object set? Yes.

(f) Name the output object: PLA_foci_in_nuclei.

13. Overlay outlines: Draw outlines for each object identified through the pipeline on top of the PLA gray image to facilitate the visual inspection of the quality of the foci modeling. (*see* **Note 40**). Suggested conditions:

(a) Display outlines on a black image? No.

(b) Select the image on which to display outlines: PLA (from color to gray step).

(c) Name the output image: Overlays_.

(d) Outline display mode: Color.

(e) How to outline: Thick.

(f) Select the object to display: Whole cell (from Identify Secondary Objects—*whole_cell_objects*).

(g) Select outline color: green is suggested as the original color for the WGA channel.

(h) Select the object to display: Nuclei (from Identify Primary Objects—*nuclear_objects*).

(i) Select the outline color: blue is suggested as the original color for the DAPI channel.

(j) Select the object to display: PLA_whole_cell (from Identify Primary Objects—PLA foci in whole cell).

(k) Select outline color: red is suggested as the original color for the PLA channel.

14. Save images: this allows you to choose which images to save for future data analysis. It is recommended to save the overlay outlines for further visual inspection of the quality of the modeling.

(a) Select the type of image to save: Image.

(b) Select the image to save: Overlays_ (from the Overlay Outlines step).

(c) Select method for constructing file names: From image filename.

(d) Select the image name for the file prefix: PLA.

(e) Append a suffix to the image file name? Yes.

(f) Text to append to the image name: _overlays.

(g) Saved file format: tiff.

(h) Image bit depth: 8-bit integer.
(i) Save with lossless compression? Yes.
(j) Output file location: Default Output Folder.
(k) Overwrite existing files without warning? No.
(l) When to save: Every cycle.
(m) Record the file and path information to the saved image? No.
(n) Create subfolders in the output folder? No.

15. Export to spreadsheet: Export all measurements to a comma-delimited file (.csv). The measurements made for the nuclei, whole cell, and foci objects will be saved to separate .csv files, in addition to the per-image .csv file. Select the comma as the column delimiter and select the Default Output Folder. Add the prefix "Calculations_" to the filename and select "No" to all other options except for the *Export all measurement types?* option.
16. Once all modeling parameters have been set, use the "Step" tab to evaluate the modeling quality by comparing the accuracy of the modeled object shape and the number of objects counted based on the image being analyzed (*see* **Note 41**).
17. Use the "*View Workspace*" tab to overlap the different shapes on the channels and check the quality of the modeling.
18. Hit the tab "*Run*" to launch the analysis of the whole image set.
19. Import the *Image.csv* file to an Excel spreadsheet.
20. Use the measurements of PLA objects per cell and per nucleus to determine the nuclear vs. cytoplasmic localization of the protein–protein interactions evaluated. The cytoplasmic PLA signals result from subtracting the PLA foci detected in the nucleus from the PLA foci detected in the whole cell.

4 Notes

1. Use the desired adherent cell line and adjust growth conditions accordingly.
2. It is highly recommended to use a freshly prepared PFA solution. Do not use solutions that are more than 1 week old.
3. A PLA experiment requires using a PLUS and a MINUS probe every time. Sigma offers anti-rabbit, anti-mouse, and anti-goat PLUS and MINUS probes. The selection of the probes will depend on the species of the primary antibodies to detect. If a primary antibody is raised in a species different from rabbit, mouse, or goat, the probemaker kits may be used to conjugate a PLA oligoprobe to any unconjugated secondary antibody.

4. These reagents are only required if commercial PLA probes are not available for the species of the primary antibodies, as described in the previous note.
5. The Duolink® In Situ detection reagents red generate PLA signals in the red channel. Make sure any other labeling to perform is not using a red fluorophore. Sigma also offers kits for detecting PLA probes using green, far-red, or orange filters.
6. The ligase buffer contains oligonucleotides that hybridize with the PLA probes. The polymerase buffer contains the oligonucleotide probes labeled with a fluorophore that hybridizes to the RCA product. Protect amplification buffer from light.
7. Using a mounting medium with DAPI for nuclear staining is also possible.
8. The ibidi Polymer Coverslip is only compatible with certain types of immersion oil. Check the manufacturer's catalog for information.
9. The μ-Slide 15 Well 3D ibiTreat ibidi plates were incorporated into the protocol, aiming for a cost-effective experiment. It has a "well-in-a-well" technology, which minimizes the required volume of PLA reagents and avoids meniscus formation. The volume of the upper well is 50 μL (used for cell culture, fixation, blocking, and washing steps) and the volume of the inner well is 10 μL. We suggest using 15 μL for incubation with the PLA reagents to guarantee full coverage of the cell monolayer. The growth area of the inner well is 0.125 cm^2. Protocol adapted with modifications from [19].
10. Keep a sterile environment for all cell culture steps. Sterile conditions are not required after cells have been fixed.
11. Poly-lysine facilitates cell adhesion to the culture surface by promoting electrostatic interactions between the positively charged lysine residues deposited in the culture surface and the negatively charged residues of the cell membranes.
12. Make sure to rinse well after coating incubation, as excess of poly-lysine solution can be toxic for the cells.
13. The seeding cell density depends on the cell type cultured and the treatments to be performed. It is essential not to allow cells to grow over 70% confluency.
14. Medium exchange may be necessary every 1–2 days, depending on the cell type. Carefully aspirate the old medium and replace it with 50 μL of fresh warm medium.
15. Fill the space between the plate's wells with sterile water or 1% agarose for less evaporation if long incubation times are required.

16. Do not use vacuum aspiration to remove liquids, as it will aspirate part of the cells. Use gel loading tips instead, locating the tip at the same edge of the well every time a liquid is aspirated. Make sure not to disturb the cell monolayer with the tip.
17. Never let the cells to dry out. Only remove buffer when the reagent for the next step is ready to be added. Make sure all reagents are at room temperature before adding to the cells, as sudden temperature changes cause the cells to detach.
18. Safe stopping points refer to steps of the protocol where it is possible to stop, store the plate at 4 °C in the dark, and continue the day after. Add sterile water around the wells to generate a humidified environment to hinder evaporation during storage.
19. WGA staining must be performed before cell permeabilization to only stain the cell surface and avoid staining of membranous intracellular organelles. WGA staining concentration may require optimization.
20. From now on, protect the plate from light to avoid photobleaching of the fluorophores.
21. Optimize the concentration and incubation times for the primary antibodies via confocal immunofluorescence experiments before performing the PLA.
22. Include one extra reaction when using master mixes.
23. Dilution optimization of homemade probes is suggested until getting a suitable signal-to-noise ratio. A 1:50–1:100 dilution usually gives suitable results.
24. Other secondary antibodies could be included in the PLA master mix if their fluorophores do not interfere with all the different fluorophores used in the experiment.
25. Keep the enzymes in the freezer until required. Use a freezing box while using them and return to the freezer immediately. Use the master mixes right after adding the enzymes.
26. Increased DNA amplification time if the PLA signals are not strong enough.
27. Labeling is stable for up to a week in the dark at 4 °C. Make sure to add water in between wells and store it in a humidity chamber to avoid evaporation. Do not let the samples to dry out. Add extra PBS if necessary.
28. Confirm the excitation and emission peaks of the fluorophores used in the experiment to set up the lasers of the confocal microscope. To avoid cross-excitation, ensure dual or triple color images are acquired by consecutive scanning with only one laser line active per scan.

29. Do not overexpose the slide to light to avoid photobleaching.
30. Focusing around 10 cells per field allows high spatial precision to discriminate between nuclear and cytoplasmic signals.
31. Make sure the pictures taken represent the overall result around the well. Avoid taking pictures at the edge of the well, as limited access of reagents to those areas may affect the efficiency of the labeling. The minimum of 20 pictures per well/condition is to have around 150–200 cells for the further quantification step, which facilitates the interpretation and statistical analysis of the results.
32. Flow cytometry could be used instead if the PLA detection using fluorescence microscopy does not offer enough sensitivity. However, that requires a different set of reagents intended for flow cytometry detection.
33. To get familiar with the basic operation of the software, please refer to the original reference (17) and watch the CellProfiler tutorial videos (https://cellprofiler.org/tutorials).
34. The conditions described here are optimal for analyzing images taken under the described conditions. Adjust CellProfiler settings to accurately identify the different signals in the different channels of the image if using an alternative set of fluorophores.
35. The pictures in the *.czi* format obtained from the microscope can be directly used in CellProfiler. Other formats are accepted, but the pipeline should be modified accordingly.
36. Example images and their corresponding CellProfiler pipelines can be downloaded from the CellProfiler webpage. Rather than adding modules one by one, it is also possible to download an example CellProfiler pipeline for foci identification from the software webpage, load it in the workspace, and modify it to mimic the conditions described here.
37. Do not leave blank spaces in names in CellProfiler, as it gives an error. Instead, use underscores (_) to connect all words in the names of images and shapes.
38. The size of the objects to identify can change from set to set of images depending on the type of cells, microscope objective, magnification used, zoom, or reagents used. Therefore, the object size ranges suggested here are examples that work for images taken under the described experimental conditions. Further optimization of diameter ranges may be required to correctly model the shapes of another set of images.
39. It is essential to set the range of nuclear-accepted diameter with a lower limit high enough to exclude unspecific DAPI stains and micronuclei and an upper limit low enough to exclude cell clumps or multinucleate cells. In the example described here,

DAPI signals smaller than 300 or larger than 600 pixels are filtered out and not considered for subsequent quantification steps.

40. Outline overlays can be automatically drawn for each shape modeled on top of each input image to evaluate the object modeling quality.

41. Carefully check the modeling quality in every step of the pipeline and modify conditions to make the drawn shapes fit as best as possible with the image analyzed. Make sure no multiple shapes are drawn per foci, cell, or nucleus. Changing parameters under the "Method to distinguish clumped objects" function could help for better fitting. It is also recommended to manually count the PLA foci, nuclei, and whole-cell numbers for a couple of images and compare them with the number of objects modeled by CellProfiler. Once confident with the modeling results, launch the analysis for the whole set of images.

References

1. Philpott CC, Patel SJ, Protchenko O (2020) Management versus miscues in the cytosolic labile iron pool: the varied functions of iron chaperones. Biochim Biophys Acta (BBA) Mol Cell Res 1867(11):118830
2. Paul VD, Lill R (2014) SnapShot: eukaryotic Fe-S protein biogenesis. Cell Metab 20(2): 384.e1
3. Paul VD, Lill R (2015) Biogenesis of cytosolic and nuclear iron–sulfur proteins and their role in genome stability. Biochim Biophys Acta (BBA) Mol Cell Res 1853(6):1528–1539
4. Rouault TA, Maio N (2017) Biogenesis and functions of mammalian iron-sulfur proteins in the regulation of iron homeostasis and pivotal metabolic pathways. J Biol Chem 292(31): 12744–12753
5. Philpott CC, Protchenko O, Wang Y, Novoa-Aponte L, Leon-Torres A, Grounds S et al (2023) Iron-tracking strategies: chaperones capture iron in the cytosolic labile iron pool. Front Mol Biosci 10:1127690
6. Robbins E, Fant J, Norton W (1972) Intracellular iron-binding macromolecules in HeLa cells. Proc Natl Acad Sci 69(12):3708–3712
7. Robinson I, Yang Y, Zhang F, Lynch C, Yusuf M, Cloetens P (2016) Nuclear incorporation of iron during the eukaryotic cell cycle. J Synchrotron Radiat 23(6):1490–1497
8. Puig S, Ramos-Alonso L, Romero A, Martínez-Pastor M (2017) The elemental role of iron in DNA synthesis and repair. Metallomics 9:1483–1500
9. Troadec M-B, Loréal O, Brissot P (2017) The interaction of iron and the genome: for better and for worse. Mutat Res Rev Mutat Res 774: 25–32
10. Gari K, León Ortiz AM, Borel V, Flynn H, Skehel JM, Boulton SJ (2012) MMS19 links cytoplasmic iron-sulfur cluster assembly to DNA metabolism. Science 337(6091): 243–245
11. Stehling O, Vashisht AA, Mascarenhas J, Jonsson ZO, Sharma T, Netz DJ et al (2012) MMS19 assembles iron-sulfur proteins required for DNA metabolism and genomic integrity. Science 337(6091):195–199
12. Netz DJ, Mascarenhas J, Stehling O, Pierik AJ, Lill R (2014) Maturation of cytosolic and nuclear iron–sulfur proteins. Trends Cell Biol 24(5):303–312
13. Weon JL, Yang SW, Potts PR (2018) Cytosolic iron-sulfur assembly is evolutionarily tuned by a cancer-amplified ubiquitin ligase. Mol Cell 69(1):113–125. e6
14. O'Halloran TV, Culotta VC (2000) Metallochaperones, an intracellular shuttle service for metal ions. J Biol Chem 275(33): 25057–25060
15. Fredriksson S, Gullberg M, Jarvius J, Olsson C, Pietras K, Gústafsdóttir SM et al (2002) Protein detection using proximity-dependent

DNA ligation assays. Nat Biotechnol 20(5): 473–477

16. Söderberg O, Gullberg M, Jarvius M, Ridderstråle K, Leuchowius K-J, Jarvius J et al (2006) Direct observation of individual endogenous protein complexes in situ by proximity ligation. Nat Methods 3(12):995–1000
17. Carpenter AE, Jones TR, Lamprecht MR, Clarke C, Kang IH, Friman O et al (2006) CellProfiler: image analysis software for identifying and quantifying cell phenotypes. Genome Biol 7:1–11
18. Schindelin J, Arganda-Carreras I, Frise E, Kaynig V, Longair M, Pietzsch T et al (2012) Fiji: an open-source platform for biological-image analysis. Nat Methods 9(7): 676–682
19. Alam MS (2018) Proximity ligation assay (PLA). Curr Protoc Immunol 123(1):e58
20. Frey AG, Palenchar DJ, Wildemann JD, Philpott CC (2016) A glutaredoxin· BolA complex serves as an iron-sulfur cluster chaperone for the cytosolic cluster assembly machinery*. J Biol Chem 291(43):22344–22356

Chapter 5

Fluorescence-Based Proteoliposome Methods to Monitor Redox-Active Transition Metal Transmembrane Translocation by Metal Transporters

Mitchell A. Pope, Rose M. Curtis, Humera Gull, Madujika A. Horadigala Gamage, Sameera S. Abeyrathna, Nisansala S. Abeyrathna, Christoph J. Fahrni, and Gabriele Meloni

Abstract

Transmembrane transition metal transporter proteins are central gatekeepers in selectively controlling vectorial metal cargo uptake and extrusion across cellular membranes in all living organisms, thus playing key roles in essential and toxic metal homeostasis. Biochemical characterization of transporter-mediated translocation events and transport kinetics of redox-active metals, such as iron and copper, is challenged by the complexity in generating reconstituted systems in which vectorial metal transport can be studied in real time. We present fluorescence-based proteoliposome methods to monitor redox-active metal transmembrane translocation upon reconstitution of purified metal transporters in artificial lipid bilayers. By encapsulating turn-on/-off iron or copper-dependent sensors in the proteoliposome lumen and conducting real-time transport assays using small unilamellar vesicles (SUVs), in which selected purified Fe(II) and Cu (I) transmembrane importer and exporter proteins have been reconstituted, we provide a platform to monitor metal translocation events across lipid bilayers in real time. The strategy is modular and expandable toward the study of different transporter families featuring diverse metal substrate selectivity and promiscuity.

Key words Proteoliposome, Iron, Copper, Fluorescence, Real time, Small unilamellar vesicles, Transporter, Transmembrane protein

1 Introduction

Transition metals are essential trace elements, and despite their low abundance, they are central to the chemistry of life [1, 2]. Serving as enzyme cofactors, structural centers, and signaling ions, biologically relevant transition metals can be broadly classified into two

Authors Mitchell A. Pope, Rose M. Curtis, Humera Gull and Madujika A. Horadigala Gamage have equally contributed to this chapter.

Oleh Khalimonchuk (ed.), *Iron Metabolism: Methods and Protocols*, Methods in Molecular Biology, vol. 2839, https://doi.org/10.1007/978-1-0716-4043-2_5,

main classes: redox-active and redox-inert metals [1–7]. Redox-inert metals, such as zinc (Zn(II)), exist and function in a single oxidation state under physiological conditions and redox potentials [1, 2]. Contrarily, redox-active metals, such as iron and copper, have the ability to cycle between their available oxidation states (Fe (II)/(IV)) and Cu(I)/(II)), allowing these metals in metalloenzyme active sites to catalyze redox chemistry, electron transfer, and/or substrate(s) activation (e.g., molecular oxygen), thus permitting challenging reactions to take place in the biological context [1–5, 8]. Reactions catalyzed through these metalloenzymes are essential to cellular metabolism, replication, and virulence in pathogenic bacteria [9, 10]. The ability of iron and copper to redox cycle in the presence of O_2 and biological reductants also necessitates the strict regulation of their concentrations to prevent the generation of free radicals, reactive oxygen and nitrogen species, and associated biomolecular and cellular damage [5, 8, 11]. Intricate pathways of metal uptake, sequestration, delivery to target metalloproteins via metallochaperones, control of intracellular concentrations via metal sensors and metal-dependent transcription factors, and cellular export across biological membranes have evolved to compensate for metal-induced toxicity while ensuring that the necessary cellular metal quotas are maintained [1–4, 12, 13].

Prokaryotic iron homeostasis has been extensively characterized in gram-negative bacteria, which all follow a generally conserved scheme for their acquisition [4, 13–15]. Unique import systems have evolved for each iron source (Fe(II) and Fe(III)) to ensure that homeostasis is maintained in a variety of environments. Transmembrane iron uptake transporters can be categorized into outer membrane (OM) and inner membrane (IM) acquisition systems [15]. Generally, Fe(II) is acquired in the periplasm via poorly characterized porin-type proteins and Fe(III) is acquired via transporter-mediated uptake of Fe(III)-associated complexes (e.g., siderophore-Fe(III) complexes by the OM receptor FhuA) [16–19]. Translocation across the IM of the periplasmic Fe(II) pool is mediated by a variety of import systems such as *Feo*, *ZupT*, and *MntH* transporters, all of which allow the uptake of Fe(II) [12, 20–23]. After iron import through the OM Fe(III)-complex TonB-dependent receptors, Fe(III) is transferred to periplasmic binding proteins for translocation to the cytoplasm via the unique IM transport systems, such as Fe(III)-siderophore ABC-transporters (e.g., FhuABCD and FecABCDE complexes which are selective for ferrichrome and ferric citrate, respectively) [24–28]. Irrespective of the acquisition system, intracellular iron is primarily stored in bacterioferritin and is utilized in numerous systems for a variety of crucial cellular functions [1, 2, 29]. Export of iron in prokaryotes, in contrast to import, is less widely investigated [30, 31]. While deletion of cation diffusion facilitator (CDF) proteins, such as yiiP, are associated with excess iron-related phenotypes, Yiip and its homologs are Zn(II)-transporting proteins with no demonstrated capacity to transport iron [31, 32]. Due to its redox properties, complex homeostatic pathways to maintain appropriate cellular Cu (I) concentrations have also evolved [1–4, 33–36]. In contrast to iron, bacterial copper exporter systems have been extensively studied, and the ATP-dependent P_{1B-1}-type ATPase Cu(I)-pumps, often referred to as CopAs, are highly conserved throughout prokaryotes and eukaryotes (ATP7A/B), while pathways for Cu (I) uptake in bacteria remain yet to be fully characterized [37–45].

Clearly, membrane-associated iron and copper transporters are essential to the survival and pathogenicity of prokaryotic organisms [9, 46, 47]. Despite their crucial role in metal homeostasis, these transporter systems remain poorly characterized due to a variety of challenges in conducting structural, biophysical, and biochemical investigations, including overexpression, solubilization, and purification in functional forms [48]. In particular, characterization of cargo translocation events in detergent-solubilized, purified proteins is only possible through proxy methods, such as substrate stimulation of catalytic ATP hydrolysis (e.g., in ABC transporters and P-type ATPases) [49]. To overcome this challenge, strategies to study substrate translocation events by purified transmembrane transporters reconstituted in artificial lipid bilayers have been developed, in particular the use of small unilamellar vesicles (SUVs), also termed proteoliposomes (PLs), to characterize the selectivity and kinetics of substrate translocation of a variety of cargo molecules [50, 51]. However, the study of transporter-mediated translocation events of redox-active transition metals in artificial lipid bilayers is underrepresented in the literature. We have developed a methodology to reconstitute transmembrane metal transporters into SUVs followed by encapsulation of metal-dependent turn-on/−off fluorescent probes (e.g., FluoZin-3, Calcein, and CTAP-3) to detect real-time translocation of metal substrates into the liposomal lumen by fluorescence spectroscopy (*see* Fig. 1) [30, 52, 53]. This methodology allows for the study of translocation kinetics and selectivity of a variety of transporters, in an isolated system, providing methodologies that facilitate insights into metal translocation across biological membranes of both prokaryotic and eukaryotic systems.

Presented herein, proteoliposome-based protocols are described for the characterization of three different metal transmembrane transporters using fluorescent probes encapsulated in PL: i) *Legionella pneumophila* IroT/MavN (Iron Transporter/ More regions allowing vacuolar colocalization N protein), ii) *Pseudomonas aeruginosa* AitP (Alternative iron transport Protein), and iii) *Escherichia coli* CopA (Cu(I) P-type ATPase). *Lp*IroT is a secondary active transporter that facilitates the intravacuolar Fe (II) uptake from the host organism in the macrophage *Legionella*-containing vacuole (LCV) [30, 54–56]. *Pa*AitP is a

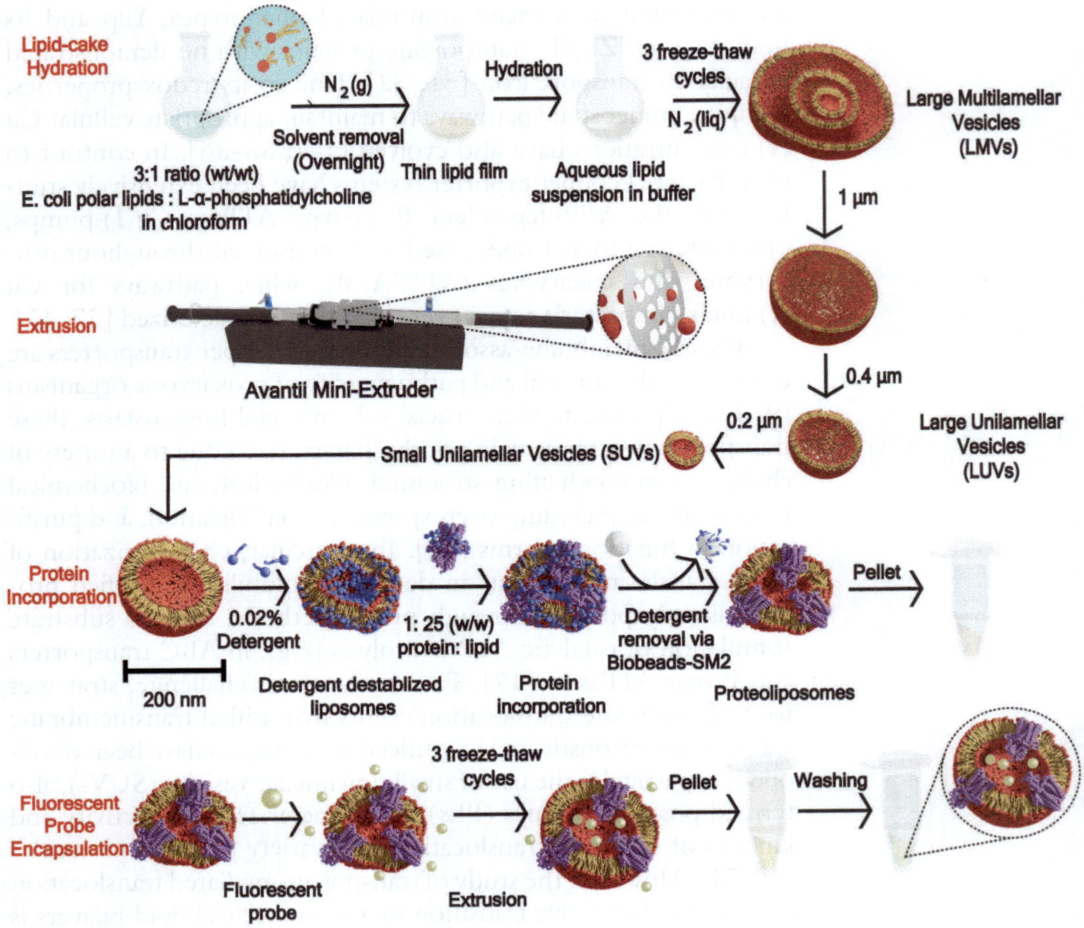

Fig. 1 Scheme summarizing the preparation of artificial lipid bilayer systems (proteoliposomes) to study transition metal translocation across membranes by purified transmembrane transporter proteins. The following steps are showcased: lipid cake preparation, hydration, and preparation of multilamellar vesicles by freeze-thaw cycles, extrusion of SUVs with controlled size distribution, transmembrane protein incorporation in SUVs, and fluorescent probe encapsulation for transition metal translocation studies

CDF believed to be involved in the export of Fe(II), and potentially Co(II), through a poorly characterized mechanism [32, 57, 58]. *Ec*CopA is a primary active P_{1B-1}-type ATPase involved in the export of Cu(I), thereby playing a key role in controlling intracellular Cu(I) concentrations [3, 33, 37, 41, 59, 60]. Together, these proteins represent only a few of the diverse metal transporter families involved in the translocation of metal substrates across lipid bilayers. Nevertheless, the presented methods can be easily extended to the identification and characterization of a variety of other iron and copper transporters and transporter families that mediate the translocation of other essential and toxic transition metals.

2 Materials

1. Lipids: *E. coli* polar lipid extract and chicken egg *L*-α-phosphatidylcholine were purchased as nitrogen-purged stocks in chloroform ($CHCl_3$) at a lipid concentration of 25 mg/mL (*see* **Note 1**).
2. Chelex-treated Milli-Q H_2O: Mix 50 g of Chelex 100 Molecular Biology Grade Resin with 1.0 L of Milli-Q H_2O in a 2.0 L beaker. Stir for at least 60 min with a stir bar and then filter through a 0.2 μm filter into a glass bottle for long-term storage (*see* **Note 2**).
3. 1.0 mM tris(2-carboxyethyl)phosphine hydrochloride (TCEP): Weigh 2.9 mg of TCEP into a 15 mL Falcon tube and resuspend to 10 mL using Chelex-treated Milli-Q H_2O (*see* **Note 3**).
4. 1.0 mM *L*-ascorbic acid: Weigh 8.8 mg of *L*-ascorbic acid into a 50 mL Falcon tube and resuspend to 50 mL using Chelex-treated Milli-Q H_2O.
5. 2.0% (w/v) 7-Cyclohexyl-1-heptyl-β-*D*-maltoside (CYMAL-7) in Milli-Q H_2O: Weigh 20 mg of CYMAL-7 (Anatrace Products) into a 1.5 mL Eppendorf tube and add 1.0 mL of Chelex-treated Milli-Q H_2O. Gently mix until it is dissolved (*see* **Note 4**).
6. 2.0% (w/v) n-Dodecyl-β-*D*-Maltoside (DDM) in Milli-Q H_2O: Weigh 20 mg of DDM into a 1.5 mL Eppendorf tube and add 1.0 mL of Chelex-treated Milli-Q H_2O. Gently mix until it is dissolved.
7. Lipid resuspension buffer: 200 mM 3-(*N*-morpholino)propanesulfonic acid (MOPS)-NaOH pH 7.0, 1.0 M NaCl, and 1.0 mM TCEP. Add 10 mL of a 5.0 M NaCl stock, 10 mL of a 1.0 M MOPS-NaOH pH 7.0 stock, and 14.5 mg of TCEP to a 50 mL Falcon tube and gently mix before adding 2.5 g of Chelex 100 resin. Gently mix for 60 min and adjust the pH to 7.0 before filtering through a 0.2 μm filter. Dilute to 50 mL with Chelex-treated Milli-Q H_2O.
8. *Lp*IroT purification buffer: 20 mM MOPS-NaOH pH 7.0, 500 mM NaCl, 1.0 mM dithiothreitol (DTT), and 0.05% (w/v) CYMAL-7. Weigh 2.093 g of MOPS, 14.61 g of NaCl, 77.1 mg of DTT, and 0.25 g of CYMAL-7 and transfer them to a 1.0 L beaker. Add approximately 400 mL of Milli-Q H_2O and mix until dissolved. Adjust the pH to 7.0 using NaOH and dilute the buffer to 500 mL using Milli-Q H_2O. Filter through a 0.2 μm filter to remove particulates and store in a sealed bottle at 4 °C.

9. *Pa*AitP purification buffer: 20 mM MOPS-NaOH pH 7.0, 500 mM NaCl, 1.0 mM DTT, 10% (v/v) glycerol, and 0.05% (w/v) CYMAL-7. Prepare as described for the *Lp*IroT purification buffer by adding 50 mL of 100% glycerol before diluting to the final volume.
10. *Ec*CopA purification buffer: 20 mM MOPS-NaOH pH 7.0, 500 mM NaCl, 1.0 mM L-ascorbic acid, and 0.05% (w/v) DDM. Prepare as described for the *Lp*IroT purification buffer by substituting 88.1 mg of *L*-ascorbic acid for DTT and 0.25 g of DDM for CYMAL-7.
11. *Lp*IroT and *Pa*AitP transport buffer: 20 mM MOPS-NaOH pH 7.0, 100 mM NaCl, and 1.0 mM TCEP. Weigh 0.419 g of MOPS, 0.584 g of NaCl, and 28.6 mg of TCEP and transfer them to a 250 mL beaker. Add 50 mL of Chelex-treated Milli-Q H_2O and mix until they are dissolved. Adjust the pH to 7.0 using NaOH. Weigh 20 g of Chelex 100 resin and transfer it to the beaker. Stir the solution for at least 60 min before adjusting to pH 7.0. Dilute the buffer to 100 mL using Chelex-treated Milli-Q H_2O and filter through a 0.2 μm filter to remove the Chelex resin. Store in a sealed bottle wrapped in aluminum foil at 4 °C for immediate usage (*see* **Notes 5** and **6**).
12. *Ec*CopA transport buffer: (20 mM MOPS-NaOH pH 7.0, 100 mM NaCl, and 1.0 mM *L*-ascorbic acid). Prepare as described for the previous transport buffer by substituting 17.6 mg of *L*-ascorbic acid for TCEP and reducing the Chelex 100 resin to 5.0 g.
13. Activated Bio-Beads: Weigh 5.0 g of Bio-Beads SM-2 Adsorbents (Bio-Rad Laboratories) into a 250 mL beaker and wash them by diluting to 200 mL with >95% methanol. Gently stir for 5–10 min and filter to separate the Bio-Beads. Transfer the Bio-Beads back into the beaker and repeat the washing procedure two more times using >95% ethanol. Wash the Bio-Beads one final time with at least 500 mL of Milli-Q H_2O and leave to dry on the filter under vacuum for 30 min before transferring them to a 50 mL Falcon tube for storage.
14. 10 mM FluoZin-3 in ultrapure H_2O: FluoZin-3 tetrapotassium salt (cell impermeant; Invitrogen, Thermo Fisher Scientific) is spun for 60 s at 18,000× *g* to pellet solids before resuspension. Carefully open the tube immediately after centrifugation and add 59 μL of Ultrapure H_2O to resuspend the entire contents of the tube (500 μg). Gently mix for several minutes until it is completely dissolved. Store the FluoZin-3 stock at −20 °C as 10 μL aliquots wrapped in aluminum foil (*see* **Note 7**).
15. 10 mM Calcein in 100% DMSO: Weigh 20 mg of high-purity calcein (cell impermeant; Invitrogen, Thermo Fisher Scientific)

into a 15 mL Falcon tube. Dissolve calcein by adding 3.22 mL of pure DMSO and gently mix until it is fully dissolved. Aliquot calcein into 50 μL fractions in PCR tubes and store at −20 °C protected from light, avoiding freeze–thaw cycles.

16. 10 mM CTAP-3 in ultrapure H_2O [52]: Weigh 15 mg of CTAP-3 into a 15 mL Falcon tube. Dissolve CTAP-3 by adding 3.06 mL of ultrapure H_2O and gently mix until it is fully dissolved. Aliquot and store as described for calcein.
17. 10 mM Fe(II) in Milli-Q H_2O: Weigh 27.8 mg of iron(II) sulfate heptahydrate ($FeSO_4 \cdot 7H_2O$) into a 15 mL Falcon tube. Transfer the tube into a constant flow N_2-purged glovebox and dissolve with 10 mL of Chelex-treated Milli-Q H_2O that has been made O_2 free by three vacuum/nitrogen cycles on a Schlenk line.
18. 100 mM Cu(I) in CH_3CN: Weigh 37.3 mg of tetrakis(acetonitrile)copper(I) hexafluorophosphate ($[Cu(CH_3CN)_4]PF_6$) into a 1.5 mL Eppendorf tube inside a glove box. Dissolve it with the addition of 1.0 mL of argon-purged CH_3CN and gently mix inside the glove box (*see* **Notes 8** and **9**).
19. 100 mM ATP in Milli-Q H_2O: Weigh 0.551 g of adenosine-5′-triphosphate disodium salt hydrate into a 15 mL Falcon tube and dissolve it with 10 mL of Chelex-treated Milli-Q H_2O (*see* **Note 10**).
20. 1.0 M $MgCl_2$ in Milli-Q H_2O: Weigh 2.03 g of magnesium chloride hexahydrate ($MgCl_2 \cdot 6H_2O$) into a 15 mL Falcon tube and dissolve it with 10 mL of Chelex-treated Milli-Q H_2O.

For the preparation of oxygen-free stock solutions, use a standard Schlenk line degassing protocol.

3 Methods

3.1 Liposome Preparation

3.1.1 Day 1

1. Dry lipids into a thin-film cake to remove $CHCl_3$ by transferring 3.0 mL of 25 mg/mL *E. coli* polar lipid extract and 1.0 mL of 25 mg/mL *L*-α-phosphatidylcholine, a 3:1 (w/w) ratio, respectively, to a 50 mL or 100 mL round bottom flask. Flow a steady stream of N_2 gas over the lipids, controlled by inserting a 200 μL pipette tip into the N_2 tube and suspending the tip approximately 1–2 cm above the round bottom of the flask. Gently swirl the round bottom flask or tilt the top of the flask 30° and spin the entire flask under N_2 flow until $CHCl_3$ has fully evaporated and the lipids have formed a thin film or "cake" on the walls of the flask (*see* **Note 11**) (*see* Fig. 1).

2. Take a piece of aluminum foil and puncture it 10–15 times with a thumb tack in a circular pattern. Place the circular hole pattern over the opening of the round bottom flask and seal the foil around the neck of the flask. Cover any exposed areas of the flask with foil to protect it from light. Desiccate the flask overnight under vacuum to complete the removal of $CHCl_3$.

3.1.2 Day 2

1. Remove the lipids from the desiccator and resuspend using 3.6 mL of 1.0 mM TCEP by gently pipetting the solution over the lipid cake (*see* **Note 12**).
2. Once the lipids are fully resuspended, add 400 μL of lipid resuspension buffer to the flask, yielding a 20 mM MOPS-NaOH pH 7.0, 100 mM NaCl, 1.0 mM TCEP buffer, and 25 mg/mL lipids.
3. Transfer the resuspended lipids to a 15 mL Falcon tube and rupture the liposomes by shock-freezing for ~10 min in liquid N_2, followed by thawing in a room temperature water bath. Repeat this process for a total of three freeze–thaw cycles.
4. Assemble an Avanti mini extruder as described by the manufacturer (Avanti Polar Lipids), starting with a 1.0 μm pore size filter. Extrude the lipids at least 11 times to achieve SUVs. Disassemble the extruder and dry with a Kim Wipe before reassembling using the next smallest filter pore size, 0.4 μm. Extrude the lipids at least 11 times to reduce to the size of the SUVs. Repeat this process again for the smallest filter, 0.2 μm (*see* **Notes 13** and **14**).
5. Add 40 μL of 2% (w/v) CYMAL-7 (*Lp*IroT/*Pa*AitP) or DDM (*Ec*CopA) in Milli-Q H_2O to 4.0 mL of liposome suspension for a final concentration of 0.02% (w/v) detergent (~2× critical micellar concentration (CMC) for CYMAL-7 and DDM) to destabilize the unilamellar vesicles.
6. Gently mix the lipids at room temperature for 60 min by tilting or rotating.
7. Incubate the lipids at 4 °C for 15 min to cool them before the addition of a target protein.
8. Separate the lipids into two 2.0 mL aliquots in 15 mL Falcon tubes.
9. Add the purified protein stock (*Lp*IroT, *Pa*AitP, or *Ec*CopA), yielding a final concentration of 1.0 mg protein/25 mg lipids (using 3.0–5.0 mg/mL protein stocks) to one fraction of 2.0 mL of 25 mg/mL detergent-destabilized SUVs, now termed PL. Add an equal volume of the corresponding protein purification buffer to the other 2.0 mL fraction of detergent-destabilized SUVs, now termed control liposomes (CL).
10. Gently mix both tubes by tilting or rotating at 4 °C for 60 min.

11. Weigh activated Bio-Beads into eight 15 mL Falcon tubes at a final concentration of 60 mg/mL Bio-Beads (based on the total volume to be added to each tube; lipids + protein/buffer). Label four tubes as PL and the remaining four as CL.
12. Transfer the PLs and CLs into a Bio-Bead-containing tube and incubate for 60 min at 4 °C, while gently mixing by tilting or rotating, to remove detergent.
13. Collect the supernatant from both tubes by aspirating the solution with a 1000 μL pipette tip against the bottom of the Falcon tube, transferring it to fresh Bio-Beads. Repeat this process until no liquid remains. Incubate overnight as described in the previous step (*see* **Note 15**).

3.1.3 Day 3

1. Transfer solutions to fresh Bio-Beads, as previously described, and incubate for 2 h at 4 °C while gently mixing by tilting or rotating.
2. Transfer solutions to fresh Bio-Beads for a final time, as previously described, and incubate for 2 h at 4 °C while gently mixing by tilting or rotating.
3. Carefully separate the supernatant from the Bio-Beads by transferring it into fresh tubes to completely remove the Bio-Beads (*see* **Note 16**).
4. Transfer PLs and CLs to ultracentrifuge tubes balancing to ±1.0 mg and centrifuge at 160,000× *g* for 45–60 min at 4 °C (*see* **Note 17**).
5. Carefully separate the supernatant from the pellet, which contains the PLs and CLs, and store it for analysis via SDS-PAGE.
6. Resuspend the PL and CL pellets to a final concentration of 50 mg/mL lipids (2.0 mg/mL protein) by gently pipetting with 1000 μL of the respective transport buffer. Separate into 250 μL aliquots and freeze in liquid N_2 for storage at −80 °C.

3.2 Liposome Characterization

3.2.1 SDS-PAGE

To verify the encapsulation of protein in the lipid-bilayer (proteoliposome) and estimate the incorporation efficiency from the amount of protein remaining unincorporated (in the supernatant) and incorporation amount from a standard curve (*see* Fig. 2a):

1. Mix 5.0 μL of PLs with 5.0 μL of 4× Laemmli protein sample reducing buffer (Laemmli) and load into an SDS-PAGE well (*see* **Note 18**).
2. Mix 10 μL of PL supernatants, obtained from liposome preparation **step 20**, with 10 μL of 4× Laemmli buffer and load into an SDS-PAGE well.
3. Mix 5.0 μL of the corresponding protein standard at a range of concentrations spanning ~0.2x–2x encapsulated protein

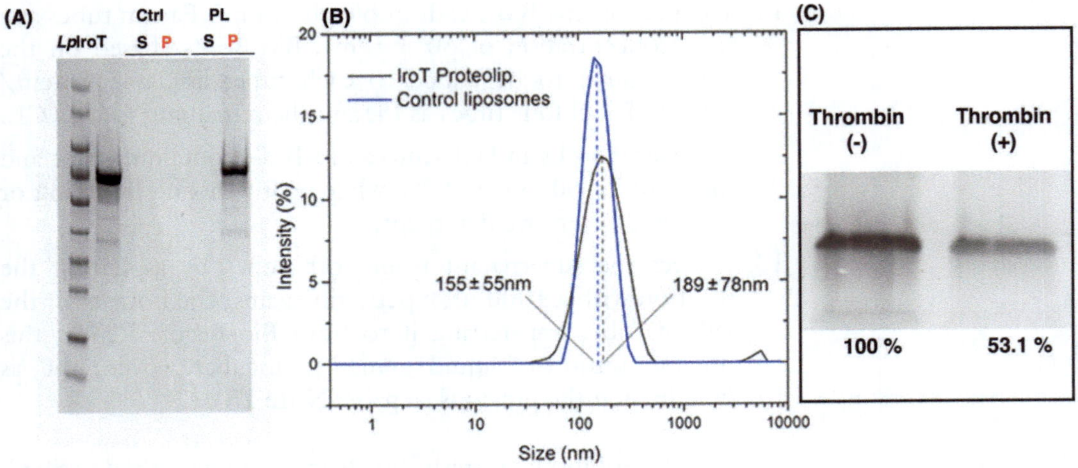

Fig. 2 Qualitative and quantitative analysis of generated proteoliposomes, exemplified for *Lp*IroT. (**a**) SDS-PAGE demonstrating successful incorporation of *Lp*IroT in SUVs. Lane 1: protein EZ-Run™ pre-stained standard (Thermo Fisher Scientific), lane 2: purified *Lp*IroT, lane 3: supernatant from CL, lane 4: pellet from CL, lane 5: supernatant from *Lp*IroT PL, and lane 6: pellet from *Lp*IroT PL. (**b**) Dynamic light scattering analysis of SUV average diameter size, illustrating size distribution for CL and PL. (**c**) Western Blot analysis of cleaved and uncleaved *Lp*IroT-PL to determine protein incorporation directionality in SUVs, following His_{10}-tag removal upon thrombin cleavage. (Adapted with permission from Abeyrathna S. et al., Biochemistry, 58, 4337–4342, Copyright 2019 American Chemical Society)

concentrations (1.5 mg/mL, 1.0 mg/mL, 0.75 mg/mL, 0.5 mg/mL, and 0.25 mg/mL) with 5.0 μL of 4× Laemmli buffer and load into an SDS-PAGE well to estimate protein concentration in the PL.

4. Mix 5.0 μL of CLs with 5.0 μL of 4× Laemmli buffer and load into an SDS-PAGE well to ensure control purity.
5. Mix 10 μL of CL supernatant, obtained from liposome preparation **step 20**, with 10 μL of 4× Laemmli buffer and load into an SDS-PAGE well.

3.2.2 Dynamic Light Scattering (DLS)

To confirm the average size and polydispersity of the PL/CL using DLS (*see* Fig. 2b):

1. Add 5.0 μL of PL/CL to 1.0 mL of the respective filtered transport buffer. Mix them well and transfer the mixture to a DLS micro cuvette.
2. Collect dynamic light scattering data using 633 nm wavelength and 175° scattering angle at 25 °C (*see* **Note 19**).
3. Record the PL/CL size and polydispersity index (PDI) separately.

3.2.3 Protein Directionality in PLs

To determine the directionality of protein reconstituted in PLs, that is, inward vs. outward, extraluminally exposed affinity purification tags can be selectively cleaved from the protein's N- or C-terminus and analyzed via SDS-PAGE and Western Blot (WB). Cleavage efficiency, which can be quantified by band densitometric analysis, compared to an uncleaved standard, is directly correlated with the outward facing orientation of the target protein (*see* Fig. 2c):

1. Dilute 500 μL of lipids at 2.0 mg/mL protein (50 mg/mL lipids) to 1.0 mL (1.0 mg/mL protein) using the respective transport buffer.
2. Add a prescreened optimized amount of the corresponding protease (protein-construct dependent) to the PLs and gently mix the sample under the optimized time and temperature conditions, to ensure complete cleavage, by tilting or rotating.
4. Take a fraction of both the cleaved and uncleaved PL samples for analysis by WB.
5. Prepare the protease-cleaved sample for SDS-PAGE by mixing 10 μL of cleaved PL with 10 μL of 4× Laemmli buffer.
6. Prepare the PL-uncleaved standard by mixing 5.0 μL of 2.0 mg/mL PL with 5.0 μL of 4× Laemmli buffer.
7. Load the samples onto an SDS-PAGE gel and transfer the separated proteins onto a nitrocellulose membrane for WB.
8. Probe the membrane with the protein construct-dependent antibody to detect the cleaved tag of interest.
9. Compare the band intensities of the cleaved and control (uncleaved) samples to quantify protein directionality (control 100% uncleaved; PL band intensity proportional to inward facing tag orientation).

3.3 Metal-Dependent Fluorescent Probe Encapsulation

1. Thaw a 250 μL aliquot of PL, or CL, frozen at a concentration of 50 mg/mL lipids (2.0 mg/mL protein) at 4 °C or on ice.
2. Dilute PL/CL to 1.0 mL, 12.5 mg/mL lipids, in their respective transport buffer and add the fluorescent probe to the specified concentration:
   (a) *Lp*IroT—Add 1.0 μL of previously prepared 10 mM FluoZin-3 to 1.0 mL of PL/CL (final: 10 μM).
   (b) *Pa*AitP—Add 4.0 μL of previously prepared 10 mM Calcein to 1.0 mL of PL/CL (final: 40 μM).
   (c) *Ec*CopA—Add 2.0 μL of supplied or synthesized 10 mM CTAP-3 to 1.0 mL of PL/CL (final: 20 μM).
3. Freeze-fracture the liposomes using three freeze–thaw cycles in liquid N_2 as previously described in the liposome preparation.

4. Extrude the liposomes as previously described in the liposome preparation (*see* **Note 20**).
5. Dilute the liposomes to 1.5 mL in a 1.5 mL ultracentrifuge tube and balance to ±1 mg. Remove unencapsulated dye from the liposomes (PL and CL) by pelleting via ultracentrifugation at 160,000× *g* for 45 min at 4 °C (*see* **Note 21**).
6. Immediately upon removing the ultracentrifuge tube from the rotor, gently remove the supernatant, containing the unencapsulated fluorescent probe, taking care not to disturb the liposome (PL and CL) pellets.
7. Resuspend and wash the liposomes in 1.5 mL of transport buffer and repeat the washing and ultracentrifugation process two more times.
8. After the third ultracentrifugation process, resuspend the PL/CL pellets to the desired final volume:
   (a) *Lp*IroT: Resuspend with 1.0 mL of transport buffer (12.5 mg/mL lipid: 0.5 mg/mL protein).
   (b) *Pa*AitP: Resuspend with 5 mL of transport buffer (1.25 mg/mL lipid: 0.05 mg/mL protein) (*see* **Note 22**).
   (c) *Ec*CopA: Resuspend with 500 μL of transport buffer (25 mg/mL lipid: 1 mg/mL protein).

3.4 Real-Time Fe(II) Transport by *Lp*IroT Monitored Through Fluozin-3 Fluorescence

1. Degas 20 mL of *Lp*IroT transport buffer and Chelex-treated Milli-Q H_2O using a standard Schlenk line protocol.
2. Use the 10.0 mM Fe(II) stock to generate the highest concentration of the concentration series (10x) being tested, 400 μM. Dilute 40 μL of the Fe(II) stock with 960 μL of degassed transport buffer in an Eppendorf tube inside the glove box.
3. Prepare a set of 1:1 serial dilutions to generate 200 μM, 100 μM, 50 μM, and 25 μM stocks by mixing 500 μL of the corresponding concentration with 500 μL of degassed transport buffer. To make the final 10 μM Fe(II) stock, mix 25 μL of the 400 μM Fe(II) stock with 975 μL of degassed transport buffer.
4. Equilibrate a fluorometer heat block temperature controller at 25 °C for at least 30 min prior to use.
5. For kinetic fluorescence measurements, select an excitation wavelength of 480 nm and an emission wavelength of 515 nm and a slit width of 2.5 nm. Data points should be collected with 0.5 s intervals.
6. Prior to and in between each run rinse the cuvette with 3% HNO_3, followed by an extensive wash with water and finally >70% ethanol. Dry the cuvettes under a flow of N_2 (*see* **Note 23**).

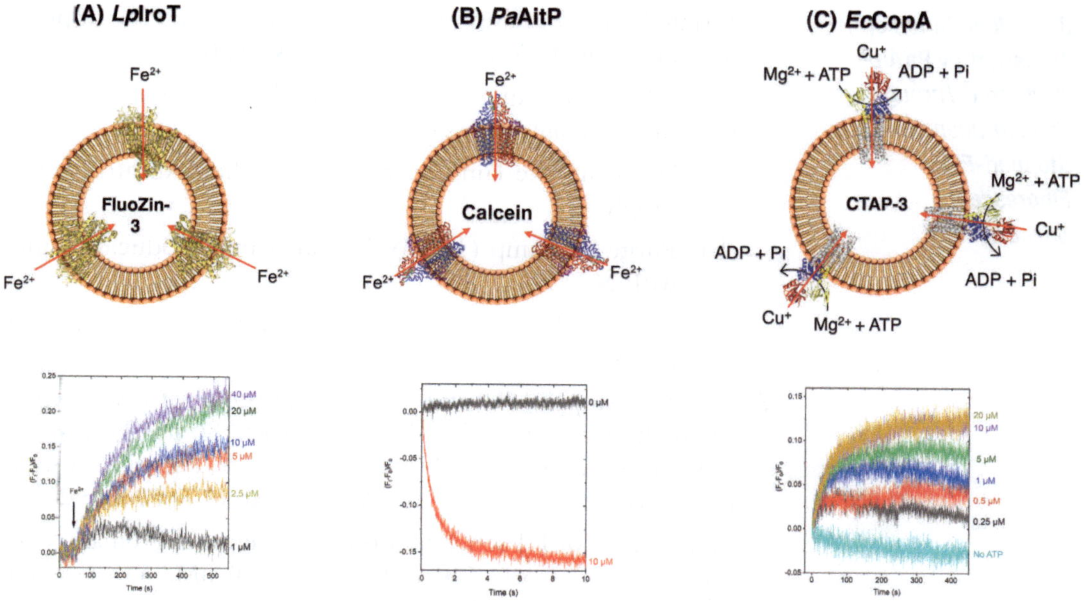

Fig. 3 Emission kinetic traces illustrating fluorescence-detected real-time metal transport in proteoliposomes: (**a**) Fe(II) transport by *Lp*IroT upon FluoZin-3 encapsulation in the proteoliposome lumen (Adapted with permission from Abeyrathna S. et al., Biochemistry, 58, 4337–4342, Copyright 2019 American Chemical Society); (**b**) Fe(II) transport by *Pa*AitP upon calcein encapsulation in the proteoliposome lumen; and (**c**) Cu (I) transport by *Ec*CopA upon CTAP-3 encapsulation in the proteoliposome lumen. (Reproduced from Abeyrathna N. et al., Dalton Trans., 2020, 49, 16,082–16,094, with permission from the Royal Society of Chemistry)

7. Transfer 108 μL of PLs or CLs into a microcuvette and equilibrate until the fluorescence baseline is stable for at least 60 s (typically 200–300 s) (*see* **Note 24**).
8. Pause the run and add 12 μL of the Fe(II) stock prepared above and rapidly restart the kinetics measurement, collecting data for 600 s.
9. Repeat **steps 7–9** until data for all Fe(II) substrate concentrations have been collected in triplicate for both PL and CL.
10. Plot the kinetics data of the fluorescence change as $(F_t—F_0)/F_0$, where F_0 is the fluorescence emission prior to Fe (II) addition and F_t is the fluorescence emission at each given time after Fe(II) addition (*see* Fig. 3a).
11. Correct the emission response of the PL transport traces by subtracting the traces obtained with the corresponding CL for each Fe(II) concentration. Kinetics data fitting can be performed subsequently (*see* **Note 25**).

3.5 Real-Time Fe(II) Transport by *Pa*AitP Monitored Through Calcein Quenching via Stopped-Flow Fluorescence Spectroscopy

1. Install the appropriate emission filter (515 nm) on the stopped-flow instrument (Applied Photophysics SX20) (*see* **Note 26**).
2. All startup procedures should be completed 60 min before the start of data collection.
   (a) Warm up the lamp by turning on the dedicated power supply.
   (b) Purge the lamp (150 W Xenon ozone-producing lamp) with N_2.
   (c) Turn on the water circulator unit and set the temperature to 25 °C.
3. While the stopped-flow instrument is purging with N_2, prepare the highest concentration 2× Fe(II)-stock, 80 μM Fe(II), in a glove box by mixing 3.97 mL of degassed *Pa*AitP transport buffer with 32 μL of the 10 mM Fe(II) stock (*see* **Note 27**).
4. Prepare a set of 1:1 serial dilutions to generate the remainder of the 2× Fe(II) stocks (40 μM, 20 μM, 10 μM, 5.0 μM, and 2.5 μM) by mixing 2.0 mL of the corresponding concentration with 2.0 mL of degassed *Pa*AitP transport buffer (experimental concentration series: 40 μM, 20 μM, 10 μM, 5.0 μM, 2.5 μM, and 1.25 μM).
5. Turn on the lamp after the system has purged with N_2 for 30 min and allow the lamp to stabilize for 30 min before starting data collection.
6. Turn on the electronics unit.
7. Open SX control software and set the excitation wavelength to 495 nm using a 10 nm slit width. Set data collection to 30 s with 0.3 ms collection intervals.
8. Remove the Chelex-treated Milli-Q H_2O from the drive syringes by manually operating the auto-stop system.
9. To clean the drive syringes, fill two syringes with Chelex-treated Milli-Q H_2O, remove any air bubbles, and connect to the drive syringes. Turn the drive valves to the loading position. Gently push down on the plungers of the water-filled syringes to load the sample into the drive syringes. Then turn the two drive valves to the drive position.
10. Repeat **step 9** with transport buffer instead of Chelex-treated Milli-Q H_2O.
11. Load one syringe with diluted PL/CL solution and another with transport buffer. Attach both syringes to the system.
12. Load the samples into the drive syringes following the steps listed in **step 9**.
13. Purge the system by driving five or more injections according to the manufacturer's recommendation and subsequently

adjust the fluorescence photomultiplier voltage using the AutoPM function until the system is fully equilibrated. Note the number of drives required to fully equilibrate the system.

14. Collect control kinetic traces with samples containing no iron.
15. Once the collection is finished, remove any remaining sample and dispose of the remaining liquid into a waste collector.
16. Rinse the drive syringes with Milli-Q H_2O, 3% HNO_3, and then Chelex-treated Milli-Q H_2O water following the steps listed in **step 9**.
17. Ensure that the prechamber of the glove box is O_2 free.
18. Inside the glove box, fill a syringe with the lowest concentration of the previously prepared 2× Fe(II) stock. Cap the syringe and seal the edge of the syringe with a parafilm.
19. Place the syringe, filled with the Fe(II) stock solution, into the prechamber of the glovebox and seal the inner door.
20. Outside the glovebox, fill a syringe with PL or CL, remove any air bubbles, and attach the syringe to the drive valve.
21. Retrieve the Fe(II)-filled syringe from the glovebox outer chamber.
22. Quickly remove any air bubbles from the sealed syringe, remove the cap, and attach the Fe(II) stock to the drive syringe.
23. Load the samples into the drive syringes as previously described and drive the system the number of times noted during **step 13** to equilibrate the system. Then click "acquire" to collect the kinetic traces.
24. Repeat **steps 16–23** until you have collected spectra for all desired Fe(II) concentrations, moving from low to high concentrations.
25. Repeat this process for the CLs until all Fe(II) stock concentrations have been collected in triplicate.
26. Plot the kinetic data using the fluorescence emission intensity (e.g., emission photomultiplier voltage, $(V_t—V_o)/V_o$), as previously described for FluoZin-3 fluorescence measurements. Kinetic data fitting can be performed subsequently (*see* Fig. 3b).

3.6 Real-Time Cu(I) Transport by EcCopA Monitored Through CTAP-3 Fluorescence

1. Prepare Cu(I) stocks using degassed buffer in the glove box as 10× stocks of the desired final concentration by diluting 1000× stocks prepared in 100% CH_3CN, from the 100 mM stock, into buffer to ensure that the final CH_3CN concentration is constant across the dilution series, generating the concentration series: 200 μM, 100 μM, 50 μM, 10 μM, 5 μM, and 2.5 μM (final Cu(I) concentrations: 20 μM, 10 μM, 5.0 μM, 1.0 μM, 0.5 μM, and 0.25 μM).

2. Equilibrate a fluorometer heat block temperature controller at 37 °C for at least 30 min prior to use.
3. For kinetic fluorescent measurements, select an excitation wavelength of 365 nm and an emission wavelength of 450 nm and a slit width of 2.0 nm. Data points should be collected every 0.1 s.
4. Wash a cuvette prior to beginning data collection and in between samples as described in the *Lp*IroT FluoZin-3 transport procedure.
5. Transfer 105.6 μL of CTAP-3-encapsulated PLs or CLs into a microcuvette and equilibrate fluorescence until the baseline is stable for 60 s.
6. Pause data acquisition to add 1.2 μL of 100 mM ATP stock (1.0 mM ATP final concentration). Mix the sample in the cuvette by gently pipetting, avoiding bubbles.
7. After ~30 s pause, start data acquisition again and add 1.2 μL of 1.0 M $MgCl_2$ (10 mM $MgCl_2$ final concentration). Gently mix and resume the kinetic measurement.
8. While collecting baseline data for another ~30 s, remove the desired 10× metal stock, sealed with the parafilm, from the glove box prechamber.
9. Pause the run and add 12 μL of the 10× metal stock to the cuvette, gently mix, and rapidly restart the measurement.
10. Collect the time-dependent fluorescence kinetic traces for 600 s before completing data collection.
12. Plot the kinetic data fluorescence change as $(F_t—F_0)/F_0$, where F_0 is the fluorescence prior to Cu(I) addition and F_t is the fluorescence emission at each given time after Cu (I) addition (*see* Fig. 3c).
11. Correct the response of the PL transport traces by subtracting the traces obtained with the corresponding CL. Kinetic data fitting can be performed subsequently.
12. An additional control is collected for P-type ATPases by performing transport assays in the absence of ATP or Mg^{2+} (both required for transport activity). Chelex-treated Milli-Q H_2O is substituted in place of those reagents.

4 Notes

1. Lipids may be supplied as lyophilized powder instead of $CHCl_3$ solutions. Resuspend lipids to the desired final concentration of lipids using 100% $CHCl_3$ and then proceed as described. All

lipids, lyophilized powder or chloroform solution, should be stored at −20 °C under an inert atmosphere to prevent lipid oxidation.

2. Fresh Chelex 100 should always be used for the purpose of removing metals from buffers and stocks. It is recommended to wash the Chelex resin prior to usage to remove non-covalently bound ligands. According to the manufacturer, the resin can be regenerated and reused. In our experience, however, the metal binding capacity of regenerated resin is reduced and no longer suitable for use in PL preparation.
3. Reducing agents (e.g., TCEP, DTT, and *L*-ascorbic acid) should be prepared fresh to ensure consistency across the preparations.
4. Detergents should be purchased at a minimum purity of Sol-Grade (≥ 98%).
5. For *Lp*IroT Fe(II)-transport, all solutions are treated in large excess of Chelex due to zinc contamination in salts and isolated lipids from native sources. Enough zinc can be carried over to the fluorescent probe encapsulation step to potentially saturate FluoZin-3 fluorescence. This results in quenching signal from the formation of FluoZin3-Fe(II), which has lower fluorescence than FluoZin3-Zn(II). Zinc contamination will vary from preparation to preparation, thus extensive treatment of solutions with Chelex is necessary.
6. Buffers may be stored at −20 °C for up to one month if needed, but it is always preferable to use freshly prepared buffers.
7. All fluorescent probes should be purchased as cell impermeant forms, otherwise, the probe can freely diffuse across the lipid bilayer and will not discriminate between metals in the intra- and extra-luminal spaces.
8. Argon is preferred in place of nitrogen for purging CH_3CN. Flush the CH_3CN bottle with a constant argon gas stream on a Schlenk line.
9. Cu(I) salts should be stored in the glove box under an inert atmosphere and solutions should be prepared immediately prior to use to prevent disproportionation.
10. ATP should be purchased at the highest purity available.
11. The top border of the lipid cake should not exceed half the height of the round bottom flask. Lipids resuspended too high on the walls of the round bottom flask can be difficult to resuspend after the lipid cake is dried. If serious enough, this may result in a significantly lower lipid concentration in the resuspension.

12. For *Ec*CopA PLs, substitute 1.0 mM *L*-ascorbic acid for 1.0 mM TCEP in all buffers.
13. Lipids can be extruded more than 11 times to increase monodispersity, but extrusion should always end on an odd number of passages through the filter, that is, in the syringe opposite to the one where the initial sample was placed. This prevents contamination by remaining multilamellar vesicles and lipids that have not been successfully extruded.
14. Regardless of the desired final liposome size, lipids should be extruded in a stepwise fashion, gradually reducing the size.
15. Particular attention should be paid to not transfer Bio-Beads between Falcon tubes. If minor amounts of Bio-Beads are carried over between tubes, it will not harm the detergent removal process.
16. A thin, long needle (>30G and 1.5 inches long) connected to a syringe can facilitate and reduce sample loss from multiple transfers compared to a micropipette.
17. Centrifugation speeds and time should be adjusted to ensure complete pelleting without damaging the SUVs. The listed speed and time are sufficient to pellet ~200 nm vesicles but smaller diameter vesicles may need to be spun at higher speeds or for longer duration. The minimum speed necessary to pellet SUVs should be used to prevent mechanical damage to the membrane.
18. Extra Laemmli buffer is used according to manufacturer's recommendations to improve SDS-PAGE quality because of the high concentration of lipids in the samples.
19. Specific scattering angle and wavelength will vary depending on the instrument being used.
20. Extrude CLs to the desired final size before extruding the PLs to prevent contamination of the CLs with PLs.
21. Samples are diluted after dye incorporation into the maximum volume of the ultracentrifugation tubes to prevent tube damage and wear. After encapsulation, liposomes can be diluted significantly to increase washing.
22. *Pa*AitP can be diluted significantly to lower protein concentrations as data are collected using a filter-based stopped-flow instrument resulting in a larger fluorescent signal.
23. Cuvettes should be rinsed with 3% HNO_3 to prevent absorption of metals on the cuvette quartz, resulting in contamination of samples placed in the cuvettes.
24. The volume should be adjusted to ensure that the meniscus of the liquid falls above the top of the cuvette window to prevent additional scattering that would lead to unreliable measurements.

25. This correction will account for SUV leakiness due to poor detergent removal and/or unencapsulated fluorescent probes from insufficient washing during probe encapsulation.
26. If the instrument being used features an emission monochromator, this step is not necessary. In general, all procedures listed are specific to the instruments used in our laboratory and should be adjusted to fit the available instrumentation.
27. Metal stocks are prepared at 2× desired experimental concentrations for stopped-flow experiments because metal is mixed in a 1:1 (v/v) ratio with PL/CL for measurement, yielding the desired 1× concentration.

Acknowledgments

We thank Dr. M. Thomas Morgan (Georgia Institute of Technology) for synthetizing the CTAP-3 Cu(I) fluorescent probe. This work was supported by the National Institute of General Medical Sciences (NIH, R35GM128704 to G. M. and R35GM136404 to C. J. F.) and the Robert A. Welch Foundation (AT-2073-20210327 and AT-2073-20240404 to G. M.). The contents of this publication are solely the responsibility of the authors and do not necessarily represent the official views of NIGMS or NIH.

References

1. Bertini I (2007) Biological inorganic chemistry: structure and reactivity. University Science Books, Sausalito, CA
2. Lippard SJ, Berg JM (1994) Principles of bioinorganic chemistry. University Science Books, Mill Valley, CA
3. Amie K, Rosenzweig AC (2009) Structural biology of copper trafficking. Chem Rev 109: 4760–4779
4. Ma Z, Jacobsen FE, Giedroc DP (2009) Coordination chemistry of bacterial metal transport and sensing. Chem Rev 109:4644–4681
5. Kohen R, Nyska A (2002) Oxidation of biological systems: oxidative stress phenomena, antioxidants, redox reactions, and methods for their quantification. Toxicol Pathol 30:620–650
6. D'Ambrosi N, Rossi L (2015) Copper at synapse: release, binding and modulation of neurotransmission. Neurochem Int 90:36–45
7. Koppenol W, Bounds P (2011) Redox-active metals: iron and copper. Princ Free Radic Biomed 1:91–111
8. Hrbac J, Kohen R (2000) Biological redox activity: its importance, methods for its quantification and implication for health and disease. Drug Dev Res 50:516–527
9. Klein JS, Lewinson O (2011) Bacterial ATP-driven transporters of transition metals: physiological roles, mechanisms of action, and roles in bacterial virulence. Metallomics 3:1098
10. Huynh C, Sacks DL, Andrews NW (2006) A leishmania amazonensis ZIP family iron transporter is essential for parasite replication within macrophage phagolysosomes. J Exp Med 203: 2363–2375
11. Weidinger A, Kozlov AV (2015) Biological activities of reactive oxygen and nitrogen species: oxidative stress versus signal transduction. Biomol Ther 5:472–484
12. Sestok AE, Lee MA, Smith AT (2022) Prokaryotic ferrous iron transport: exploiting pools of reduced iron across multiple microbial environments. In: Hurst CJ (ed) Microbial metabolism of metals and metalloids. Springer, Cham, pp 299–357
13. Andrews SC, Robinson AK, Rodríguez-Quiñones F (2003) Bacterial iron homeostasis. FEMS Microbiol Rev 27:215–237

14. Parrow NL, Fleming RE, Minnick MF (2013) Sequestration and scavenging of iron in infection. Infect Immun 81:3503–3514
15. Brown JB, Lee MA, Smith AT (2021) Ins and outs: recent advancements in membrane protein-mediated prokaryotic ferrous iron transport. Biochemistry 60:3277–3291
16. Létoffé S, Heuck G, Delepelaire P et al (2009) Bacteria capture iron from heme by keeping tetrapyrrol skeleton intact. Proc Natl Acad Sci 106:11719–11724
17. Kammler M, Schön C, Hantke K (1993) Characterization of the ferrous iron uptake system of Escherichia coli. J Bacteriol 175:6212–6219
18. Ganz T, Nemeth E (2015) Iron homeostasis in host defence and inflammation. Nat Rev Immunol 15:500–510
19. Locher KP, Rees B, Koebnik R et al (1998) Transmembrane signaling across the ligand-gated FhuA receptor: crystal structures of free and ferrichrome-bound states reveal allosteric changes. Cell 95:771–778
20. Lau CKY, Krewulak KD, Vogel HJ (2016) Bacterial ferrous iron transport: the feo system. FEMS Microbiol Rev 40:273–298
21. Sestok AE, Linkous RO, Smith AT (2018) Toward a mechanistic understanding of feo-mediated ferrous iron uptake. Metallomics 10:887–898
22. Roberts CS, Ni F, Mitra B (2021) The zinc and iron binuclear transport center of ZupT, a ZIP transporter from Escherichia coli. Biochemistry 60:3738–3752
23. Makui H, Roig E, Cole ST et al (2000) Identification of the Escherichia coli K-12 Nramp orthologue (MntH) as a selective divalent metal ion transporter. Mol Microbiol 35: 1065–1078
24. Chu BC, Garcia-Herrero A, Johanson TH et al (2010) Siderophore uptake in bacteria and the battle for iron with the host; a bird's eye view. Biometals Int J Role Met Ions Biol Biochem Med 23:601–611
25. Mosbahi K, Wojnowska M, Albalat A et al (2018) Bacterial iron acquisition mediated by outer membrane translocation and cleavage of a host protein. Proc Natl Acad Sci 115:6840–6845
26. Arnold FM, Weber MS, Gonda I et al (2020) The ABC exporter IrtAB imports and reduces mycobacterial siderophores. Nature 580:413–417
27. Ferguson AD, Chakraborty R, Smith BS et al (2002) Structural basis of gating by the outer membrane transporter FecA. Science 295: 1715–1719
28. Ogierman M, Braun V (2003) Interactions between the outer membrane ferric citrate transporter FecA and TonB: studies of the FecA TonB box. J Bacteriol 185:1870–1885
29. Rivera M (2017) Bacterioferritin: structure, dynamics, and protein–protein interactions at play in iron storage and mobilization. Acc Chem Res 50:331–340
30. Abeyrathna SS, Abeyrathna NS, Thai NK et al (2019) IroT/MavN Is a *Legionella* transmembrane Fe(II) transporter: metal selectivity and translocation kinetics revealed by *in Vitro* real-time transport. Biochemistry 58:4337–4342
31. Sankari S, O'Brian MR (2014) A bacterial iron exporter for maintenance of iron homeostasis. J Biol Chem 289:16498–16507
32. Lu M, Fu D (2007) Structure of the zinc transporter YiiP. Science 317:1746–1748
33. Argüello JM, Raimunda D, Padilla-Benavides T (2013) Mechanisms of copper homeostasis in bacteria. Front Cell Infect Microbiol 3:73
34. Chen J, Jiang Y, Shi H et al (2020) The molecular mechanisms of copper metabolism and its roles in human diseases. Eur J Phys 472:1415–1429
35. Espinoza A, Le Blanc S, Olivares M et al (2012) Iron, copper, and zinc transport: inhibition of divalent metal transporter 1 (DMT1) and human copper transporter 1 (hCTR1) by shRNA. Biol Trace Elem Res 146:281–286
36. Solioz M (2018) Copper homeostasis in gram-negative bacteria. In: Solioz M (ed) Copper and bacteria: evolution, homeostasis and toxicity. Springer, Cham, pp 49–80
37. Sitsel O, Grønberg C, Autzen HE et al (2015) Structure and function of Cu(I)- and Zn(II)-ATPases. Biochemistry 54:5673–5683
38. Telianidis J, Hung YH, Materia S et al (2013) Role of the P-type ATPases, ATP7A and ATP7B in brain copper homeostasis. Front Aging Neurosci 5:44
39. Buiakova OI, Xu J, Lutsenko S et al (1999) Null mutation of the murine ATP7B (Wilson disease) gene results in intracellular copper accumulation and late-onset hepatic nodular transformation. Hum Mol Genet 8:1665–1671
40. Tümer Z (2013) An overview and update of *ATP7A* mutations leading to menkes disease and occipital horn syndrome. Hum Mutat 34: 417–429
41. Rensing C, Fan B, Sharma R et al (2000) CopA: an Escherichia coli Cu(I)-translocating P-type ATPase. Proc Natl Acad Sci 97:652–656

42. Barry AN, Shinde U, Lutsenko S (2010) Structural organization of human Cu-transporting ATPases: learning from building blocks. J Biol Inorg Chem 15:47–59
43. Kaplan JH, Lutsenko S (2009) Copper transport in mammalian cells: special care for a metal with special needs. J Biol Chem 284:25461–25465
44. Salustros N, Grønberg C, Abeyrathna NS et al (2022) Structural basis of ion uptake in copper-transporting P_{1B}-type ATPases. Nat Commun 13:5121
45. Abeyrathna N, Abeyrathna S, Morgan MT et al (2020) Transmembrane Cu(I) P-type ATPase pumps are electrogenic uniporters. Dalton Trans Camb Engl 49:16082–16094
46. Ding C, Festa RA, Sun TS et al (2014) Iron and copper as virulence modulators in human fungal pathogens: iron and copper as virulence modulators in human fungal pathogens. Mol Microbiol 93:10–23
47. Palmer LD, Skaar EP (2016) Transition metals and virulence in bacteria. Annu Rev Genet 50: 67–91
48. Carpenter EP, Beis K, Cameron AD et al (2008) Overcoming the challenges of membrane protein crystallography. Curr Opin Struct Biol 18:581–586
49. Repen B, Schneider E, Alexiev U (2012) Optimization of a malachite green assay for detection of ATP hydrolysis by solubilized membrane proteins. Anal Biochem 426:103–105
50. Knowles AF, Kandrach A, Racker E et al (1975) Acetyl phosphatidylethanolamine in the reconstitution of ion pumps. J Biol Chem 250:1809–1813
51. Geertsma ER, Nik Mahmood NAB, Schuurman-Wolters GK et al (2008) Membrane reconstitution of ABC transporters and assays of translocator function. Nat Protoc 3: 256–266
52. Morgan MT, McCallum AM, Fahrni CJ (2016) Rational design of a water-soluble, lipid-compatible fluorescent probe for Cu (I) with sub-part-per-trillion sensitivity. Chem Sci 7:1468–1473
53. Abeyrathna SS, Abeyrathna NS, Basak P et al (2022) Plastic recognition and electrogenic uniport translocation of 1st-, 2nd-, and 3rd-row transition and post-transition metals by primary-active transmembrane P_{1B-2}-type ATPase pumps. Chem Sci 14:6059–6078
54. Cianciotto NP (2015) An update on iron acquisition by Legionella pneumophila: new pathways for siderophore uptake and ferric iron reduction. Future Microbiol 10:841–851
55. Christenson ET, Isaac DT, Yoshida K et al (2019) The iron-regulated vacuolar legionella pneumophila MavN protein is a transition-metal transporter. Proc Natl Acad Sci 116: 17775–17785
56. Portier E, Zheng H, Sahr T et al (2015) *IroT/ mavN* , a new iron-regulated gene involved in *L egionella pneumophila* virulence against amoebae and macrophages: *IroT*, a new iron-regulated gene. Environ Microbiol 17:1338–1350
57. Salusso A, Raimunda D (2017) Defining the roles of the cation diffusion facilitators in Fe^{2+}/ Zn^{2+} homeostasis and establishment of their participation in virulence in pseudomonas aeruginosa. Front Cell Infect Microbiol 7:84
58. Mihelj P, Abreu I, Moreyra T et al (2023) Functional characterization of the Co^{2+} transporter AitP in sinorhizobium meliloti: a new player in Fe^{2+} homeostasis. Appl Environ Microbiol 89:e01901–e01922
59. Abeyrathna N, Abeyrathna S, Morgan M et al (2021) Mechanistic dissection of Cu(I)-translocating P-type ATPase pumps by *In-vitro* real-time transport in artificial lipid bilayer vesicles. FASEB J. 35:fasebj.2021.35.S1.03160
60. Argüello JM, González-Guerrero M, Raimunda D (2011) Bacterial transition metal P_{1B}-ATPases, transport mechanism and roles in virulence. Biochemistry 50:9940–9949

Chapter 6

Metal Uptake by Mitochondrial Carrier Family Proteins Using *Lactococcus lactis*

Xinyu Zhu, Laura E. Oldfather, and Paul A. Cobine

Abstract

Metal ion homeostasis in mitochondria is essential to maintaining proper cellular physiology. However, the ability of metals to bind off target or form complexes with multiple metabolites presents major challenges to understanding the mechanisms that govern this homeostasis. Adding further to the complexity, some of the major mitochondrial transporters have shown substrate promiscuity. In many cases, mitochondrial metals are found in the matrix compartment that is surrounded by the impermeable inner membrane. Four major classes of transporters facilitate the movement of solute across the inner membrane. These are mitochondrial carrier family, ATP-binding cassette transporters, mitochondrial pyruvate carriers, and sideroflexins. For iron, the matrix is the site of iron-sulfur clusters and heme synthesis and therefore transport must occur in a coordinated fashion with the cellular needs for these critical cofactors. Iron could be transported in numerous forms as it has been shown to form complexes with abundant metabolites such as citrate, nucleotides, or glutathione. Here, we describe assays to study iron (or any metal) transport by mitochondrial carrier family proteins expressed in *Lactococcus lactis* using a nisin-controlled expression system.

Key words Mitochondria, Mitochondrial inner membrane, Mitochondrial carrier family, Iron, Transport

1 Introduction

Solutes are translocated to the mitochondrial matrix through the impermeable inner membrane (IM) via four major classes of transporters: the mitochondrial carrier family (MCF/SLC25), ATP-binding cassette transporters, mitochondrial pyruvate carriers, and sideroflexins [1]. MCF proteins comprise the most abundant family with 53 members in humans and 30 members in *Saccharomyces cerevisiae* that are collectively responsible for the transport of numerous substrates for cellular processes including TCA cycle, nucleoside for energy metabolism, nucleotide for replication, and amino acids for degradation or maintenance of the urea cycle [2]. The MCF proteins have a conserved fold with three repeats that contain two transmembrane helices connected by a short

Oleh Khalimonchuk (ed.), *Iron Metabolism: Methods and Protocols*, Methods in Molecular Biology, vol. 2839,
https://doi.org/10.1007/978-1-0716-4043-2_6,

α-helical loop [3–5]. Structural studies of the MCF proteins show that they can adopt two distinct conformational states: cytoplasmic state (c-state) and matrix state (m-state). In the c-state, the salt bridges form on the matrix side of the protein, make the carrier open to the IMS, and give the MCF access to metabolites and other substrates present in this compartment, while the m-state is open to the matrix side [6, 7]. Currently, no substrates have been definitively identified for 16 out of 53 human MCFs, and it has been recently demonstrated that some MCF members may transport multiple solutes, suggesting promiscuity in their substrates [1, 8].

Recruitment of transition metals such as copper, iron, zinc, and manganese to the matrix is required for optimal mitochondrial function; however, the transporters that move these ions across the IM are under investigation. There are at least four distinct experimental approaches to identify and characterize a mitochondrial metal transporter: (1) characterization of null mutant phenotype that results from loss of metal in the compartment, (2) direct detection of a defect in substrate accumulation in the organelle or compartment, (3) in vitro purification and reconstitution of the transporter followed by assays in combination with a specific inhibitor of transport as proof of principle for folding fidelity, and (4) heterologous expression in an organism that can fold the protein and insert in the membrane. Mitochondrial iron has been studied by all four approaches.

The MCF proteins MITOFERRIN1 (MFRN1/SLC25A37) and MITOFERRIN2 (MFRN2/SLC25A28) have been characterized as iron transporters. Deletion of MFRN1 in zebrafish and mouse results in decreased heme production in developing erythroid cells that gives rise to severe anemias, which phenotypically demonstrated the role of MFRN1 as an iron transporter for heme production. Although MFRN2 partially overlaps with MFRN1 in iron transport, it failed to suppress the anemia phenotype because of its relative low abundance in erythroid cells [9]. The *S. cerevisiae* homologs of MITOFERRIN1/2 are Mrs3/4. Deletion of both *MRS3* and *MRS4* causes a severe growth defect in yeast grown under iron-depleted conditions [10]. Additional phenotypic evidence supports the idea that total mitochondrial iron availability is reduced in the *mrs3Δ* including the activation of iron-dependent transcription factor in yeast AFT1 and impaired loading of an heterologously expressed Fe-SOD that was localized to the matrix [11, 12]. In addition to Mrs3 and Mrs4 in yeast, Rim2 was shown to mediate transport of nucleotide-bound iron across the IM [13]. Finally, deletion of *MTM1* results in mismetallation of manganese Sod2 with iron that could be reversed by *MRS3* deletion [14]. These examples all support the phenotypic characterization of Mrs3 as an iron transporter. Biochemically, proteoliposomes containing purified MFRN1 and IM-derived vesicles with overexpressed MFRN1 demonstrate the iron uptake capability of this

MCF protein [15–17]. In fact, MFRN1 in proteoliposomes can transport iron, copper, manganese, and zinc but not nickel [17], suggesting some substrate promiscuity.

The combination of assays demonstrated the role of MITOFERRIN1/MRS3 in iron transport. However, the exact mechanisms and the potential role of interacting partners are under investigation [18, 19]. In general, disrupted mitochondrial iron homeostasis results in accumulation of insoluble iron in mitochondria, which is observed in numerous Fe-related disease states. Mutation of phosphate carrier in yeast which limits the availability of this iron-binding metabolite helps reverse some iron accumulation defects in a *YFH1* mutant [20]. Therefore, a possible function of a to date unidentified endogenous iron ligand(s) could be to maintain soluble matrix metal pools and prevent accumulation of insoluble iron in mitochondria [21, 22]. Identification of such a ligand(s) would have a major impact on determining the mechanism of the iron transporters.

Here, we describe the use of *Lactococcus lactis* as a tool for studying iron transport in mitochondria. *L. lactis* is a gram-positive bacterium that can fold and insert eukaryotic proteins into its cytoplasmic membrane [23]. A commercially available nisin-controlled gene expression system (NICE) in which nisin serves as an inducer of protein production has been developed [24]. The NICE system has been utilized for expressing various prokaryotic and eukaryotic membrane proteins including MCFs [25]. This system has potential to help expand our understanding of iron transport as it allows rapid production of stably folded proteins that can be tested without purification or refolding. This allows for experimental design that can test multiple redox states, possible exchange of substrates, and multiple buffering/ligand components (Fig. 1). The method has been adapted from the successful assays of metabolites such as ATP/ADP, oxaloacetate, citrate, and metals including copper and silver. In this chapter, we discuss the use of heterologous expression in *L. lactis* to characterize iron transport by mitochondrial carrier family proteins.

2 Materials

All solutions should be prepared in ultrapure water with analytical grade reagents at room temperature unless noted otherwise.

2.1 Construction of Lactococcus lactis *Expression Plasmid*

1. *Lactococcus lactis* expression plasmid pNZ8148 carries a chloramphenicol resistance cassette and a limited range of usable restriction sites (*see* **Notes 1** and **2**).
2. *Escherichia coli* hosting strain MC1061, chemically competent. Store in glycerol 20% (v/v) in a -80 °C freezer (*see* **Note 3**).

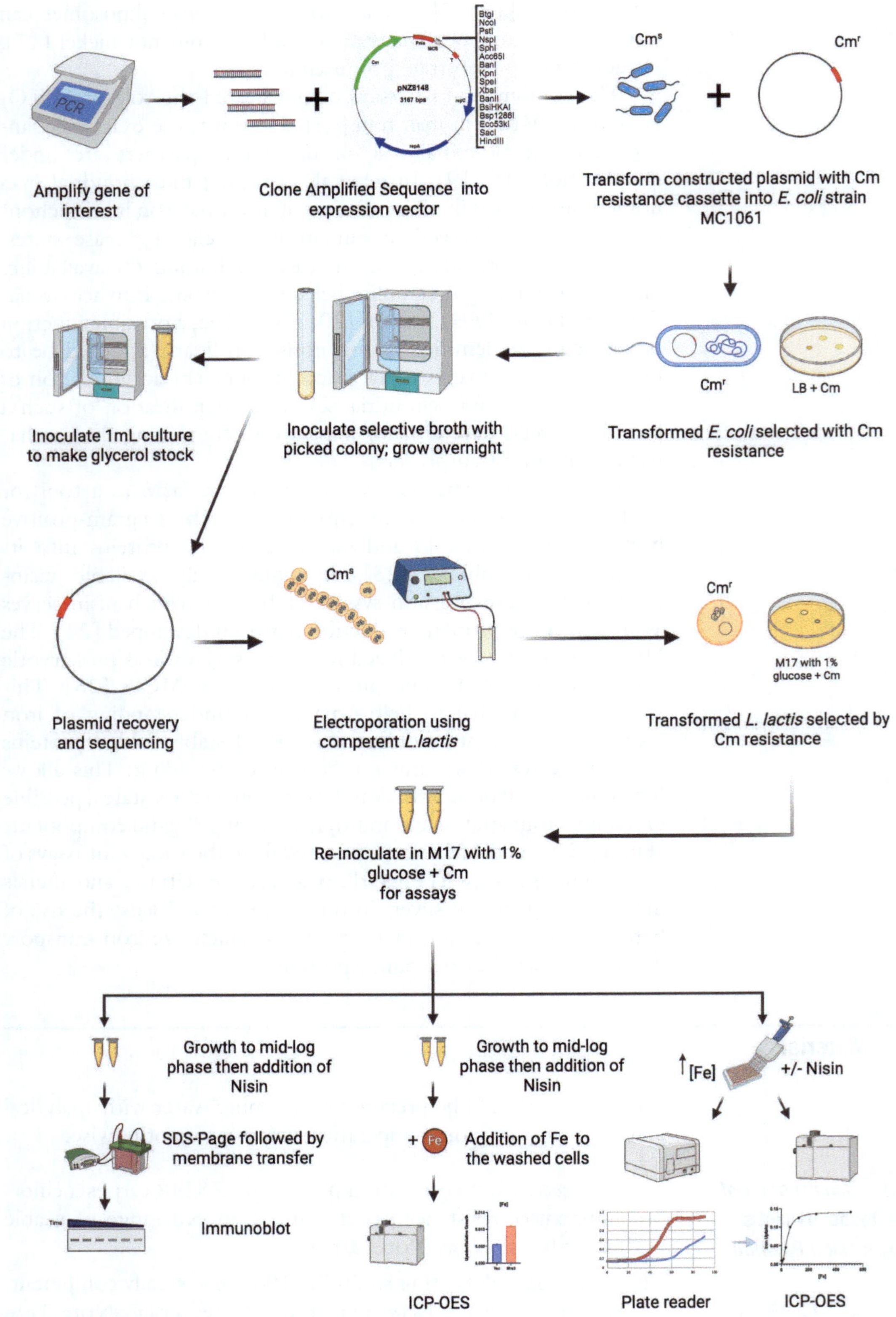

Fig. 1 *Workflow of iron uptake assay.* The target gene is amplified from genomic DNA (yeast) or cDNA (other eukaryotes) and cloned into the *L. lactis* expression plasmid. The constructed plasmid is then transformed into

3. Lysogeny broth (LB) liquid media: tryptone 1% (w/v), NaCl 1% (w/v), and yeast extract 0.5% (w/v).
4. Chloramphenicol: dissolve chloramphenicol powder in ethanol absolute (≥ 99.9%) to make a 25 mg/mL stock. Store at −20 ° C for up to 1 year.
5. SOC medium: yeast extract 0.5% (w/v), tryptone 2% (w/v), 20 mM glucose, 10 mM NaCl, 2.5 mM KCl, 10 mM $MgCl_2$, and 10 mM $MgSO_4$. Adjust pH to 7 with NaOH.

2.2 Transformation of Lactococcus lactis

1. Strain for heterologous expression: *L. lactis* NZ9000.
2. Growth Media: M17 premix 3.725% (w/v) (M17 premix contains pancreatic digest of casein 0.5% (w/v), soy peptone 0.5% (w/v), beef extract 0.5% (w/v), yeast extract 0.25% (w/v), ascorbic acid 0.05% (w/v), magnesium sulfate 0.025% (w/v), disodium-β-glycerophosphate 1.9% (w/v)), glucose 1% (w/v), sucrose 17% (w/v), and glycine 2% (w/v)).
3. Washing Buffer: Sucrose 17% (w/v) and glycerol 10% (v/v).
4. Recovery Media: M17 premix 3.725% (w/v), glucose 1% (w/v), sucrose 17% (w/v), 20 mM $MgCl_2$, and 2 mM $CaCl_2$.
5. Selective Plate: Tryptone 1% (w/v), yeast extract 0.5% (w/v), sucrose 20% (w/v), glucose 1% (w/v), gelatin 2.5% (w/v), 2.5 mM $MgCl_2$, 2.5 mM $CaCl_2$, and 1.5% agar, pH 6.8 [26].
6. Electroporation cuvette with 2 mm gap.

2.3 Expression of Target Protein in Lactococcus lactis

1. Nisin: Make the nisin stock at 1 mg/mL in 0.05% acetic acid. Store aliquoted vials at −20 °C for up to 1 year.
2. 5× SDS-PAGE loading dye: 250 mM Tris-HCl, pH 6.8, SDS 8% (w/v), bromophenol blue 0.1% (w/v), glycerol 40% (v/v), and 100 mM dithiothreitol. Store aliquoted vials at −20 °C for up to 1 year.

2.4 Iron Uptake Assay

1. Nitric acid: 67–70%, ARISTAR® ULTRA, Ultrapure for trace metal analysis, VWR Chemicals BDH®.

Fig. 1 (continued) E. coli. Transformants are grown on selective broth overnight. The plasmid is extracted, sequenced, and electroporated into *L. lactis*. The transformants are used to assay iron uptake. Before the assay, protein expression should be confirmed by immunoblot. Direct iron uptake assay is carried out by adding variable concentrations of Fe salts/complexes to washed bacterial cells after protein induction. Growth-based assays of iron uptake in medium are performed by monitoring optical density (to determine possible inhibitory concentration of the iron substrates) and direct measurement of the total iron uptake

3 Methods

3.1 Transforming pNZ8148 with Cloned Gene into MC1061

1. Thaw 50 μL chemically competent MC1061 cells on ice.
2. Add 50–100 ng plasmid DNA (the volume should be 1–5 μL) to the tube containing cells.
3. Gently mix by inverting tubes four to five times. Do not vortex.
4. Place the DNA/cell mixture on ice for 30 min.
5. Heat-shock the DNA/cell mixture at exactly 42 °C for 45 s. Do not mix.
6. Place on ice for 5 min.
7. Add 950 μL of SOC medium into the tube and shake at 250 rpm and 37 °C for 60 min (*see* **Note 4**).
8. Prepare several dilutions (e.g., 1:10, 1:20, and 1:100) of the transformed cells and plate 50–100 μL of each dilution on an LB agar plate containing 25 μg/mL chloramphenicol.
9. Incubate the plates at 37 °C for 16–24 h.
10. Pick individual colonies from the selection plate to inoculate separate 1 mL LB media containing 25 μg/mL chloramphenicol in sterile microcentrifuge tubes.
11. Incubate the cultures at 37 °C for 16–24 h.
12. Make cell stock by centrifugation at 10,000x *g* for 1 min, removing media and suspending cell pellets in 100 μL 20% (v/v) glycerol.
13. Store the *E. coli* stocks at −80 °C.

3.2 Making Electrocompetent *Lactococcus lactis* Cells

1. Grow *L. lactis* in 1 mL of M17 with 1% glucose at 30 °C to reach OD_{600} of 0.5–0.8.
2. Dilute the culture 100-fold in Growth Media.
3. Incubate at 30 °C until OD_{600} reaches 0.7.
4. Wash twice with ice cold Washing Buffer.
5. Suspend electrocompetent cells in 1/100 volume of Washing Buffer to be used or stored at −80 °C.

3.3 Electroporation of *Lactococcus lactis*

1. Add 1 ng–1 μg of pNZ8148 carrying cloned gene in water to 50 μL of electrocompetent cells. Gently mix by pipetting several times (*see* **Note 5**).
2. Transfer the mixture into an ice cold 2 mm gap cuvette.
3. Electroporate at 2 kV for 5 ms.
4. Immediately after electroporation, add 1 mL Recovery Media and transfer to a sterile microcentrifuge tube.
5. Incubate the cultures for 2 h at 30 °C.

6. Pellet *L. lactis* cells by centrifugation at 5000× *g* for 10 min.
7. Without disturbing the cell pellet, remove 900 μL media from the tube.
8. Resuspend the cell pellet in the rest of the media and plate all the cells on a Selective Plate, incubate for 2–3 days at 30 °C.
9. Take several isolated colonies from the Selective Plate to inoculate separate 1 mL Growth Media containing 25 μg/mL chloramphenicol.
10. Incubate for 24 h at 30 °C.
11. Make stocks of transformants by centrifugation at 10,000× *g* for 1 min, removing media and suspending cell pellets in 100 μL 20% glycerol.
12. Store at −80 °C.

3.4 Induction of the Target Protein in Lactococcus lactis

1. Inoculate 1 mL of M17 media containing 1% glucose with *L. lactis* transformant.
2. Incubate at 30 °C overnight.
3. Inoculate new M17 media containing 1% glucose with the overnight culture until an OD_{600} of ~0.1 is reached. The number of new cultures depends on the number of downstream applications (e.g., immunoblot, direct iron uptake, and growth curve).
4. Incubate tubes at 30 °C until OD_{600} reaches 0.6 (*see* **Note 6**).
5. Prepare fresh nisin solution and add to each tube for a final concentration of 10 ng/mL (*see* **Note 7**).
6. Incubate cultures at 30 °C for 3 h, and invert the tubes periodically.
7. Harvest cells by centrifugation at 5000× *g* for 10 min.
8. Resuspend the cells for downstream assays.

3.5 Immunoblot to Verify Protein Expression

2. 1 mL of induced *L. lactis* is centrifuged at 10,000× *g* for 1 min.
3. Resuspend the cell pellets in 100 μL 1% SDS.
4. Sonicate the cell suspension with 70% power for 30 s and put on ice to rest for 30 s. Repeat the cycle three times (*see* **Note 8**).
5. Add SDS loading dye and incubate at 95 °C for 5 min (*see* **Note 9**).
6. Run SDS-PAGE on 10% acrylamide SDS gel and transfer onto the nitrocellulose membrane.
7. The expression of the protein can be validated by western blot (*see* **Note 10**).

3.6 Iron Uptake Assay: Direct Measurement in Washed Cells

8. Wash freshly induced *L. lactis* cells with water.
9. Place the cells on ice for at least 30 min before assay.
10. Resuspend the pellets in 100 μL solutions of chosen iron salts at various concentrations (*see* **Note 11**).
11. Incubate the cells at 30 °C for different time points at 1 min interval.
12. Harvest the cells by centrifugation at 17,000× *g* for 30 s.
13. Wash twice with water.
14. Remove all the water before acid digestion.
15. Digest the cell pellets by adding 100 μL nitric acid to each microcentrifuge tube and incubate on a 100 °C heat block for 30 min.
16. Dilute the samples with water (usually 900 μL) (*see* **Note 12**).
17. Measure the total metals by ICP.
18. Assess the rate of iron uptake in the linear portion of the uptake curves, normally over the first 5–10 min.

3.7 Iron Uptake Assay: In Medium with Optical Density and Direct Measurement

1. Inoculate 1 mL of M17 media containing 1% glucose with *L. lactis* transformant.
2. Incubate at 30 °C overnight.
3. Inoculate 10 mL of M17 media containing 1% glucose with the overnight culture until an OD_{600} of ~0.1 is reached.
4. Prepare fresh nisin solution with a final concentration of 100 ng/mL.
5. Prepare up to eight sets of media for assays. For media 1, add sterile water in place of nisin; For other media, add a 1 in 10 dilution of nisin and then add iron (using salt or ligated iron source) at a range of different concentrations (*see* **Notes 13** and **14**).
6. Set up a 96-well plate with each row containing 90 μL of media with various additives (*see* **Note 15**) and each column containing one strain isolated as a single colony on a transformant plate (Fig. 2, *see* **Note 16**). For example, Row A: M17 with 1% glucose without nisin, Row B: M17 with 1% glucose plus nisin without iron, Row C: M17 1% glucose plus nisin with 10 μM iron, and Row D–H with increasing iron.
7. Inoculate each well with 10 μL of freshly grown transformant from Subheading 3.7, **step 3** (*see* **Note 17**).
8. The test plate is then incubated in a microplate reader with temperature control and shaking options. The plate is incubated at 30 °C for 12 h recording OD_{600} every 20 min after shaking for 15 s (*see* **Note 18**). Growth curves can be analyzed for inhibition/enhancement of growth with increasing metal concentrations (*see* **Note 19**).

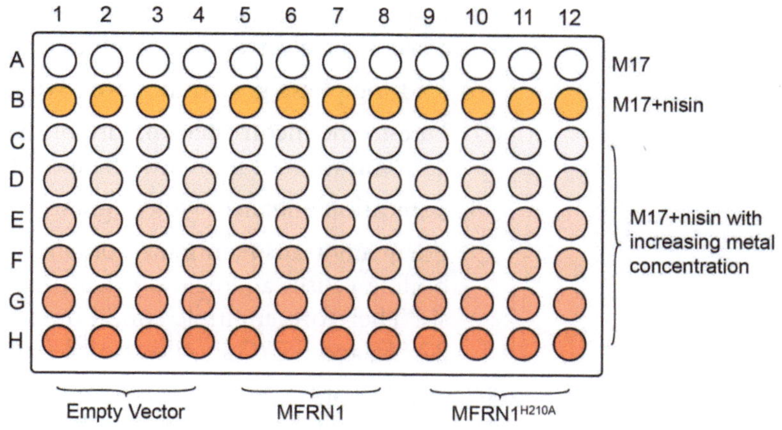

Fig. 2 *96-well plate experiment.* Each row (**a**–**h**) contains variants of the media with increasing concentrations of the metal. Each column (1–12) contains individual transformants. In this example, columns 1–4 are inoculated with *L. lactis* with empty vector; columns 5–8 are inoculated with *L. lactis* with wild type MITOFERRIN1; and columns 9–12 are inoculated with *L. lactis* with non-functional variant MITOFERRIN1^{H210A} [17]

9. After incubation, the final plate can be used for ICP analysis. The cells are removed from the plate, pelleted by centrifugation at 10,000× *g* for 30 s followed by two wash steps with sterile water. Cells are then digested with nitric acid as described in Subheading 3.6 for analysis by ICP (*see* **Note 20**).

4 Notes

1. Multiple cloning sites include NcoI, PstI, SphI, KpnI, SpeI, XbaI, SacI, and HindIII. If any restriction site other than NcoI at 5′ end is used to clone the gene of interest, make sure the inserted gene is in frame with the plasmid. An alternative approach is to use the Gibson cloning method to avoid the potential problem.
2. The codon usage of *L. lactis* should be considered for maximal expression of the eukaryotic gene of interest based on differences in the G/C content of the target gene.
3. *Escherichia coli* MC1061 strain genotype: *araD139*, *Δ(ara, leu) 7697*, *ΔlacX74*, *galU*$^-$, *galK*$^-$, *hsr*$^-$, *hsm*$^+$, and *strA* [27] are used to clone and amplify the plasmid DNA of diverse gram-positive bacteria. pNZ8148 plasmids need to be maintained in MC1061 cells to prevent potential plasmid rearrangement.
4. SOC gives two-fold higher transformation efficiency than the LB medium.

5. Commercial elution buffers from DNA extraction kits can disrupt the electroporation efficiency.
6. Mid-log phase growth requires 3–5 h.
7. Due to the inconsistent nature of the nisin stock, the working concentration of nisin and the induction time can vary.
8. Sonication enhances lysis of gram-positive bacteria.
9. For many transmembrane-containing mitochondrial proteins, boiling causes aggregation that prevents proper electrophoresis. Therefore, multiple conditions, that is, room temperature, 37 °C, and 55 °C should also be tested.
10. Validity of the individual commercial antibodies vary. Preferably, use antibody validated with knockout cell lines/strains.
11. Iron solutions should be prepared in degassed sterile filtered water freshly to minimize the change of oxidation state and solubility. Micro-precipitates are common in iron solutions. Therefore, the uninduced control is imperative to determine the background.
12. The diluted samples should be within the ICP detectable range.
13. The set-up of the media to demonstrate the activity of a variant of a protein (WT or mutant) is performed with media in each row at a particular iron concentration and different transformants in each column.
14. Variable oxidation states and concentrations should be used to assess the iron uptake.
15. Using a large-volume 96-well plate, media 1–8 can be placed in the row A-H column 1. Then, 90 μL is transferred to the test plate using a multi-channel pipette.
16. Any combination of media can be used in the 96-well plate, such as increasing Fe in each row (Row B: 0 μM; Row C: 10 μM; Row D: 20 μM; and Row E: 30 μM) or repeat concentrations (Row B–D: 0 μM and Row E–G: 25 μM Fe).
17. With a deep-well 96-well plate, the individual transformants can be placed across the rows and then transferred to the media in the experiment.
18. Shaking is only required for "even" readings in a plate reader, and the plates should be flat bottomed.
19. The toxicity of iron may not be apparent under most conditions; therefore, iron can be substituted with a toxic metal that can be transported (e.g., cobalt and cadmium) or media can contain additional components to induce enhanced toxicity (e.g., stressors).

20. The final uptake of iron per cell can be determined based on the internal denominator such as sulfur content or optical density determined using the plate reader.

References

1. Cunningham CN, Rutter J (2020) 20,000 picometers under the OMM: diving into the vastness of mitochondrial metabolite transport. EMBO Rep 21:e50071. https://doi.org/10.15252/embr.202050071
2. Palmieri F, Scarcia P, Monné M (2020) Diseases caused by mutations in mitochondrial carrier genes SLC25: a review. Biomol Ther 10:655. https://doi.org/10.3390/biom10040655
3. Robinson AJ, Overy C, Kunji ERS (2008) The mechanism of transport by mitochondrial carriers based on analysis of symmetry. Proc Natl Acad Sci 105:17766–17771. https://doi.org/10.1073/pnas.0809580105
4. Ruprecht JJ, Kunji ERS (2020) The SLC25 mitochondrial carrier family: structure and mechanism. Trends Biochem Sci 45:244–258. https://doi.org/10.1016/j.tibs.2019.11.001
5. Crichton PG, Lee Y, Kunji ERS (2017) The molecular features of uncoupling protein 1 support a conventional mitochondrial carrier-like mechanism. Biochimie 134:35–50. https://doi.org/10.1016/j.biochi.2016.12.016
6. Ruprecht JJ, King MS, Zögg T et al (2019) The molecular mechanism of transport by the mitochondrial ADP/ATP carrier. Cell 176:435–447.e15. https://doi.org/10.1016/j.cell.2018.11.025
7. Pebay-Peyroula E, Dahout-Gonzalez C, Kahn R et al (2003) Structure of mitochondrial ADP/ATP carrier in complex with carboxyatractyloside. Nature 426:39–44. https://doi.org/10.1038/nature02056
8. Zhu X, Boulet A, Buckley KM et al (2021) Mitochondrial copper and phosphate transporter specificity was defined early in the evolution of eukaryotes. eLife. https://elifesciences.org/articles/64690
9. Paradkar PN, Zumbrennen KB, Paw BH et al (2009) Regulation of mitochondrial iron import through differential turnover of mitoferrin 1 and mitoferrin 2. Mol Cell Biol 29:1007–1016. https://doi.org/10.1128/MCB.01685-08
10. Foury F, Roganti T (2002) Deletion of the mitochondrial carrier genes MRS3 andMRS4 suppresses mitochondrial iron accumulation in a yeast frataxin-deficient strain*. J Biol Chem 277:24475–24483. https://doi.org/10.1074/jbc.M111789200
11. Li L, Kaplan J (2004) A mitochondrial-vacuolar signaling pathway in yeast that affects iron and copper metabolism*. J Biol Chem 279:33653–33661. https://doi.org/10.1074/jbc.M403146200
12. Yang M, Cobine PA, Molik S et al (2006) The effects of mitochondrial iron homeostasis on cofactor specificity of superoxide dismutase 2. EMBO J 25:1775–1783. https://doi.org/10.1038/sj.emboj.7601064
13. Froschauer EM, Rietzschel N, Hassler MR et al (2013) The mitochondrial carrier Rim2 co-imports pyrimidine nucleotides and iron. Biochem J 455:57–65. https://doi.org/10.1042/BJ20130144
14. Luk E, Carroll M, Baker M et al (2003) Manganese activation of superoxide dismutase 2 in *Saccharomyces cerevisiae* requires MTM1, a member of the mitochondrial carrier family. Proc Natl Acad Sci USA 100:10353–10357. https://doi.org/10.1073/pnas.1632471100
15. Brazzolotto X, Pierrel F, Pelosi L (2014) Three conserved histidine residues contribute to mitochondrial iron transport through mitoferrins. Biochem J 460:79–89. https://doi.org/10.1042/BJ20140107
16. Froschauer EM, Schweyen RJ, Wiesenberger G (2009) The yeast mitochondrial carrier proteins Mrs3p/Mrs4p mediate iron transport across the inner mitochondrial membrane. Biochim Biophys Acta BBA Biomembr 1788:1044–1050. https://doi.org/10.1016/j.bbamem.2009.03.004
17. Christenson ET, Gallegos AS, Banerjee A (2018) In vitro reconstitution, functional dissection, and mutational analysis of metal ion transport by mitoferrin-1. J Biol Chem 293:3819–3828. https://doi.org/10.1074/jbc.M117.817478
18. Al-Karadaghi S, Hansson M, Nikonov S et al (1997) Crystal structure of ferrochelatase: the terminal enzyme in heme biosynthesis. Structure 5:1501–1510. https://doi.org/10.1016/S0969-2126(97)00299-2
19. Dietz JV, Willoughby MM, Piel RB et al (2021) Mitochondrial contact site and cristae organizing system (MICOS) machinery

supports heme biosynthesis by enabling optimal performance of ferrochelatase. Redox Biol 46:102125. https://doi.org/10.1016/j.redox.2021.102125

20. Seguin A, Santos R, Pain D et al (2011) Co-precipitation of phosphate and iron limits mitochondrial phosphate availability in *Saccharomyces cerevisiae* lacking the yeast frataxin homologue (YFH1). J Biol Chem 286:6071–6079. https://doi.org/10.1074/jbc.M110.163253
21. Hadzhieva M, Kirches E, Mawrin C (2014) Review: iron metabolism and the role of iron in neurodegenerative disorders. Neuropathol Appl Neurobiol 40:240–257. https://doi.org/10.1111/nan.12096
22. Llorens JV, Soriano S, Calap-Quintana P et al (2019) The role of iron in Friedreich's Ataxia: insights from studies in human tissues and cellular and animal models. Front Neurosci 13:75. https://doi.org/10.3389/fnins.2019.00075
23. Monné M, Chan KW, Slotboom D-J et al (2005) Functional expression of eukaryotic membrane proteins in Lactococcus lactis. Protein Sci Publ Protein Soc 14:3048–3056. https://doi.org/10.1110/ps.051689905
24. Zhou XX, Li WF, Ma GX et al (2006) The nisin-controlled gene expression system: construction, application and improvements. Biotechnol Adv 24:285–295. https://doi.org/10.1016/j.biotechadv.2005.11.001
25. Vest KE, Wang J, Gammon MG et al (2016) Overlap of copper and iron uptake systems in mitochondria in *Saccharomyces cerevisiae*. Open Biol 6:150223. https://doi.org/10.1098/rsob.150223
26. Holo H, Nes IF (1989) High-frequency transformation, by electroporation, of lactococcus lactis subsp. cremoris grown with glycine in osmotically stabilized media. Appl Environ Microbiol 55:3119–3123. https://doi.org/10.1128/aem.55.12.3119-3123.1989
27. Casadaban MJ, Cohen SN (1980) Analysis of gene control signals by DNA fusion and cloning in Escherichia coli. J Mol Biol 138:179–207. https://doi.org/10.1016/0022-2836(80)90283-1

Part II

Methods to Analyze Cellular Heme and Iron-Sulfur Cluster Species

Chapter 7

Methods for Cultivation of *Bacteroides thetaiotaomicron* and Analysis of Heme Metabolism by Mass Spectrometry and Spectroscopic Approaches

Ronivaldo Rodrigues da Silva, Victoria Adedoyin, and Jennifer L. DuBois

Abstract

Traditional studies of cellular metabolism have relied on the use of radioisotopes. These have clear disadvantages associated with safety and waste generation. Furthermore, detection of the labeled species by scintillation counting provides only a quantification of its presence or absence. The use of stable isotopes, by contrast, allows the application of powerful, orthogonal spectroscopic approaches such as nuclear magnetic resonance spectroscopy (NMR) and various mass spectrometric methods. Using stable isotope labeling to study heme metabolism requires integrating methods for (a) generating the heme in labeled forms, (b) cultivating and quantifying the organism of choice in chemically defined media, to which labeled compounds can be added, (c) recovering cellular components and/or spent growth media, and (d) analyzing these materials for the labeled species using spectroscopic and mass spectrometric methods. These methods are summarized here in the context of *Bacteroides thetaiotaomicron,* a generally nonpathogenic anaerobe and heme auxotroph.

Key words Heme metabolism, Bacteria, Anaerobe, Stable isotope, Bacteroides, Microbiome, Metabolomics, LC-MS

1 Introduction

B. thetaiotaomicron is a gram-negative anaerobic and heme auxotrophic bacterium that is often used as a model human gut symbiotic strain. Although *Bacteroides* species are described as aerotolerant, they do not grow significantly in the presence of molecular oxygen. In light of this fact, when culturing *B. thetaiotaomicron*, it is essential to supply it with heme and remove oxygen from the culture media [1]. Here, we describe the equipment, culture media, and approaches for monitoring the growth and maintenance (as a glycerol stock) of this species, preparing and characterizing modified forms of tetrapyrroles as media supplements, and monitoring the fate of the latter by spectroscopic

Oleh Khalimonchuk (ed.), *Iron Metabolism: Methods and Protocols*, Methods in Molecular Biology, vol. 2839,
https://doi.org/10.1007/978-1-0716-4043-2_7,

and mass spectrometric (MS) methods, specifically high-performance liquid chromatography (HPLC) and HPLC coupled to mass spectrometry (LC-MS). Related species can be handled in analogous ways. Parts of these protocols may be broadly adaptable for studies of other microbial species and consortia.

2 Materials

2.1 Equipment Requirements

Equipment for anaerobically cultivating this species include an anaerobic glove box (e.g., from Coy), anaerobic GasPak™ sealed canisters (Carolina), a gas cylinder (nitrogen and a mixed gas, 97.5% nitrogen/2.5% hydrogen), a sterile needle/cannula, and sealed glass vessels like Hungate tubes.

2.2 Cultivating B. thetaiotaomicron

A variety of solid and liquid media may be needed for different applications and their constituents are given here.

Tryptic Soy Agar with 5% Sheep Blood Tryptic soy broth (weigh the amount for 1 L according to the manufacturer's recommendations), resazurin solution (1 mL of 1 mg/mL stock), 15 g agar, and 950 mL deionized water. Autoclave the mixture for 20 min at 121 ° C and 1 atm. Then, cool the sterilized medium to ~47 °C and add 50 mL of defibrinated sheep blood, mix gently, and pour the medium into Petri dishes. Petri dish media are prepared aerobically and stored in a glove box to equilibrate prior to use. Place the plates in plastic bags to prevent dehydration during incubation in the glove box.

Brain Heart Infusion- Supplemented Medium (BHIS) Brain heart infusion broth powder (weigh the quantity for 1 L according to the manufacturer's recommendations), 5 g yeast extract, and 1 mL of 1 mg/mL resazurin stock (optional). Dissolve the medium components in 1 L of deionized water and autoclave for 20 min at 121 °C and 1 atm. When the autoclaved medium is cool, add 0.5 g of L-cysteine (free base) in 1 mL solution sterilized by filtration (0.22 μm pores), 10 mL dissolved hemin at 0.5 mg/mL (*see* **Note 1**), and 0.2 mL vitamin K1 solution.

Vitamin K Solution Mix 0.15 mL of vitamin K1 with 30 mL of 95% ethanol and sterilize the mixture using a 0.22 μm filter.

Defined Minimal Medium Supplemented with Hemin [5] (see ***Note 2)*** 6 mM potassium dihydrogen phosphate, 15.4 mM sodium chloride, 98 μM magnesium (II) chloride hexahydrate,

176.5 μM calcium (II) chloride dihydrate, 4.2 μM cobalt (II) chloride hexahydrate, 50.5 μM manganese (II) chloride tetrahydrate, 9.3 mM ammonium chloride, 1.75 mM sodium sulfate, 134 μM L-methionine, 23.8 mM sodium bicarbonate, 8.25 mM L-cysteine (free base), 28 mM D-glucose, and 15 μM hemin chloride. Dissolve all components in deionized water, according to their final concentrations, except the hemin chloride (*see* **Note 1** that describes how to prepare it). Then, add hemin chloride to the mixture and adjust the pH to 7.1 and sterile filter it into an autoclaved vessel using a 0.22 μm filter. If a large volume is required for this medium, it is recommended to prepare it at a concentration of 10×, as this makes it easier to filter.

***Chopped Meat Medium* (see *Note 2*)** 500 g ground beef (fat free), 1000 mL deionized water, 25 mL 1 M NaOH, 30 g tryptone, 5 g yeast extract, 5 g K_2HPO_4, 10 mL hemin solution (0.5 mg/mL, *see* **Note 1**), 1 mL resazurin solution (1 mg/mL), and 0.5 g L-cysteine (free base). Media can be dispensed into Balch tubes with caps and crimp tops for autoclaving and subsequent use.

2.3 Preparation of Isotopically Labeled Heme

The requirements for the expression and purification of this protein and for the extraction of its heme are described below.

2.3.1 Expression System

Competent *E. coli* cells (e.g., BL21 (DE3), New England Biolabs®) and plasmid (e.g., pET-11a, GenScript Biotech) harboring the gene encoding the outer membrane portion of cytochrome b5 and isopropyl-β-thiogalactoside (IPTG).

2.3.2 Growth Media

M9 Minimal Medium (per 1 L) 12.8 g $Na_2HPO_4.7H_2O$, 3 g KH_2PO_4, 1 g NH_4Cl, and 0.5 g NaCl ($^{15}NH_4Cl$ may be substituted for natural isotopic abundance NH_4Cl if desired). Autoclave the mixture for 20 min at 121 °C and 1 atm. Next, add to the 1 L autoclaved salt solution the following filter-sterilized components (0.22 μm pore filter): 20 mL of 20% dextrose (w/v), 2 mL of 1 M $MgSO_4$, 0.1 mL of 1 M $CaCl_2$, 0.5 mL of 100 mM ferric chloride, and 1 mL of 100 mg/mL ampicillin. If desired, use ^{13}C labeled dextrose. This will result in full incorporation of ^{13}C label throughout the bacterial culture. Delta-aminolevulinic acid (δ-ALA) is optional (17 mg/L) and may also be used as the means of introducing specific isotopic labels into heme if it itself is labeled.

SOC Medium 2% Tryptone (w/v), 0.5% yeast extract (w/v), 10 mM NaCl, and 2.5 mM KCl. Autoclave the mixture for 20 min at 121 °C and 1 atm.

2.3.3 Protein Purification and Heme Extraction

Buffer A (50 mM Tris-HCl, pH 8.0), Buffer B (50 mM Tris-HCl, pH 8.0, 0.3 M NaCl), Buffer C (50 mM Tris-HCl, pH 8.0, 150 mM NaCl), 1 M phenylmethylsulfonyl fluoride (PMSF), lysozyme (from NEB and other vendors, typically 50 mg/mL), DEAE resin in a suitable low-pressure glass column, and Sephacryl S-300 HR in a low-pressure glass column. Column sizes and resin bed volumes may vary depending on the scale needed for purification. Additionally, acetone and 12 M HCl are required for heme extraction, along with a rotary evaporator for removing the solvent.

2.4 Analysis of Heme

2.4.1 Pyridine Hemochromagen Assay

Solution I: 0.2 M NaOH and 40% pyridine (v/v); Solution II: 0.1 M potassium ferricyanide in deionized water; and Solution III: 0.5 M sodium dithionite in 0.5 M NaOH solution. Only a small volume is made, and dithionite is added immediately prior to use. A UV/visible spectrophotometer is needed for this procedure.

1. Using a fume hood, prepare 6 mL of 0.5 M NaOH in deionized water and mix 4 mL of this solution with 4 mL of pyridine. Add deionized water until the volume reaches 10 mL.
2. Add 50 μL of 0.1 M potassium ferricyanide (solution II) to 10 mL of solution I.
3. Prepare 1 mL of 0.5 M sodium dithionite in 0.5 M NaOH (use the remaining NaOH solution described in **Step 1**).

2.4.2 Analysis of Cell-Based Hemes by HPLC and LC-MS

One molar HCl:DMSO (1:1, v/v), 2 mL FastPrep lysis B matrix tube (MP Biomedicals™) for use in a FastPrep-24 5G instrument (MP Biomedicals™), solid-phase extraction columns (Sep-Pak® tC18 3 cc Vac Cartridge, Waters 036815), acetonitrile + 0.1% trifluoroacetic acid, a reverse phase column (e.g., PLRP-S column, Agilent), solvent A (ultrapure water + 0.1% formic acid), solvent B (acetonitrile + 0.1% formic acid), and 2 mM stock solutions of protoporphyrin IX and hemin in DMSO.

3 Methods

3.1 Maintaining an Anaerobic Environment

Options for producing anoxic atmosphere for culture growth include the use of an anaerobic glove box, anaerobic GasPak™ sealed canisters or similar systems, and sealed glass vessels like Hungate tubes where oxygen-free gases (nitrogen, carbon dioxide, argon, helium, etc.) are used to sparge solutions [2]. When culturing anaerobes in an anoxic chamber, any container (tubes, Petri dishes, etc.) can be used so long as it maintains sterility, but if the culture is made outside an anaerobic chamber, sealed containers are needed to preserve the oxygen-free space previously produced to perform the culture.

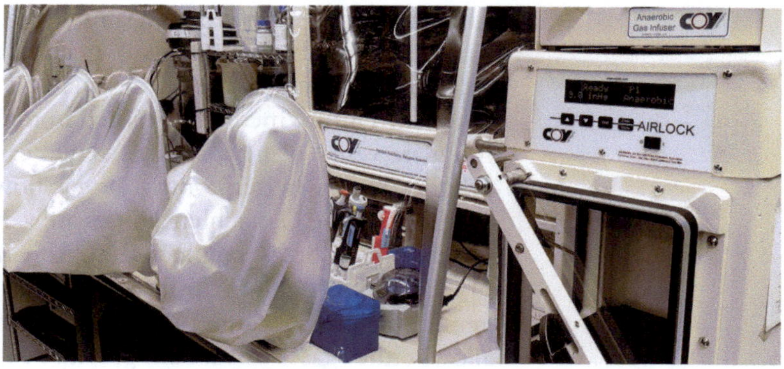

Fig. 1 Anaerobic glove box with devices inside. The antechamber for admitting materials into the main glove box is shown on the right, labeled "AIRLOCK." A double-door variable temperature incubator, suitable for incubating Petri plates at 37 °C, is shown in the back (also from Coy). The black device in the back left of the photo shows a column for removing atmospheric H_2S, which is produced as a volatile sulfur end product by sulfate-reducing organisms. Chambers holding O_2-scrubbing and platinum-containing catalysts are not shown in this image. Models sold with Coy anaerobic chambers come with small fans to facilitate circulation of the chamber gas over the catalyst surface

Anaerobic glove boxes are controlled atmosphere workspaces suitable for handling and culturing anaerobes. The sealed chamber isolates the microorganisms from the user who handles samples using hand/arm rubber gloves. To create an oxygen-free environment, inert gases are flushed into the chamber followed by evacuation, with three to four repetitions. Residual traces of O_2 react with H_2 used as a 5% component in the inert gas mixture over heated metal catalysts like palladium. A functional glovebox chamber can minimize the O_2 content to less than 2 ppm, that is, much less compared to the natural oxygen content in the air, ~2×10^5 ppm [1–3]. Inside the chamber, it is possible to place small equipment such as an incubator, spectrophotometer, centrifuge, pipettes, and other necessary devices (Fig. 1). The use of an antechamber, which is sealed off from the main chamber and which can be independently equilibrated to the inert gas, permits the introduction and removal of materials from the main workspace of the glove box.

An atmosphere of 2.5–5% H_2 and 97.5–95% N_2 gas mixture and incubation at 37 °C are common parameters for growing *B. thetaiotaomicron* in an anaerobic chamber. If the bacterial culture requires other atmospheric conditions, different combinations of gases (CO_2, H_2, and N_2) can be used. Extreme caution should be taken with hydrogen flammability at ≥10% concentration [1].

Oxygen-free solutions and head space may also be generated in glass tubes, which need to be gas-impermeable and sealed to maintain the anaerobic condition. Hungate tubes are supplied with a thick butyl stopper and a screw cap. Balch-type tubes with butyl rubber stoppers and crimped aluminum tops are also commonly used. The thick butyl rubber stopper and aluminum crimp are more able to resist the gas pressure generated during bacterial growth. Once the culture medium is placed in the tube, sparging gases into the glass vessel using a sterile needle/cannula connected to a gas cylinder or, for better control of gas flow, a Schlenck line, can be used to remove O_2. Schlenck lines are valved pairs of glass tubes in which one tube is connected to the inert gas (often with an O_2-scrubbing column in-line) and the other is connected to a gentle vacuum. Tygon tubing emanating from each glass tube allows either gases or vacuum pressure to be introduced into sealed containers via needles. Bacterial inoculation into the anaerobic medium can be performed inside the glove box, where the tubes can be opened, or using a sterile syringe and a gas-tight needle to inject the bacterial cells into the sealed tube.

Another option, especially for incubating agar plates, is the use of sealable GasPak™ canisters placed in an anaerobic chamber or placed into contact with reagents that generate H_2 and CO_2. A commonly used GasPak envelope contains sodium borohydride and sodium bicarbonate, which react with water to produce the above gases. Then, catalyzed by palladium in the chamber, O_2 reacts with H_2 to produce H_2O, creating an oxygen-free environment [3].

3.2 Choosing and Preparing Media for Cultivating Heme Auxotrophs

B. thetaiotaomicron is a heme auxotrophic anaerobe, which means it needs heme to grow but cannot produce it. Exogenous heme sources are used to supplement minimal culture media. These sources typically include sheep blood, hemin chloride, or extracted heme from proteins. Other rich media that contain heme in their composition, including beef broth and brain heart infusion (BHI) medium, have also been used for *B. thetaiotaomicron* growth.

Apart from the oxygen-free atmosphere, strict anaerobes also require chemically reduced media. Reducing agents such as L-cysteine and sodium thioglycolate are used to create a low redox potential in the media. To ensure the reducing potential of the medium or the bacterial growth chamber, resazurin, a sensitive redox dye that is not toxic to the bacterial cell, is used as a redox indicator. This dye can be added to the culture medium or introduced into the culture chamber, as with anaerobic jars or bags. Oxidized resazurin is dark blue, but when it is added to a rich anaerobic medium with excess organic compounds, this dye is irreversibly reduced to resorufin, which is pink near neutral pH. Under a reduced atmosphere between −100 mV and −51 mV, a range suitable for growth of anaerobes, there is a further reversible conversion of resorufin to hydroresorufin, which is colorless [4].

Several media recipes suitable for cultivating *B. thetaiotaomicron* are given in Materials (Subheading 2), where the choice of medium depends on the application. Tryptic soy agar is commonly used for growth of colonies on Petri plates. For liquid media, BHIS is a typical rich medium used to monitor growth via UV/visible absorbance/optical density, while chopped meat medium is reserved for cultivation prior to storage because of its opacity. Chemically defined minimal media are important for growing cells for heme/iron studies as isotopically labeled species can be added in known amounts and their uptake and fate can be monitored. Procedures for preparing these media are given in the Materials section, as little more than combining and sterilizing ingredients is required. The only exception is the chopped meat (or beef broth) medium, which has a slightly more elaborate method of preparation, given here.

Chopped Meat Medium

1. In a beaker, add 500 g of ground beef to 1 L of deionized water, and 25 mL of 1 M NaOH, keep this mixture under stirring, and boil for 5 min. Cool to room temperature or 4 °C and remove the fat from the surface.
2. Filter with a cheesecloth and retain the broth and meat particles separately.
3. Pour the broth into a volumetric cylinder and add deionized water to make up the volume to 1 L.
4. Pour this 1 L of broth into a sealable 2 L flask and add 30 g of tryptone, 5 g of yeast extract, 5 g of K_2HPO_4, 10 mL of hemin solution, and 1 mL of resazurin solution (1 mg/mL).
5. Boil this 1 L broth for 2 min and cool it while sparging N_2. Then add 0.5 g of L-cysteine, and when the broth is cold, close the flask and store it in an anaerobic glove box.
6. Inside the glove box, add pieces of ground beef in Hungate tubes and 10 mL of the broth. Seal the tubes and autoclave for 20 min at 121 °C and 1 atm.

3.3 Monitoring the Growth of B. thetaiotaomicron

Bacterial growth on liquid minimal media is monitored by optical density at 600 nm (OD_{600nm}) using ultraviolet/visible (UV/vis) absorbance spectroscopy. For sealed glass tubes, a spectrophotometer adapted for admitting the tubes themselves allows absorbance to be measured without removing the culture samples. This enables continuous monitoring of bacterial growth and avoids exposing the cells to oxygen (Fig. 2a, b). Inside the glove box, it is also possible to monitor bacterial growth in microplates using a UV/vis spectrophotometer microplate reader.

In situations where it is necessary to collect the sample, it is possible to aspirate the liquid culture from tubes sealed with a

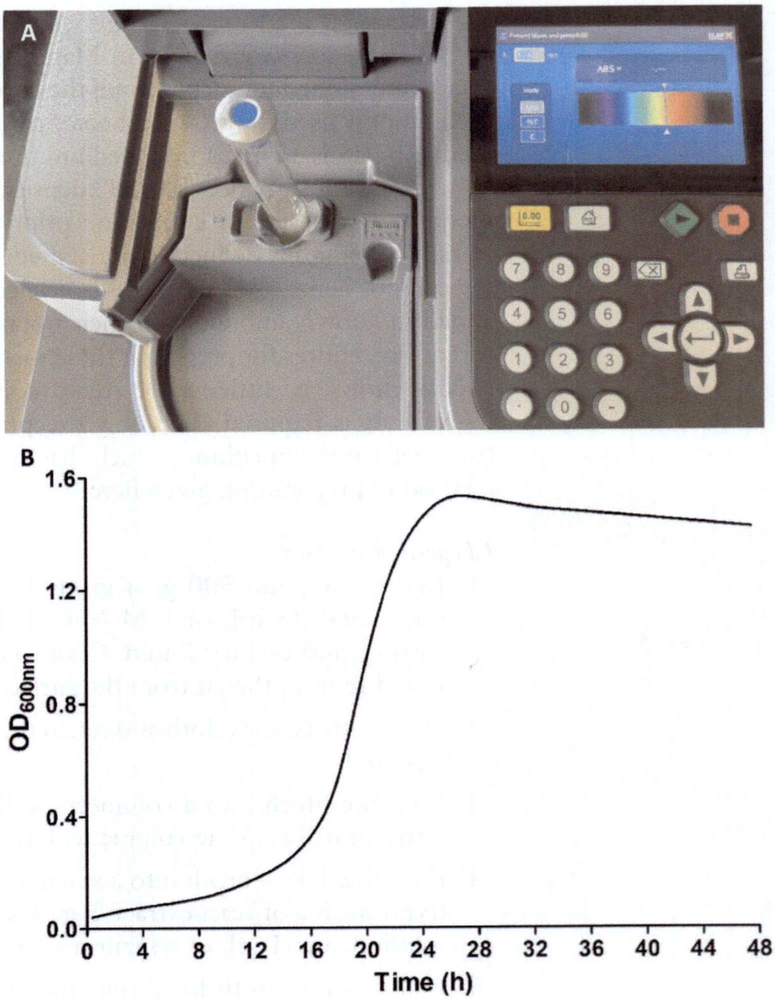

Fig. 2 Monitoring the growth of *B. thetaiotaomicron* with OD_{600nm} measured directly from the tube without taking sample aliquots. (**a**) A UV/vis spectrophotometer adapted for samples in 18 × 150 mm glass tubes. This model is made by Thermo Scientific. (**b**) *B. thetaiotaomicron* growth in a sealed glass Balch tube containing 10 mL minimal medium with supplementation of 15 μM hemin chloride was monitored by UV/vis over time. The OD_{600nm} values measured at a series of time points were plotted, and a curve was extrapolated through the points. Incubation was carried out at 37 °C under genus-optimal headspace conditions (2.5% H_2 and 97.5% N_2 atmosphere)

rubber stopper (Hungate and Balch-type tubes) using a sterile gas-tight syringe (Fig. 3). First, it is important to sterilize the rubber stopper by dripping ethanol on it. Then, turning the tube upside down and injecting a sterile needle into the rubber stopper allows the sample removal without introducing O_2 into the culture [4].

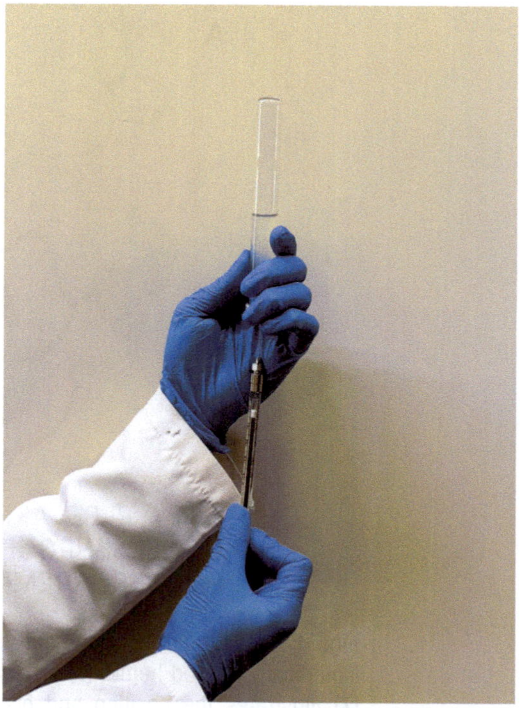

Fig. 3 Anoxic removal of the sample from a tube sealed with a butyl rubber stopper and an aluminum crimp (18 × 150 mm)

3.4 Cryopreservation of Anaerobic B. thetaiotaomicron

Cryoprotectants need to be used for long-term preservation of the *B. thetaiotaomicron* cell at −70/−80 °C. Usually, preservative solutions contain glycerol concentrations of 15–30% (v/v, final concentration), and once the cell suspension is prepared, it is flash-frozen at −70 °C or below. Skim milk is another alternative. Once the fresh bacterial culture is ready:

1. Prepare a solution of 50% glycerol in deionized water (v/v) and sterilize it in an autoclave at 121 °C for 20 min. Then sparge it with oxygen-free inert gas.
2. Autoclave 2 mL cryotubes with screw caps. After that, label them (*see* **Note 3**).
3. Using a glove box and sterile materials, gently mix 500 μL of a 50% glycerol solution with 500 μL of a bacterial culture. Remove the storage vials from the anaerobic chamber before rapidly freezing them in liquid nitrogen and store them in a freezer at −80 °C.

3.5 Expression and Purification of Cytochrome b5 (Soluble Fragment)

When monitoring heme metabolism by bacteria is a research goal, multidimensional ^{13}C, ^{15}N, or ^{1}H magnetic resonance spectroscopy (NMR) and mass spectrometry (MS) methods can be used. Heterologous overexpression of heme-containing proteins, such as cytochrome b5, allows for the production of milligram quantities of

A

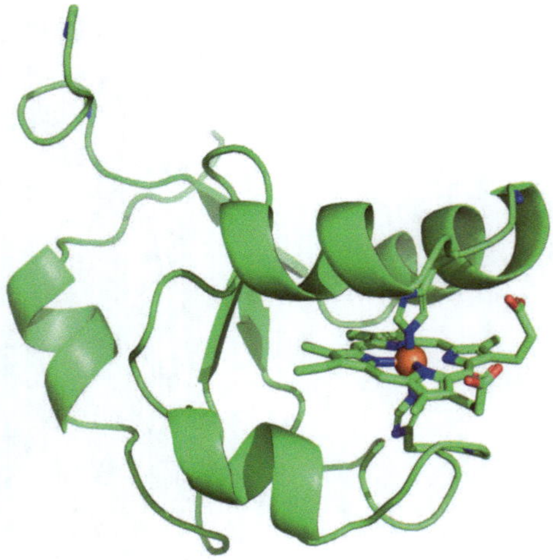

B

AVTYYRLEEVAKRNTAEETWMVIHGRVY
DITRFLSEHPGGEEVLLEQAGADATESFE
DVGHSPDAREMLKQYYIGDVHPNDLKP

Fig. 4 Cytochrome b5 fragment expressed as a vehicle for overproducing isotopically labeled heme. (**a**) An image of the crystal structure (PDB ID 1B5M) [7] illustrates the position and bis-histidine ligation of the heme within the expressed protein fragment. (**b**) FASTA sequence of the 84-amino-acid protein. The expressed protein has a computed monoisotopic molecular weight of 9876.76 daltons and a theoretical isolectric point of 4.69

labeled heme. Methods for the expression of the soluble fragment of this protein containing labeled heme to be used to supplement bacterial cultures are presented below. These methods are adapted from published reports which used cloned cytochrome b5 from *Rattus norvegicus* [6, 7]. The procedure given here begins with a synthetic gene fragment encoded in the pET-11a vector. The pET vector has a T7 viral promoter sequence that is suitable for use with an *E. coli* strain that expresses T7 RNA polymerase. This includes any of the lambda DE3 lysogens sold by a variety of biological suppliers. T7 promoters are positively regulated (derepressed) by addition of the nonmetabolizable galactose analog isopropyl-β-thiogalactoside (IPTG). The FASTA sequence of the expressed protein and an image of the heme-bound crystal structure (Protein Data Bank ID 1B5M) are given in Fig. 4.

3.5.1 Transforming Expression Strain with Plasmid

This methodology for expressing and purifying cytochrome b5 was adapted from Rivera and Walker [6].

1. Place tubes containing BL21 (DE3)-competent *E. coli* cells and the plasmid harboring the gene encoding the outer membrane portion of cytochrome b5 on ice for 10 min.
2. Transfer 50 μL of cells to an empty tube on ice.

3. Pipette 5 μL of plasmid DNA (~2 ng/μL) into the 50 μL of cells.
4. Flick the tube carefully four to five times to ensure the proper mixing of cells and DNA.
5. Place the mixture on ice for 30 min.
6. Carry out heat shock at 42 °C for 10 s and immediately allow to sit on ice for 5 min.
7. Pipette 950 μL of room temperature SOC medium and transfer it into the mixture.
8. Incubate the mixture at 37 °C for 60 min under vigorous shaking (250 rpm).
9. Pipette 50–100 μL of the sample onto Petri dishes with Luria-Bertani (LB) medium (*see* **Note 4**) and incubate overnight at 37 °C. Note that pET-11a contains a gene encoding an enzyme conferring ampicillin resistance; hence, all successful transformants should be ampicillin resistant. LB-agar plates supplemented with 50–100 μg/mL of ampicillin (added to the plate medium immediately before plates are poured) will consequently select for the desired transformants.
10. A glycerol frozen stock of the transformed bacterial strain can now be made as described above. This is typically made from a single colony picked from the Petri plate and restreaked onto a selective growth medium to ensure that the stock comes from a single parent cell.

3.5.2 Expression of Heme-Bound Cytochrome b5 in E. coli

E. coli grows in a wide variety of media. M9 minimal medium contains a defined set of nutrients, including inorganic salts, a carbon source, and a nitrogen source, required for bacterial growth. The inorganic salts include sodium chloride, potassium phosphate, and magnesium sulfate. Use of a single carbon source, such as glucose or glycerol, and a single nitrogen source, such as ammonium chloride or ammonium sulfate, allows for the cell-wide incorporation of ^{13}C or ^{15}N into biomolecules, respectively, including heme. Alternatively, various isotopically labeled forms of δ-aminolevulinic acid (ALA), a biosynthetic precursor of tetrapyrrole biosynthesis, can be used to introduce specific isotopes at particular locations in the heme. Similarly, iron supplementation to the growth medium is used to promote production of metalated heme, and it can also be a means for introducing ^{57}Fe into the tetrapyrrole for experiments requiring spectroscopic (e.g., Mössbauer) or mass spectrometric monitoring. While bacterial cultivation in minimal media typically results in slower growth and poorer yields of expressed proteins, it is the method of choice for producing isotopically labeled metabolites and proteins.

For 1 L culture:

1. Prepare M9 salts and autoclave for 20 min at 121 °C and 1 atm in a suitable wide-bottomed growth vessel (Fernbach flask, Erlenmeyer, etc.). Instructions:
   (a) Transfer 800 mL of deionized water to a clean 1 L glass bottle.
   (b) Weigh and add Na_2HPO_4, KH_2PO_4, NH_4Cl, and NaCl to the water.
   (c) Swirl the bottle gently to dissolve the salts.
   (d) Use distilled water to make the final volume of 1 L.
   (e) Sterilize the solution by autoclaving at 121 °C for 30 min at 1 atm.
2. Prior to inoculation, supplement the autoclaved solution with 20 mL of 20% dextrose (w/v), 2 mL of 1 M $MgSO_4$, 0.1 mL of 1 M $CaCl_2$, 0.5 mL of 100 mM ferric chloride, and 1 mL of 100 mg/mL ampicillin. These materials should be sterilized by filtration through a 0.22 μm pore filter. If desired, use ^{13}C labeled dextrose. This will result in full incorporation of ^{13}C label throughout the bacterial culture.
3. Seed the medium with a 1/100 v/v dilution of an overnight starter culture in LB.
4. Incubate the cultures at 37 °C with shaking at 180 rpm until $OD_{600nm} = 0.7–1.0$.
5. At this point, add supplements if used (e.g., δ-ALA at 17 mg/L final concentration) and induce the cytochrome b5 expression with IPTG (0.4 mM final concentration).
6. Next, incubate the culture at 25 °C and evaluate the best time to harvest the cells by sampling at different time points. To do this, monitor protein production via SDS-polyacrylamide gel electrophoresis (SDS-PAGE) using a 12% acrylamide gel, selecting for a harvest time that produces the maximum amount of targeted protein. (This will vary depending on the culturing vessel used and the exact shaking speed/orbital rotation.)
7. Harvest cells by gentle centrifugation. Collect the cell pellet using a silicone spatula, and freeze it in wax paper at −80 °C until ready for use. Wet cell paste can be weighed either before freezing or immediately on removal from the freezer.

3.5.3 Isolation of Cytochrome b5

1. Resuspend the cell paste (~10 g) containing expressed cytochrome b5 in 100 mL of Buffer A (*see* Subheading 2.3.3), thawing the cells if necessary.
2. Add PMSF to 1 mM and 1 mg/mL chicken egg white lysozyme (final concentrations), respectively.

3. Incubate the solution at room temperature with stirring until the cell paste forms a uniform opaque suspension.
4. Sonicate the cell suspension with stirring while incubating in an ice bath for 20 min. The time and pulse sequences appropriate for shearing cells and their genomic DNA will vary according to the device. For example, a Branson ultrasonifier successfully lyses cells with a pulse sequence of 10 s on and 25 s off at 45% amplitude.
5. Centrifuge the cell lysate for clarification (18,000 × g, 30 min, 4 °C).
6. Load the supernatant onto a DEAE column that was preequilibrated in Buffer A.
7. Wash the column with Buffer A and elute cytochrome b5 with a linear gradient from 0% to 100% Buffer B (*see* Subheading 2.3.3).
8. Analyze eluted fractions by 12% SDS-PAGE and select fractions enriched in proteins with the appropriate molecular weight (circa 9 kDa). Pool, concentrate using a centrifugal concentrator MWCO 5 kDa, and load the collected fractions onto Sephacryl S-300 HR using an isocratic gradient with Buffer C (*see* Subheading 2.3.3). Run SDS-PAGE of the eluted fractions and select fractions of >90% purity (*see* **Note 5**).

3.6 Extraction and Quantification of Heme from Cytochrome b5

3.6.1 Acetone Extraction of Heme from Cytochrome b5

This procedure is adapted from the methods described by Lewis [8] and Fronticelli and Bucci [9]. Acidified acetone will precipitate many proteins, leaving hydrophobic small molecules in the solution phase. An acid-acetone extraction of cytochrome b5 therefore results in precipitation of the unwanted protein and retention of heme in the liquid fraction. To extract, first mix 1 mL of cytochrome b5 with 100 mL of acidic acetone (prepared by mixing 1 mL of 12 M hydrochloric acid with 99 mL of acetone) and centrifuge the mixture for 10 min at 4500 × g and 4 °C. Then discard the white pellet and take the light pink supernatant for rotary evaporation at 45 °C to remove the acetone solvent. Alternatively, the acidic acetone can be evaporated in a fume hood overnight under a light stream of nitrogen. Once dried, the heme can be resuspended in a small volume of basic water or DMSO (**Note 1**) and the concentration of the stock is determined as indicated below.

3.6.2 Determination of Heme Concentration Using the Pyridine Hemochromagen Assay

This assay takes advantage of the well-known UV/visible spectrum and extinction coefficient for heme bound by two axial pyridine ligands. Spectra are measured in the ferrous and ferric states, and the difference between the two (ferrous minus ferric) is typically used to quantify the heme.

According to the method presented by Barr and Guo [10], prepare solutions I, II, and III (*see* Subheading 2.4.1).

To quantify heme, an extinction coefficient of 34.7 mM^{-1} cm^{-1} at 557 nm can be used for the reduced protein [11]. More precisely, for a difference spectrum (reduced—oxidized), 23.98 mM^{-1} cm^{-1} at 557 nm can be used [12].

1. Set the spectrophotometer to read the absorbance between 300 nm and 700 nm.
2. For the spectrophotometer blank, mix 500 μL of solution I with 500 μL of the buffer/solvent in which the sample of interest is dissolved.
3. Mix 500 μL of solution I with 500 μL of the sample of interest (diluted if necessary) and measure the spectrum. This is the oxidized sample. It should be relatively colorless.
4. After measuring the oxidized sample, add 10 μL of solution III to it. This is the reduced sample, which should be red in color. Measure spectra from minute to minute until the absorbance does not increase.
5. Use Beer's law to calculate the heme concentration: $A = \varepsilon\ c\ l$, where A = absorbance, ε = extinction coefficient, c = concentration of the sample, and l = pathlength. It is possible to use either the reduced spectrum or the difference spectrum [reduced—oxidized] as described above.

3.7 Analysis of Porphyrins from Bacterial Cells

An isotopically labeled heme may be prepared as described above and then supplied to a bacterial culture, equilibrated with the apo form of a heme-binding protein, or monitored in the context of an animal or consortium. In the simple case of heme used as a source of iron or cofactor to a bacterial culture (e.g., *B. thetaiotaomicron*), the cell culture would first be prepared as described above and then harvested. Many analyses could be carried out. This procedure for recovering and quantifying the various porphyrins, capable of distinguishing various isotopomers (i.e., chemically identical but isotopically distinct forms of the same compound), is described according to Meslé et al. [13] and adapted from Knippel et al. [14].

3.7.1 Extraction of Porphyrins from a Lysate of Bacterial Cells

A small cell pellet is collected from a bacterial culture by centrifugation and lysed mechanically. The soluble fraction may be analyzed directly or the intracellular porphyrins (e.g., hemin and protoporphyrin IX) can first be extracted and concentrated as described here.

1. Resuspend the cell pellet in 1 mL of 1 M HCl:DMSO (1:1, v/v) and vortex vigorously.
2. Transfer the resuspension to a 2 mL FastPrep lysis B matrix tube.

3. Lyse the cells in the FastPrep-24 5G instrument for 6.0 m/s, 40 s total, twice.
4. Centrifuge the cell lysate (10,000 rpm, 5 min, 4 °C) and collect the supernatant.
5. Re-resuspend the cell pellet in 1 mL of 1 M HCl:DMSO (1:1, v/v), vortex vigorously, and centrifuge as above.
6. Pool the supernatants (2 mL total).

For 1.0 mL of supernatant:

1. Dilute the supernatant with ultrapure Milli-Q H_2O to 25 mL (*see* **Note 6**).
2. Equilibrate a solid-phase extraction column (Sep-Pak® tC18 3 cc Vac Cartridge, Waters 036815, works well for porphyrins) with 2 mL of the solution consisting of acetonitrile + 0.1% trifluoroacetic acid (TFA) followed by 4 mL of ultrapure Milli-Q H_2O.
3. Using a vacuum manifold to draw the sample into the column matrix, load the supernatant onto the column.
4. Wash the column with 5–10 mL of ultrapure Milli-Q H_2O.
5. For porphyrin elution, add to the column 2 mL of acetonitrile + 0.1% TFA followed by 2 mL of methanol. Collect the eluate.
6. Dry the sample in a vacuum centrifuge concentrator at 45 °C until complete evaporation.
7. Store the sample protected from light at −20 °C until analysis.

3.7.2 Sample Analysis by LC-MS

Heme and porphyrins can be definitively identified and quantified by HPLC, LC-MS, or both. When HPLC is used, visible absorbance detection is the means by which heme and any of its precursors (such as protoporphyrin IX) or its degradation products (e.g., bilirubin) are identified. The UV/visible spectra and extinction coefficients for these standards are published [15].

1. Resuspend evaporated samples in 100–1000 μL of acetonitrile + 0.1% TFA. Note that this solvent, also adopted as an elution solvent above, is intended to be optimal for downstream analysis by LC-MS. A different solvent may be appropriate for another application.
2. Add 100 μL of the sample to replicate MS sample vials.
3. Equilibrate a reverse-phase column (e.g., PLRP-S column) to an 85:15 ratio of solvent A to solvent B (*see* Subheading 2.4.2), with a 600 μL/minute flow rate at 50 °C. A linear gradient elution, from 15% to 95% solvent B over 6 min followed by a 2 min hold at 95% B, can be used to separate porphyrins if desired. Separation is not necessary if the mass analyzer of a mass spectrometer is used as the primary means of identifying

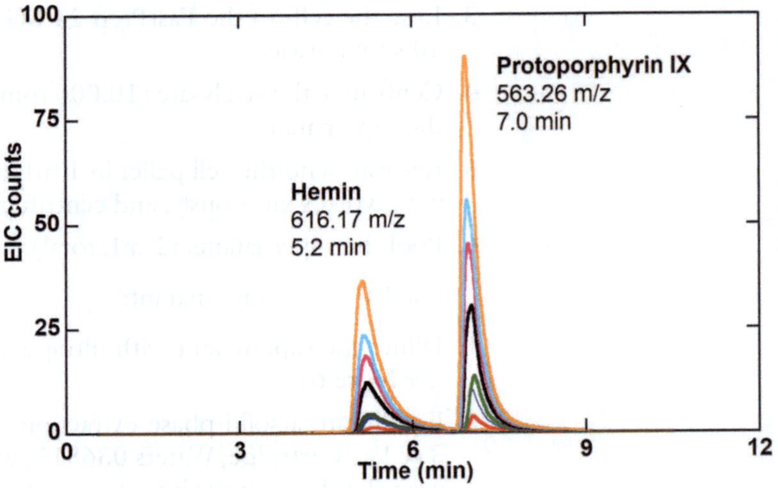

Fig. 5 Representative total ion chromatogram of heme and PPIX at different concentrations. Elution times of 5.2 and 7 min for heme (616.17 *m/z*) and PPIX (563.26 *m/z*), respectively. The colored lines represent the different concentrations of porphyrins: 0.25 μM (red), 0.5 μM (blue), 1 μM (green), 2 μM (black), 3 μM (pink), 4 μM (aqua), and 6 μM (orange). Peak integration can be used to generate standard curves of area versus concentration and to determine concentrations of unknowns in turn

the analytes because the analytes will have distinct masses, whereas good separation is necessary if UV/vis absorbance (e.g., from a standard HPLC instrument) is used to detect and quantify the analytes.

4. The column must be reequilibrated at 15% solvent B for 2 min. It is also recommended to perform two blank runs between the samples to ensure that no analytes are retained on the column.

5. For standard curves, prepare 2 mM stock solutions of porphyrin standards individually in DMSO. Next, mix and dilute them to make a 100 μM master mix (final concentration of hemin and protoporphyrin IX) in DMSO and use this stock solution to make standard curves from 0.25 to10 μM in acetonitrile + 0.1% TFA (500 μL final). Based on that, *m/z* values from MS can be determined empirically for standards of heme and protoporphyrin IX (Fig. 5).

4 Notes

1. To prepare a stock solution of hemin, dissolve 50 mg of hemin chloride in 1 mL of basic water (1 M NaOH) and bring the volume up to 100 mL with deionized water. Sterilize it using a 0.22 μm filter. Alternatively, stocks of hemin can be prepared in dimethylsulfoxide (DMSO). Throughout this chapter,

"hemin" (the chloro complex of ferric-protoporphyrin IX) is used interchangeably with "heme" (ferrous-protoporphyrin IX) for the sake of simplicity. If handled in air and in the dissolved, pH 7 state, the iron complex of protoporphyrin IX will generally be predominantly Fe (III). Ligation of ferric iron by Cl- prevents coordination by water, which can in turn lead to polymerization of the iron-protoporphyrin IX species via oxo (O^{2-}) or hydroxo (OH^{-}) bridging ligands.

2. Adapted from Varel and Bryant [5], the defined medium is suitable for routine cultivation of *B. thetaiotaomicron*. The clear appearance of this medium is favorable for monitoring optical density of cultures (OD) during bacterial growth. In contrast, since chopped meat medium is a rich and turbid medium with meat particles, it is recommended to use it to grow *B. thetaiotaomicron* for short- and long-term storage.
3. When autoclaving tubes and other vessels, do not seal them. Leave the cap on loosely to allow steam to flow during sterilization.
4. Usually, Luria-Bertani (LB) agar is used to grow transformed *E. coli*. According to the selectable marker present in the plasmid harboring the target gene, antibiotics are added to the medium (e.g., ampicillin, chloramphenicol, and kanamycin).
5. To purify proteins, different routes with alternating resin columns can be used. Here, we describe anionic and size exclusion chromatography steps to purify cytochrome b5, but other resin columns can be tested to obtain the pure protein.
6. Milli-Q water (Millipore) is used throughout the procedures described here. When metal-free or defined metal conditions are required, the use of deionized water-rinsed plasticware, or nitric acid-washed glassware, is necessary. Nitric acid can be purchased at high-purity metal-free grade for this purpose.

References

1. Bacic MK, Smith CJ (2008) Laboratory maintenance and cultivation of *Bacteroides* species. Curr Protoc Microbiol Chapter 13:Unit 13C.1. https://doi.org/10.1002/9780471729259.mc13c01s9
2. Strobel HJ (2009) Basic laboratory culture methods for anaerobic bacteria. In: Mielenz J (ed) Biofuels. Methods in molecular biology, vol 581. Humana Press, Totowa. https://doi.org/10.1007/978-1-60761-214-8_16
3. Madigan MT, Bender KS, Buckley DH, Sattley WM, Stahl DA (2019) Brock biology of microorganisms, 15th edn. Pearson Education, London
4. Leibniz Institute DSMZ-German Collection of Microorganisms and Cell Cultures GmbH. Cultivation of anaerobes. https://www.dsmz.de/collection/catalogue/microorganisms/culture-technology. Accessed 30 Apr 2023
5. Varel VH, Bryant MP (1974) Nutritional features of *Bacteroides fragilis* subsp. *fragilis*. Appl Microbiol 28:251–257. https://doi.org/10.1128/am.28.2.251-257.1974
6. Rivera M, Walker FA (1995) Biosynthetic preparation of isotopically labeled heme. Anal Biochem 230:2295–2302. https://doi.org/10.1006/abio.1995.1477

7. Rodríguez-Marañón MJ, Qiu F, Stark RE, White SP, Zhang X, Foundling SI, Rodríguez V, Schilling CL, Bunce RA, Rivera M (1996) 13C NMR spectroscopic and X-ray crystallographic study of the role played by mitochondrial cytochrome b5 heme propionates in the electrostatic binding to cytochrome c. Biochemist 35:16378–16390. https://doi.org/10.1021/bi961895o
8. Lewis UJ (1954) Acid cleavage of heme proteins. JBC 206:109–120
9. Fronticelli C, Bucci E (1963) Acetone extraction of heme from myoglobin and hemoglobin at acid pH. Biochim Biophys Acta 78:530–531
10. Barr I, Guo F (2015) Pyridine hemochromagen assay for determining the concentration of heme in purified protein solutions. Bio-protocol 5(18):e1594. https://doi.org/10.21769/BioProtoc.1594
11. Paul KG, Theorell H, Åkeson A (1953) The molar light absorption of pyridine ferroprotoporphyrin (pyridine haemochromogen). Acta Chem Scand 7:1284–1287
12. Berry EA, Trumpower BL (1987) Simultaneous determination of hemes a, b, and c from pyridine hemochrome spectra. Anal Biochem 161(1):1–15. https://doi.org/10.1016/0003-2697(87)90643-9
13. Meslé MM, Gray CR, Dlakić M, DuBois JL (2023) *Bacteroides thetaiotaomicron*, a model gastrointestinal tract species, prefers heme as an iron source, yields protoporphyrin IX as a product, and acts as a heme reservoir. Microbiol Spectr 11(2):e0481522. https://doi.org/10.1128/spectrum.04815-22
14. Knippel RJ, Zackular JP, Moore JL, Celis AI, Weiss A, Washington MK, DuBois JL, Caprioli RM, Skaar EP (2018) Heme sensing and detoxification by HatRT contributes to pathogenesis during *Clostridium difficile* infection. PLoS Pathog 14(12):e1007486. https://doi.org/10.1371/journal.ppat.100748
15. Sullivan SA, Streit BR, Ferguson EL, Jean PA, Mcnett DA, Llames LT, Dubois JL (2015) Mass-spectrometric profiling of porphyrins in complex biological samples with fundamental, toxicological, and pharmacological applications. Anal Biochem 478:82–89. https://doi.org/10.1016/j.ab.2015.03.004

Chapter 8

Heme *o* Isolation and Analysis Using Heme *o* Synthase Heterologously Expressed in *Escherichia coli*

Elise D. Rivett, Minh Anh Dang, and Eric L. Hegg

Abstract

Heme *o* is an Fe–porphyrin involved in the majority of aerobic respiration pathways found in all three domains of life. In eukaryotes and most aerobic prokaryotes, heme *o* functions solely as the precursor for the synthesis of heme *a*, a necessary cofactor for most heme–copper terminal oxidases. In some prokaryotes, such as *Escherichia coli* (*E. coli*), heme *o* can serve as a cofactor for heme–copper oxidases instead of heme *a*. Given its role as a key substrate or cofactor, purified heme *o* promises to be a valuable resource for the study of heme–copper oxidase assembly and activity. However, commercially available heme *o* is sold in limited quantities at a relatively high cost (compared to the prototypical heme *b*), making the use of heme *o* purchased from suppliers unfeasible for such studies. In this chapter, we present step-by-step methods both for heme *o* isolation from *E. coli* overexpressing heme *o* synthase and for HPLC analysis of cellular hemes (i.e., heme *o* and heme *b*).

Key words CtaB, *Escherichia coli*, Heme *b*, Heme extraction, Heme *o*, High-performance liquid chromatography (HPLC), Heterologous expression, Reverse-phase chromatography

Abbreviations

B. subtilis	*Bacillus subtilis*
CH_3CN	Acetonitrile
CtaB	Heme *o* synthase
CyoE	Heme *o* synthase
DMSO	Dimethyl sulfoxide
E. coli	*Escherichia coli*
HCl	Hydrochloric acid
HPLC	High-performance liquid chromatography
IPTG	Isopropyl β-D-1-thiogalactopyranoside
LB	Luria–Bertani
OD_{600}	Optical density at 600 nm
sh	Shoulder
TFA	Trifluoroacetic acid

Oleh Khalimonchuk (ed.), *Iron Metabolism: Methods and Protocols*, Methods in Molecular Biology, vol. 2839, https://doi.org/10.1007/978-1-0716-4043-2_8,

1 Introduction

Heme *o* is a biologically relevant Fe–porphyrin involved in aerobic respiration. Structurally, the only difference between heme *o* and heme *b* (Fe–protoporphyrin IX, the most common type of heme in the cell) is the identity of the substituent at the C2 position of the porphyrin ring. In heme *b*, the C2 substituent is a vinyl group, while in heme *o*, this substituent is a hydroxyethylfarnesyl moiety (Fig. 1). Functionally, heme *b* plays a wide variety of roles in the cell,

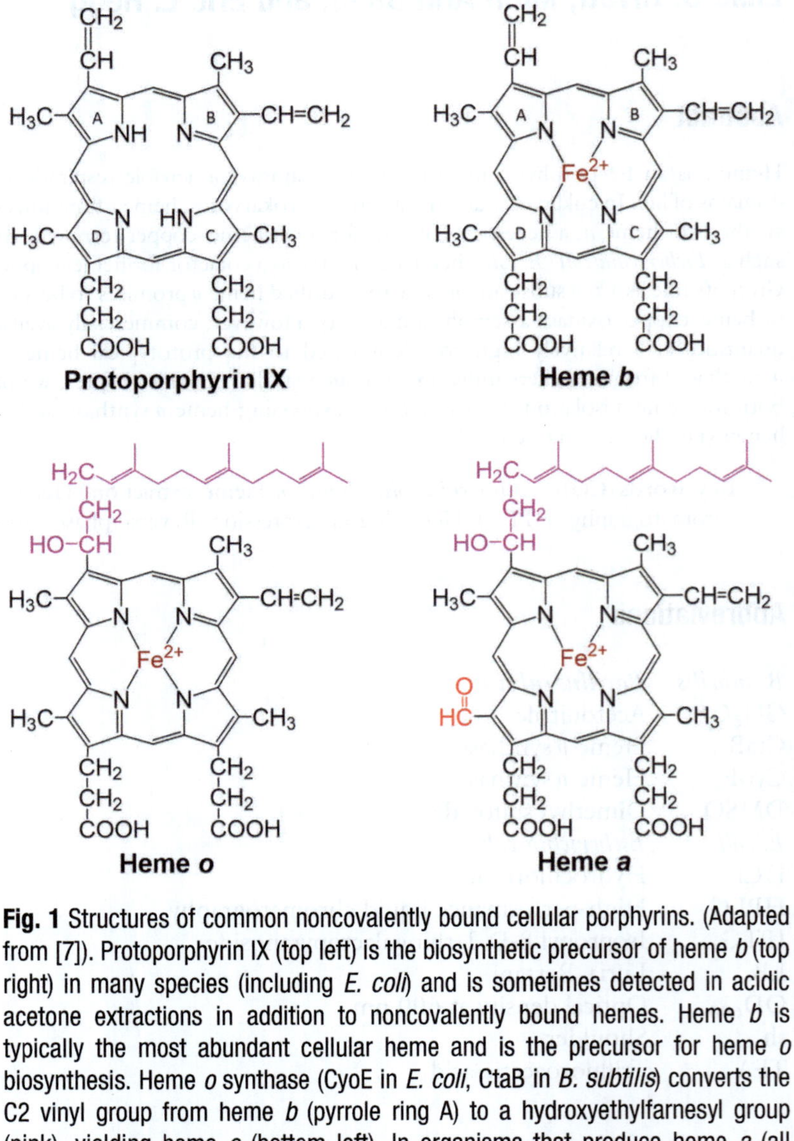

Fig. 1 Structures of common noncovalently bound cellular porphyrins. (Adapted from [7]). Protoporphyrin IX (top left) is the biosynthetic precursor of heme *b* (top right) in many species (including *E. coli*) and is sometimes detected in acidic acetone extractions in addition to noncovalently bound hemes. Heme *b* is typically the most abundant cellular heme and is the precursor for heme *o* biosynthesis. Heme *o* synthase (CyoE in *E. coli*, CtaB in *B. subtilis*) converts the C2 vinyl group from heme *b* (pyrrole ring A) to a hydroxyethylfarnesyl group (pink), yielding heme *o* (bottom left). In organisms that produce heme *a* (all eukaryotes and many aerobic prokaryotes, but not *E. coli*), heme *o* serves as a precursor for heme *a* biosynthesis. In heme *a* (bottom right), the C8 substituent (pyrrole ring D) is a formyl group (red)

including acting as a prosthetic group for diverse proteins [1, 2], serving as a regulatory molecule [3–5], and acting as the precursor for the synthesis of specialized heme types (including heme *o*) [6, 7]. In contrast, the only known functions of heme *o* are related to aerobic respiration. Most commonly, heme *o* serves as a precursor for the synthesis of heme *a* (Fig. 1), a cofactor required for all eukaryotic cytochrome *c* oxidases and most prokaryotic heme–copper oxidases (i.e., quinol oxidases and cytochrome *c* oxidases). In contrast, some prokaryotes incorporate heme *o* into heme–copper oxidases. *Escherichia coli* (*E. coli*), for example, lacks the gene for heme *a* synthase and exclusively uses heme *o* as the active-site (high-spin) heme cofactor for its bo_3 quinol oxidase. Other prokaryotes possess heme–copper oxidases that can incorporate either heme *o* or heme *a* into the active site to produce a functional enzyme [8–12]. Thus, purified heme *o* is a valuable resource for studying heme–copper oxidase assembly and activity and for catalytic studies of heme *a* synthase. However, very few chemical suppliers sell heme *o*, and the cost of heme *o* is substantially higher than that of heme *b* on a per mg basis. In this chapter, we present methods both for isolating heme *o* from *E. coli* that is heterologously overexpressing heme *o* synthase and for HPLC analysis of cellular hemes. Broadly, this method consists of four parts: (1) growth of an *E. coli* strain overexpressing heme *o* synthase, (2) small-scale heme extraction to verify high heme *o* production, (3) large-scale heme extraction and chromatographic purification of heme *o*, and (4) HPLC analysis of hemes.

When grown at high O_2 concentrations, *E. coli* synthesizes heme *o* using its native heme *o* synthase (CyoE), allowing for bo_3 quinol oxidase assembly [13]. Under aerobic growth conditions, heme *o* comprises approximately 10% of the total noncovalently bound heme in *E. coli* BL21(DE3) [14]. However, a much higher heme *o* yield can be achieved if native [14] or nonnative [15] heme *o* synthase is overexpressed in *E. coli*. For example, we have found that when *E. coli* BL21(DE3) coexpresses the T7 lysozyme from pLysS and *Bacillus subtilis* heme *o* synthase (*Bs*CtaB) from a pET vector, the percentage of heme *o* increases to 40–70% of noncovalently bound cellular heme (i.e., heme *b* and heme *o*) (Figs. 2 and 3). As detailed below, this coexpression system can be leveraged to isolate heme *o* from *E. coli* in microgram quantities.

Small-scale or large-scale extraction of heme from *E. coli* can be accomplished by mixing an aqueous cell suspension with acidic acetone. Acetone causes cellular proteins to precipitate, and the addition of acid ensures that the heme propionate groups are protonated so that heme will remain in the solution. Following the removal of the precipitate by centrifugation, the extracted hemes can be separated on a reverse-phase C18 column. Hemes extracted from small cell pellets are applied to an analytical HPLC column to confirm that cellular heme *o* production is high, while hemes

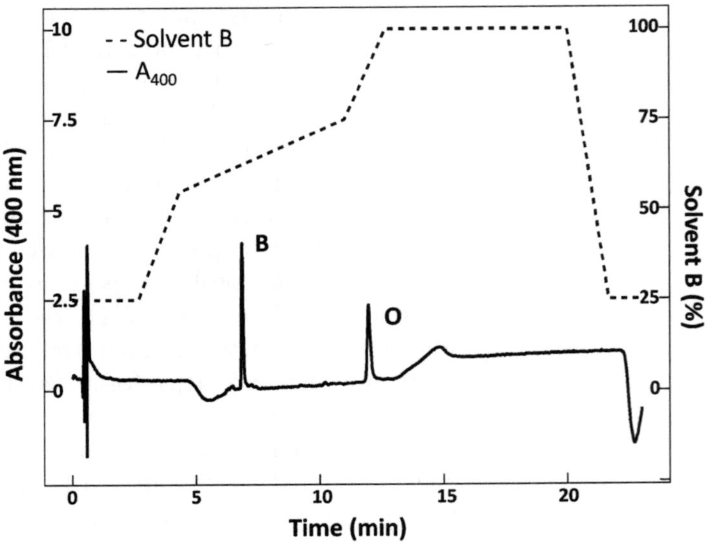

Fig. 2 Chromatographic profile of hemes extracted from *E. coli* BL21(DE3) pLysS pET3a T7-*Bs*CtaB and separated on an InfinityLab Poroshell 120 EC-C18 column. Hemes were separated using a gradient of Solvent A (water, 0.1% (v/v) TFA) and Solvent B (acetonitrile, 0.1% (v/v) TFA): 0–2.67 min: 25% B, 2.67–4.33 min: gradient from 25% B to 55% B, 4.33–11.00 min: gradient from 55% B to 75% B, 11.00–12.67 min: gradient from 75% B to 100% B, 12.67–20.00 min: 100% B, 20.00–23.00 min: gradient to return to starting conditions of 25% B, with a constant 1.18 mL/min flow rate. Dashed line: Solvent B (%), solid line: absorbance at 400 nm (A_{400}). The peaks corresponding to heme *b* (6.9 min, ~62% Solvent B) and heme *o* (12 min, ~90% Solvent B) are labeled

obtained from a large-scale heme extraction are applied to a Sep-Pak C18 cartridge to allow for the collection of purified heme *o*. While several variations of these protocols have been reported [16–18], the procedural details are often sparse. Additionally, our method of heme purification is simpler than previously reported protocols that require additional extractions with diethyl ether or ethyl acetate [16–18].

HPLC analysis is useful for both the classification and quantification of extracted hemes. Flowing the extracted hemes through an HPLC equipped with a C18 column and coupled to a UV-vis detector enables one to uniquely identify different hemes based on their retention time (Fig. 3) and Soret absorbance maximum (i.e., the absorbance maximum at ~400 nm) (Fig. 4). Thus, HPLC analysis can be used for relatively rapid identification of each heme type in a mixture (provided the appropriate heme standards are available for comparison). HPLC data can also be used to determine approximate heme concentrations using a standard curve prepared with heme *b* standards. Notably, we have found that HPLC-based concentration measurements of both pure heme *b* and purified heme *o* are similar to the values determined by the

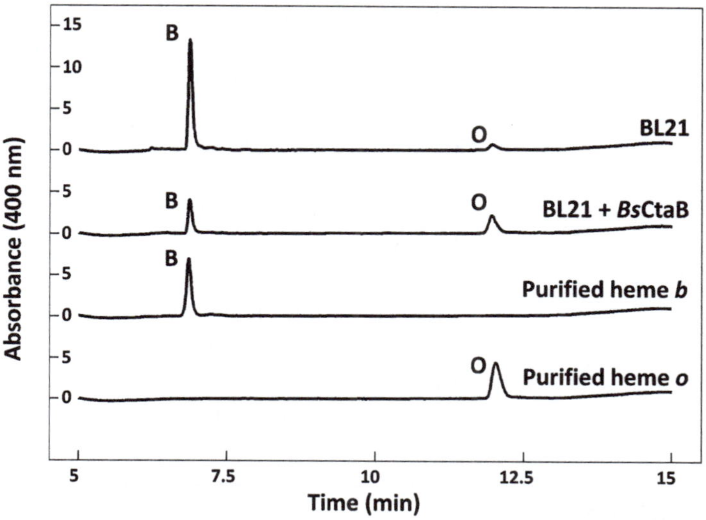

Fig. 3 HPLC chromatograms of hemes from small-scale heme extractions from *E. coli* or purified hemes. Top two traces: HPLC chromatograms of hemes extracted from *E. coli* BL21(DE3) grown aerobically with (BL21 + *Bs*CtaB) or without (BL21) heterologously overexpressed heme *o* synthase. Bottom two traces: HPLC chromatograms of pure heme *b* (hemin) and heme *o* (isolated from *E. coli* expressing *B. subtilis* heme *o* synthase). The concentrations of heme *b* and heme *o* are both approximately 1 μM

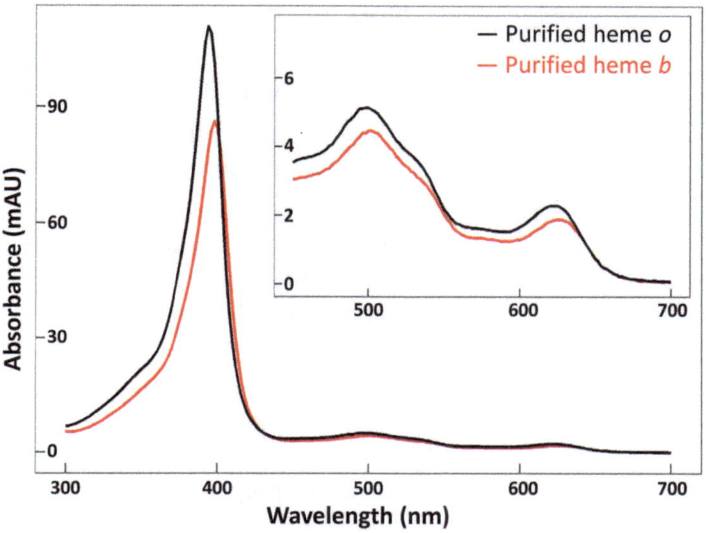

Fig. 4 Optical spectra of heme *b* and heme *o* (in HPLC solvent). Optical spectra of purified heme *b* (hemin) and purified heme *o* (isolated from *E. coli* expressing *B. subtilis* heme *o* synthase). Heme *b* (red trace, approximately 10 μM) elutes at ~62% Solvent B (acetonitrile, 0.1% TFA). Heme *o* (black trace, approximately 13 μM) elutes at ~90% Solvent B. The inset shows the α- and β-bands of each heme

well-established but more labor-intensive pyridine hemochrome method [11, 19]. As it is difficult to use the pyridine hemochrome approach to quantify individual hemes in mixtures of heme *b* and heme *o* (due to their overlapping α-bands) [20], HPLC analysis is preferable for measuring heme concentrations in such samples. Another advantage of HPLC analysis is that the detection limit (≤ 0.5 μM heme *b*) for this method is lower than that of the pyridine hemochrome method (10 μM heme *b*) [21], making HPLC analysis a convenient, sensitive method for analyzing heme mixtures and purified heme. Overall, we hope that the step-by-step methods for heme *o* overexpression and extraction and subsequent HPLC analysis outlined here will prove useful for researchers studying heme *o* or other prenylated hemes.

2 Materials

Unless otherwise specified, prepare all reagents in Milli-Q water and store at room temperature.

2.1 Growth of E. coli

1. Vector: expression plasmid pET3a T7-*Bs*CtaB [15]. This vector encodes the gene for *B. subtilis* CtaB with an N-terminal T7 epitope tag (*see* **Note 1**).
2. *E. coli* BL21(DE3) pLysS chemically competent cells (*see* **Note 2**).
3. 1000× Ampicillin Solution: 100 mg/mL ampicillin in Milli-Q water (*see* **Note 3**). Filter-sterilize and store at −20 °C.
4. 1000× Chloramphenicol Solution: 25 mg/mL chloramphenicol in 95% ethanol. Store at −20 °C.
5. Solid Luria–Bertani (LB) Medium: 10 g/L NaCl, 10 g/L tryptone, 5 g/L yeast extract, and 1.5–2% (w/v) agar. Autoclave and supplement with ampicillin (100 μg/mL) and chloramphenicol (25 μg/mL) using the previously prepared stock solutions. Store at 4 °C.
6. Liquid LB Medium: 10 g/L NaCl, 10 g/L tryptone, and 5 g/L yeast extract. Autoclave and supplement with ampicillin (100 μg/mL) and chloramphenicol (25 μg/mL) using the previously prepared stock solutions.
7. Sterile Milli-Q water. Sterilize by autoclaving.
8. Lysis buffer [15]: 30 mM Tris–HCl, pH 8.0, 20% (w/v) sucrose, and 10 mM EDTA. Filter-sterilize and store at 4 °C.
9. Liquid nitrogen.
10. Culture tubes (15-mL), 2-L baffled Erlenmeyer flasks, 1.7-mL microcentrifuge (Eppendorf) tubes, 500-mL centrifuge tubes, and 15-mL and 50-mL conical tubes.

11. Mini-centrifuge (for microcentrifuge tubes; maximum speed ~17,000 × *g*).
12. Centrifuge for 15-mL and 50-mL conical tubes (maximum speed ~5000 × *g*).
13. Centrifuge for 500-mL centrifuge tubes (maximum speed 5000 × *g*).

2.2 Small-Scale Extraction of Noncovalently Bound Hemes

1. Benzonase nuclease (≥ 250 units/μL) or similar endonuclease (*see* **Note 4**).
2. Milli-Q water.
3. HCl–Acetone: 0.6 M HCl, 95% (v/v) acetone (*see* **Note 5**).
4. Microliter syringe (*see* **Note 6**).
5. 1.7-mL microcentrifuge (Eppendorf) tubes, HPLC vials, and vial inserts (*see* **Note 7**).
6. Vortex mixer.
7. Mini-centrifuge (for microcentrifuge tubes; max speed ~17,000 × *g*).

2.3 Large-Scale Heme Extraction and Isolation of Heme o

1. Lysis Buffer [15]: 30 mM Tris–HCl, pH 8.0, 20% (w/v) sucrose, and 10 mM EDTA. Filter-sterilize and store at 4 °C.
2. Benzonase nuclease or similar endonuclease (*see* **Note 4**).
3. HCl–Acetone: 0.6 M HCl, 95% (v/v) acetone (*see* **Note 5**).
4. HPLC-grade acetonitrile.
5. Solvent A: Milli-Q water, 0.1% (v/v) trifluoroacetic acid (TFA) (*see* **Note 8**).
6. Solvent B: HPLC-grade acetonitrile, 0.1% (v/v) TFA (*see* **Note 8**).
7. Dimethyl sulfoxide (DMSO).
8. Vortex mixer.
9. Sonicator (with microtip).
10. Centrifuge for 50-mL centrifuge tubes (maximum speed ~30,000 × *g*) (*see* **Note 9**).
11. Waters Sep-Pak C18 cartridge (3 cc, 500 mg sorbent, 55–105 μm particle size) (*see* **Note 10**).
12. Extra reservoir for C18 column: 50-mL serological pipette with the top cut off (*see* **Note 11**).
13. Compressed air source (*see* **Note 12**).
14. Ring stand and clamps.
15. UV-vis spectrophotometer and glass or quartz cuvettes.
16. 1.7-mL microcentrifuge (Eppendorf) tubes, 50-mL centrifuge tubes, 15-mL and 50-mL conical tubes, and 100-mL or 250-mL beakers or Erlenmeyer flasks.

2.4 HPLC Analysis of Noncovalently Bound Hemes

1. HPLC reverse-phase InfinityLab Poroshell 120 EC-C18 column (4.6 × 50 mm, 2.7 μm particle size, 120 Å pore size) or similar column.
2. HPLC system equipped with a diode array detector (*see* **Note 13**) and capable of flow rates of 1.2 mL/min.
3. HPLC-grade acetonitrile.
4. Solvent A: Milli-Q water, 0.1% (v/v) trifluoroacetic acid (TFA) (*see* **Note 8**).
5. Solvent B: HPLC-grade acetonitrile, 0.1% (v/v) TFA (*see* **Note 8**).
6. Dimethyl sulfoxide (DMSO).
7. Hemin (from bovine, ≥ 90%) (ferriprotoporphyrin IX chloride).
8. 5-mL volumetric flask (*see* **Note 14**).
9. Vortex mixer.
10. Mini-centrifuge (for microcentrifuge tubes; max speed ~17,000 × *g*).
11. 1.7-mL microcentrifuge (Eppendorf) tubes, HPLC vials, and vial inserts (*see* **Note 7**).

3 Methods

3.1 Growth of *E. coli* BL21(DE3) pLysS pET3a T7-*Bs*CtaB

1. Transform pET3a T7-*Bs*CtaB into chemically competent BL21 (DE3) pLysS (*see* **Notes 1** and **2**) and plate the transformation mixture onto LB agar plates supplemented with ampicillin (100 μg/mL) (*see* **Note 3**) and chloramphenicol (25 μg/mL). Incubate overnight (12–16 h) at 37 °C.
2. Use an isolated transformant colony (BL21(DE3) pLysS pET3a T7-*Bs*CtaB) to inoculate 10 mL of LB supplemented with ampicillin (100 μg/mL) and chloramphenicol (25 μg/mL). Grow overnight (12–16 h), shaking, at 37 °C.
3. Use 10 mL of the overnight culture to inoculate 1 L of LB supplemented with ampicillin (100 μg/mL) and chloramphenicol (25 μg/mL) (*see* **Note 15**). Grow the culture at 37 °C, shaking at 180 rpm, until the OD_{600} reaches 1–1.2 (*see* **Note 1**).
4. Collect a sample for small-scale heme extraction from intact cells by harvesting the equivalent of 20 mL of culture at an OD_{600} of 1 (*see* **Note 16**): Centrifuge the appropriate culture volume for 7 min at 5000 × *g*, 4 °C. Discard the supernatant (*see* **Note 17**).

5. Resuspend the small-scale cell pellet in 1 mL of Milli-Q water, transfer the suspension to a microcentrifuge tube, and centrifuge for 30 s at maximum speed (~17,000 × *g*) to pellet the cells. Remove the supernatant by pipetting. The cell pellet can be used immediately for heme extraction (Subheading 3.2) or flash-frozen in liquid nitrogen and stored at −80 °C.
6. Harvest the remaining culture (~1 L) by centrifuging for 7 min at 5000 × *g*, 4 °C. Discard the supernatant (*see* **Note 17**).
7. Resuspend the cell pellet in 10 mL of lysis buffer, transfer to a 15-mL or 50-mL conical tube, and centrifuge for 7 min at 5000 × *g*, 4 °C. Discard the supernatant. Flash freeze the cell pellet in liquid nitrogen and store at −80 °C (for use in Subheading 3.3).

3.2 Small-Scale Extraction of Noncovalently Bound Hemes

1. Thaw the cell pellet on ice, if necessary. Resuspend the cell pellet (equivalent to 20 mL at $OD_{600} = 1$; *see* **Note 16**) in 200 μL of Milli-Q water (*see* **Notes 4** and **18**).
2. Add 150 μL of HCl–acetone to a new microcentrifuge tube (*see* **Notes 5** and **6**).
3. Transfer 100 μL of the cell suspension to the microcentrifuge tube containing HCl–acetone. Vortex briefly (~15 s) to mix.
4. Incubate on ice for approximately 10 min.
5. Vortex briefly (~15 s).
6. Incubate on ice for an additional 10 min. Vortex briefly (~15 s) at the end of the incubation.
7. Centrifuge at maximum speed (~17,000 × *g*) 10 min at room temperature to pellet precipitated protein/cell debris (*see* **Note 19**).
8. Transfer the supernatant to a new microcentrifuge tube.
9. Centrifuge at maximum speed (~17,000 × *g*) for 10 min at room temperature to pellet any remaining insoluble material.
10. Transfer the supernatant to an HPLC vial (*see* **Note 7**). Perform HPLC analysis promptly (*see* **Note 20**).

3.3 Large-Scale Heme Extraction and Isolation of Heme o

1. Resuspend each cell pellet (harvested from 1 L of culture) in 5 mL of lysis buffer (approximately 5 mL of lysis buffer per 2 g wet cells). Combine two 5-mL suspensions and transfer to one 50-mL conical tube (*see* **Notes 4** and **21**).
2. Lyse the cells by sonication (*see* **Note 22**).
3. Add 30 mL of HCl–acetone to 10 mL of cell suspension (*see* **Note 5**). Vortex the sample for 1–2 min to break up cell aggregates and then sonicate again to mix thoroughly (*see* **Notes 22** and **23**).

4. Divide each sample between two 50-mL centrifuge tubes (~20 mL sample/tube). Centrifuge 10 min at 30,000 × *g*, 4 °C to pellet the cell debris (*see* **Note 9**). Combine the supernatant from all four samples (~80 mL total).
5. Add 20 mL of lysis buffer to the supernatant to bring the final water/HCl–acetone ratio to 2:3 (*see* **Note 24**).
6. Prepare a Waters Sep-Pak C18 cartridge (3 cc, 500 mg sorbent, 55–105 μm particle size) by applying 15 mL of pure, HPLC-grade acetonitrile (a strong solvent) to the column via gravity flow.
7. Wash the column with 15 mL of Solvent B applied via gravity flow.
8. Wash the column with 15 mL of 75:25 (v/v) Solvent A:Solvent B applied via gravity flow.
9. Apply the heme-containing supernatant from **Step 5** to the C18 cartridge (*see* **Notes 11** and **12**). A red band should form at the top of the column as hemes bind.
10. Wash the column with 15 mL of 75:25 (v/v) Solvent A: Solvent B.
11. Wash the column with 25 mL of 20:80 (v/v) Solvent A: Solvent B to remove heme *b*. Check the UV-vis absorbance of the wash periodically to monitor heme *b* removal (*see* **Note 25**).
12. Elute heme *o* using 4 mL of 100% DMSO. Collect the concentrated heme *o* band as one fraction (~ 0.7 mL) (*see* **Note 26**). Store isolated heme *o* at −80 °C.
13. Dilute a small volume of isolated heme *o* with acetonitrile (*see* **Note 27**) starting with a dilution factor of 0.1 and diluting further if necessary. Analyze the UV-vis absorbance spectrum. For pure heme *o*, the absorbance maximum of the Soret peak should be ~395 nm.
14. Dilute a small volume of isolated heme *o* with acetonitrile (*see* **Note 27**) prior to HPLC analysis (Subheading 3.4) using the height of the sample's Soret peak (measured in **Step 13**) to gauge an appropriate dilution factor. Typically, a dilution factor of 0.1 works well.

3.4 HPLC Analysis of Noncovalently Bound Hemes

3.4.1 Preparation of Heme b (hemin) Standards for HPLC Analysis

1. Prepare a 1 mM hemin stock solution in DMSO: Weigh ~0.004 g of hemin and record the weight to the nearest 0.0001 g. Transfer the hemin to a 5-mL volumetric flask (*see* **Note 14**), add 5 mL DMSO, and stir at room temperature for 1–2 h (*see* **Note 28**).
2. Calculate the concentration of the hemin stock solution using the measured hemin weight and the formula weight of hemin (651.96 g/mol).

3. Prepare a 0.1 mM hemin solution by diluting the 1 mM hemin stock solution 1:10 with HPLC-grade acetonitrile (*see* **Notes 27** and **29**). Vortex briefly to mix.
4. Calculate the precise concentration of the diluted (~0.1 mM) heme solution using the concentration of the stock hemin solution, the density of DMSO (1.10 g/mL), and the density of acetonitrile (0.786 g/mL) (*see* **Note 29**).
5. Prepare hemin standards (0.5, 1, 2, 4, 6, 8, and 10 μM) by diluting 0.1 mM hemin with acetonitrile (*see* **Note 27**). Measure the 0.1 mM hemin solution and acetonitrile by weight and use these weight measurements to calculate the precise concentration of each standard (*see* **Note 30**). Vortex briefly to mix.
6. Centrifuge at maximum speed (~17,000 × *g*) for 10 min at room temperature to pellet any insoluble material. Transfer the supernatant (~900 μL) to an HPLC vial.

3.4.2 Heme Separation and Analysis by HPLC

1. Apply heme to the InfinityLab Poroshell 120 EC-C18 column (4.6 × 50 mm, 2.7 μm particle size, 120 Å pore size) by injecting 10 μL of the sample (*see* **Note 31**) at a flow rate of 1.18 mL/min.
2. Separate the hemes using a gradient of Solvent A and Solvent B: 0–2.67 min: 25% B, 2.67–4.33 min: gradient from 25% B to 55% B, 4.33–11.00 min: gradient from 55% B to 75% B, 11.00–12.67 min: gradient from 75% B to 100% B, 12.67–20.00 min: 100% B, 20.00–23.00 min: gradient to return to starting conditions of 25% B, with a constant 1.18 mL/min flow rate (Fig. 2) (*see* **Note 32**). Monitor heme elution by plotting absorbance at 400 nm.
3. For each heme peak, record the peak retention time, area, and absorbance maximum of the Soret peak. Identify each heme based on its retention time and absorbance maximum (Figs. 3 and 4, and Table 1) (*see* **Note 33**) [15].
4. Quantify each heme using the standard curve for hemin (heme *b*). To generate the standard curve, plot heme concentration on the x-axis and area under the chromatogram curve on the y-axis and determine the equation of the best-fit line by linear regression (*see* **Notes 31** and **34**).

4 Notes

1. Protein expression from the pET3a vector is controlled by a T7 promoter, which is specifically recognized by T7 RNA polymerase. The *E. coli* strain BL21(DE3) has a chromosomal copy of the gene encoding T7 RNA polymerase under the control of the strong, isopropyl β-D-1-thiogalactopyranoside (IPTG)-

Table 1
Characteristics of common cellular porphyrins eluted from an HPLC C18 column

Porphyrin	Approximate % solvent B at elution	Soret A_{max} (nm)	α/β bands, shoulders (sh) (nm)	Notes	References
Heme *b*	62	398	501, 535 (sh), 626	Typically the most abundant cellular heme	[15]
Protoporphyrin IX	69	402	540 (sh), 562, 602 (sh)	Heme *b* precursor in many organisms; occasionally detected in *E. coli*	This study
Heme *a*	74	406	494, 551, 604, 655 (sh)	Most common active-site cofactor for heme–copper oxidases; not found in *E. coli* unless a heterologous heme *a* synthase is expressed	[15]
Heme *o*	90	394	498, 532 (sh), 621	Cofactor for heme–copper oxidase in *E. coli*, other prokaryotes; intermediate in heme *a* biosynthetic pathway in many organisms	[15], this study

Hemes were separated using a gradient of Solvent A (water, 0.1% (v/v) TFA) and Solvent B (acetonitrile, 0.1% (v/v) TFA). The optical absorbance properties listed are specific for each heme's elution solvent [15]

inducible *lacUV5* promoter. For this reason, protein expression from pET vectors requires the use of BL21(DE3) (or a similar strain). Typically, high protein expression from a pET vector is achieved by adding IPTG to a BL21(DE3) pET culture to induce expression from the T7 promoter (by increasing expression of T7 RNA polymerase). Indeed, IPTG induction of BL21 (DE3) pLysS pET3a T7-*Bs*CtaB increases steady-state levels of the *Bs*CtaB protein. However, we have found that *uninduced* cells generally produce a higher level of heme *o* than induced cultures even though the level of *Bs*CtaB is lower. While the underlying cause of this phenomenon is not completely clear, it could be that IPTG-driven overexpression of the integral membrane protein *Bs*CtaB strains cellular metabolism, causing overall heme production to decrease over the course of extended incubations with IPTG. Along the same lines, another group has observed that overexpression of *Bs*CtaB or the native CyoE from *E. coli* causes a decrease in the ratio of total heme (heme *b* + heme *o*) to total protein [14].

2. As explained above (*see* **Note 1**), BL21(DE3) (or a similar strain) is required for expression from pET3a. We have also found that using BL21(DE3) transformed with pET3a

T7-*Bs*CtaB ***and*** pLysS is necessary to achieve high heme *o* yields. The vector pLysS produces relatively low levels of T7 lysozyme, which inhibits uninduced, low-level expression of T7 RNA polymerase and thus decreases "leaky" expression from pET vectors. Without pLysS, BL21(DE3) pET3a T7-*Bs*CtaB sometimes has an extended lag phase (4–6 h) and produces a low percentage of heme *o* (relative to heme *b*), probably because higher expression of T7-*Bs*CtaB is detrimental to the cells. If BL21(DE3) pLysS is unavailable, chemically competent BL21(DE3) may also be co-transformed with pLysS and pET3a T7-*Bs*CtaB via simultaneous or stepwise transformations. Many protocols have been published for the preparation and transformation of chemically competent *E. coli*; see, for example, the excellent paper by Chang et al. [22].

3. We recommend using ampicillin sodium salt (CAS# 69-52-3), which is water-soluble, allowing stock solutions to be prepared in water. Anhydrous ampicillin (CAS# 69-53-4) is only slightly soluble in water; if this form of the antibiotic is used, an acid or base solution is typically used as the diluent.
4. Some *E. coli* suspensions are very viscous, and the addition of a small volume of benzonase nuclease can help decrease the viscosity sufficiently such that the sample can be pipetted. For small-scale heme extraction (Subheading 3.2), we typically add 0.5 μL of benzonase nuclease (≥ 250 units/μL) to 200 μL of cell suspension and incubate 5 min on ice. If the cell suspension is still too viscous after 5 min, we add another 0.5 μL of benzonase nuclease and incubate for an additional 2–3 min. For large-scale heme extraction (Subheading 3.3), we add 1–2 μL of benzonase nuclease to 5–10 mL of cell suspension and incubate 20–30 min at 4 °C.
5. HCl–acetone should be prepared fresh (the same day it is used) by combining concentrated (~12 M or ~ 37% (w/w)) HCl and acetone in a 5:95 (v/v) ratio. To make 10 mL of HCl–acetone, mix 0.5 mL concentrated HCl with 9.5 mL acetone. The final concentration of HCl will be ~0.6 M.
6. Glass microliter syringes (e.g., a Hamilton 500-μL syringe) can be used to accurately measure small volumes of HCl–acetone. While micropipettes can be used to transfer HCl–acetone, we do not recommend this because (1) the measurements will be inaccurate as micropipettes are calibrated for water, not acetone, and (2) HCl–acetone can damage the O-rings in the pipette over time.
7. For small volumes (≤ 200 μL) of HPLC samples, we recommend using an HPLC vial insert.

8. To prepare solvents with 0.1% (v/v) TFA, add 1 mL of TFA to 1 L of the appropriate solvent (Milli-Q water or HPLC-grade acetonitrile).
9. For large-scale heme extraction, a centrifuge rotor for 50-mL centrifuge tubes is needed (maximum speed ~30,000 × *g*). While this maximum speed (30,000 × *g*) is low enough that a carbon fiber rotor could be used, we used a titanium rotor because titanium is more resistant to acetone than carbon fiber.
10. We have found that unused Waters Sep-Pak C18 cartridges can last at least 10 years.
11. To apply a rather large volume (~100 mL) of extracted hemes to a 3 cc Sep-Pak C18 cartridge, we use a homemade 50-mL reservoir. A reservoir can be made by cutting off the end of a polystyrene 50-mL serological pipette, producing a large opening at one end of the pipette (where a pipette bulb is normally attached). (This end of the pipette is heated over a Bunsen burner to soften it before cutting.) The other end of the pipette (the end with the tip) is then placed firmly inside the reservoir of the C18 column and held in place with parafilm and electrical tape to prevent leakage. The extracted hemes can then be loaded into the reservoir.
12. To expedite the process of applying extracted hemes to a C18 column during gravity chromatography, a tube attached to a compressed air source can be inserted into the top of the column (or column reservoir, *see* **Note 11**) and secured with parafilm. Assuming the junction is reasonably airtight, the pressure will force the sample through the C18 column faster than gravity alone. Alternatively, a Sep-Pak C18 cartridge can be used with a vacuum manifold, although we have not tested this method of loading extracted hemes.
13. Typically, absorbance at 400 nm is monitored to detect hemes. A diode array detector is particularly useful because it records the full UV-vis spectrum of the eluate, allowing for heme identification based on each heme's absorbance maximum (ranging from 394 to 406 nm) (Table 1). For more concentrated (approximately 10 μM or greater) samples, the maximum absorbances for the α- and β-bands of each heme can also be determined from the optical absorbance spectrum.
14. To remove any heme or other metals from the volumetric flask, we recommend washing the volumetric flask with acid (4 M HCl) and drying thoroughly prior to use.
15. We recommend growing a total of at least 4 L of culture. Typically, 1-L cultures grown in a 2-L baffled Erlenmeyer flask at 37 °C and shaken at 180 rpm reach an OD_{600} of 1–1.2 in 3.5–4 h. Under these conditions, heme *o* typically comprises 40–60% of the total extractable heme. A similar

percentage of heme *o* is produced (~50% of total extractable heme) by growing 0.5 L of BL21(DE3) pLysS pET3a T7-*Bs*CtaB in a 2.8-L nonbaffled Fernbach flask at a lower temperature (24 °C) for 48 h.

16. The culture volume equivalent to 20 mL at an OD_{600} of 1 can be calculated using the equation $v_1 = c_2v_2/c_1$. The variables c_1 and v_1 are the initial concentration and volume, respectively, while c_2 and v_2 are the final concentration and volume. In this case, $c_2 = 1$ OD_{600}, $v_2 = 20$ mL, and $c_1 =$ the final OD_{600} value of the culture.

17. The collection and washing steps for small-scale cell pellets (Subheading 3.1, **Steps 4–5**) and large-scale cell pellets (Subheading 3.1, **Steps 6–7**) are listed separately for clarity. However, these steps are often performed simultaneously to maximize efficiency.

18. In addition to performing whole-cell heme extraction with *E. coli* cell pellets, we have used very similar procedures to extract noncovalently bound hemes from other samples, including cell pellets from different bacteria (*Shewanella oneidensis* and *Rhodobacter sphaeroides*), crude mitochondrial extracts [23] from *Saccharomyces cerevisiae*, and purified protein. For *S. cerevisiae* mitochondrial extracts, 0.25–0.5 mg of total protein is pelleted by centrifugation and then resuspended in a small volume (20–50 μL) of buffer (600 mM sorbitol, 20 mM HEPES buffer, pH 7.4). The appropriate volume of HCl–acetone is then added to achieve a final buffer/HCl–acetone ratio of 2:3.

19. It is critical to remove all insoluble/particulate matter from the sample before injection into the HPLC. Because heme (especially the very hydrophobic heme *o*) adsorbs onto many common types of filter material, such as mixed cellulose ester (MCE), we have found that it is preferable to remove precipitate by multiple rounds of centrifugation instead of filtration. Installing a stainless steel in-line filter (0.2 μm pore size) on the HPLC system can also help protect the HPLC column from insoluble material.

20. Heme samples should be analyzed the same day they are prepared. In our experience, hemes in aqueous solutions (including water–HCl–acetone) tend to decrease in concentration over time, presumably because the hemes aggregate or adsorb to the container. This occurs regardless of whether the hemes are stored in plastic or glass containers.

21. This protocol is written for extracting heme from a total of 4 L of culture, which, in our experience, yields 50–80 μg of purified heme *o*. This procedure can also be scaled up for larger culture volumes. For example, to isolate heme *o* from a combined cell

pellet harvested from 11 L of culture, the volumes of lysis buffer and HCl–acetone were increased slightly, and a larger (12 cc) Sep-Pak C18 cartridge was used. The specific volumes used in this case were as follows: 30 mL of lysis buffer to initially resuspend the cells (Subheading 3.3, **Step 1**), 90 mL of HCl–acetone (Subheading 3.3, **Step 3**), and an additional 30 mL of lysis buffer (Subheading 3.3, **Step 5**).

22. Sonicate the cells 10 times using 10-s intervals. Incubate the mixture on ice between sonication steps. We use a Tekmar High Intensity Ultrasonic Processor (Model 501) equipped with a microtip, with the output control set at 4.
23. At this point, the ratio of water (aqueous cell suspension) to HCl–acetone should be approximately 1:3.
24. The solution may become cloudy at this point. This should not affect heme isolation.
25. Periodically (approximately every 5 mL), collect a small fraction of the wash and measure its UV–vis absorbance spectrum. As heme *b* is removed from the column, the height of the Soret peak (peak in the ~400 nm region due to the combination of heme *b* and heme *o*) should decrease significantly. The maximum absorbance (A_{max}) of the Soret peak should also shift from ~399 nm (heme *b* A_{max}) to ~395 nm (heme *o* A_{max}). After observing two consecutive fractions with an A_{max} of ~395 nm, it can be assumed that all heme *b* has been washed from the column. At this point, the majority of heme *o* should still be bound to the column (visible as a brown band).
26. If using an extra solvent reservoir (*see* **Note 11**), remove it before adding DMSO. Allow almost all of the wash solvent to enter the column bed before adding DMSO.
27. As with HCl–acetone, acetonitrile and DMSO cannot be measured accurately with micropipettes. For this reason, when working with these solvents, we recommend using microsyringes or glass serological pipettes for more accurate dilutions. When preparing heme standards, we measure solvent weights for even higher accuracy (*see* **Notes 29** and **30**).
28. Concentrated hemin solutions (1 mM or greater) are very dark, making it difficult to visually verify complete hemin dissolution. Therefore, the hemin stock solution is stirred for at least an hour to ensure that all of the hemin has dissolved.
29. To dilute a 1 mM hemin stock solution (dissolved in DMSO) with acetonitrile by weight: Transfer ~100 μL of 1 mM hemin to a microcentrifuge tube and record the weight. Add 1 mL (0.786 g) of acetonitrile (CH_3CN) to the tube and record the final total weight. Calculate the weight of added acetonitrile: weight CH_3CN = total weight—weight hemin stock.

Calculate the final concentration of the diluted hemin solution using the following equation:

$$[\text{final hemin (mM)}] = \frac{(\text{weight hemin stock (g)}) * [\text{hemin stock (mM)}]/\left(1.10\frac{\text{g}}{\text{mL}}\right)}{\frac{(\text{weight hemin stock (g)})}{1.10\frac{\text{g}}{\text{mL}}} + \frac{(\text{weight CH}_3\text{CN (g)})}{0.786\frac{\text{g}}{\text{mL}}}}$$

where "weight hemin stock" is the weight of 1 mM hemin (in g), "hemin stock" is the precise concentration of the 1 mM hemin stock (in mM), and "weight CH_3CN" is the weight of acetonitrile (in g).

30. To dilute the 0.1 mM hemin solution to 0.5–10 μM with acetonitrile (CH_3CN): Transfer the appropriate volume of 0.1 mM hemin to a microcentrifuge tube and record the weight. Add acetonitrile to dilute the hemin stock to a final volume of 1 mL (target final weight = 0.786 g). Record the final weight. Calculate the final concentration of each standard solution using the following equation:

$$[\text{final hemin } (\mu\text{M})] = \frac{(\text{weight} \sim 0.1 \text{ mM hemin (g)}) * [\text{hemin } (\mu\text{M})]}{(\text{total weight (g)})}$$

where "weight ~0.1 mM hemin" is the weight of the 0.1 mM hemin solution (in g), "[hemin]" is the precise concentration of the 0.1 mM hemin solution (*in μM*), and "total weight" is the final total weight (in g) after the addition of acetonitrile.

31. To quantify hemes accurately using an HPLC standard curve, it is critical to use the same injection volume for the standards and samples. We have found that the area under the curve does not scale linearly with injection volume when the injection volume is increased. Injection volumes can be increased if hemes are not going to be quantified.

32. Flow rates and the duration of each step may need to be adjusted for different C18 columns/HPLC systems. The most common types of extractable heme typically elute between ~62% Solvent B (heme *b*) and 90% Solvent B (heme *o*) (Table 1). (Heme ***a*** elutes at ~74% Solvent B.) However, retention times can shift (usually increasing slightly in each consecutive sample) as nonpolar molecules from successive samples adsorb to the C18 column. For this reason, we included a rather long wash with 100% Solvent B to ensure that all hemes were eluted and to wash other hydrophobic compounds from the C18 column.

33. For properties of prenylated hemes produced during heme *a* biosynthesis (usually present at very low concentrations), see Brown et al. [15].

34. Because the extinction coefficients of heme *o* and heme *a* are typically similar to the extinction coefficient for heme *b* [11, 24], we have used a heme *b* standard curve to estimate heme *b* concentrations and to estimate concentrations of heme *o* and heme *a* in heme mixtures and preparations of purified heme. Alternatively, once HPLC analysis has been used to confirm that a sample contains only one type of heme, the pyridine hemochrome method [19, 21] can be used to determine heme concentrations using previously determined extinction coefficients and the appropriate peak/trough wavelengths. For heme *b*, $\varepsilon_{\text{red-ox,557–540}} = 23.98\ \text{mM}^{-1}\ \text{cm}^{-1}$ [19]; for heme *a*, $\varepsilon_{\text{red-ox,587–620}} = 25.02\ \text{mM}^{-1}\ \text{cm}^{-1}$ [19]; and for heme *o*, $\varepsilon_{\text{red-ox,551–535}} = 10.3\ \text{mM}^{-1}\ \text{cm}^{-1}$ [11]. Note that using the pyridine hemochrome method to determine individual heme concentrations in mixtures of hemes *b* and *o* is difficult because the α-bands of these hemes only differ by ~4 nm and cannot be resolved [20].

References

1. Li T, Bonkovsky HL, Guo JT (2011) Structural analysis of heme proteins: implications for design and prediction. BMC Struct Biol 11:13
2. Smith LJ, Kahraman A, Thornton JM (2010) Heme proteins: diversity in structural characteristics, function, and folding. Proteins 78: 2349–2368
3. Brewitz HH, Hagelueken G, Imhof D (2017) Structural and functional diversity of transient heme binding to bacterial proteins. Biochim Biophy Acta (BBA) 1861:683–697
4. Zhang L, Hach A (1999) Molecular mechanism of heme signaling in yeast: the transcriptional activator Hap1 serves as the key mediator. Cell Mol Life Sci 56:415–426
5. Zhang X, Guo J, Wei X et al (2018) Bach1: function, regulation, and involvement in disease. Oxidative Med Cell Longev 2018: 1347969
6. Kranz RG, Richard-Fogal C, Taylor JS et al (2009) Cytochrome *c* biogenesis: mechanisms for covalent modifications and trafficking of heme and for heme-iron redox control. Microbiol Mol Biol Rev 73:510–528
7. Rivett ED, Heo L, Feig M et al (2021) Biosynthesis and trafficking of heme *o* and heme *a*: new structural insights and their implications for reaction mechanisms and prenylated heme transfer. Crit Rev Biochem Mol Biol 56:640–668
8. Peschek GA, Alge D, Fromwald S et al (1995) Transient accumulation of heme O (cytochrome *o*) in the cytoplasmic membrane of semi-anaerobic *Anacystis nidulans*: evidence for oxygenase-catalyzed heme O/A transformation. J Biol Chem 270:27937–27941
9. Sone N, Fujiwara Y (1991) Haem O can replace haem A in the active site of cytochrome *c* oxidase from thermophilic bacterium PS3. FEBS Lett 288:154–158
10. Matsushita K, Ebisuya H, Ameyama M et al (1992) Change of the terminal oxidase from cytochrome a_1 in shaking cultures to cytochrome *o* in static cultures of *Acetobacter aceti*. J Bacteriol 174:122–129
11. Matsushita K, Ebisuya H, Adachi O (1992) Homology in the structure and the prosthetic groups between two different terminal ubiquinol oxidases, cytochrome a_1 and cytochrome *o*, of *Acetobacter aceti*. J Biol Chem 267:24748–24753
12. Contreras-Zentella M, Mendoza G, Membrillo-Hernández J et al (2003) A novel double heme substitution produces a functional bo_3 variant of the quinol oxidase aa_3 of *Bacillus cereus*: purification and partial characterization. J Biol Chem 278:31473–31478

13. Borisov VB, Verkhovsky MI (2015) Oxygen as acceptor. EcoSal Plus 6. https://doi.org/10.1128/ecosalplus.ESP-0012-2015
14. Mogi T (2009) Effects of replacement of low-spin haem *b* by haem O on *Escherichia coli* cytochromes *bo* and *bd* quinol oxidases. J Biochem 145:599–607
15. Brown KR, Allan BM, Do P et al (2002) Identification of novel hemes generated by heme A synthase: evidence for two successive monooxygenase reactions. Biochemistry 41:10906–10913
16. Weinstein JD, Beale SI (1983) Separate physiological roles and subcellular compartments for two tetrapyrrole biosynthetic pathways in *Euglena gracilis.* J Biol Chem 258:6799–6807
17. Lübben M, Morand K (1994) Novel prenylated hemes as cofactors of cytochrome oxidases: archaea have modified hemes A and O. J Biol Chem 269:21473–21479
18. Puustinen A, Wikstrom M (1991) The heme groups of cytochrome *o* from *Escherichia coli.* Proc Natl Acad Sci USA 88:6122–6126
19. Berry EA, Trumpower BL (1987) Simultaneous determination of hemes *a*, *b*, and *c* from pyridine hemochrome spectra. Anal Biochem 161:1–15
20. Puustinen A, Morgan JE, Verkhovsky M et al (1992) The low-spin heme site of cytochrome *o* from *Escherichia coli* is promiscuous with respect to heme type. Biochemistry 31: 10363–10369
21. Barr I, Guo F (2015) Pyridine hemochromagen assay for determining the concentration of heme in purified protein solutions. Bio-protocol 5:e1594
22. Chang A, Chau V, Landas J et al (2017) Preparation of calcium competent *Escherichia coli* and heat-shock transformation. JEMI Methods 1:22–25
23. Diekert K, De Kroon AI, Kispal G et al (2001) Isolation and subfractionation of mitochondria from the yeast *Saccharomyces cerevisiae.* Methods Cell Biol 65:37–51
24. Zhuang J, Reddi AR, Wang Z et al (2006) Evaluating the roles of the heme *a* side chains in cytochrome *c* oxidase using designed heme proteins. Biochemistry 45:12530–12538

Chapter 9

Fluorometric Methods to Measure Bioavailable and Total Heme

Iramofu M. Dominic, Mathilda M. Willoughby, Abigail K. Freer, Courtney M. Moore, Rebecca K. Donegan, Osiris Martinez-Guzman, David A. Hanna, and Amit R. Reddi

Abstract

Heme *b* (iron protoporphyrin IX) is an essential but potentially cytotoxic cofactor, signaling molecule, and nutritional source of iron. Its importance in cell biology and metabolism is underscored by the fact that numerous diseases, including various cancers, neurodegenerative disorders, infectious diseases, anemias, and porphyrias, are associated with the dysregulation of heme synthesis, degradation, trafficking, and/or transport. Consequently, methods to measure, image, and quantify heme in cells are required to better understand the physiology and pathophysiology of heme. Herein, we describe fluorescence-based protocols to probe heme bioavailability and trafficking dynamics using genetically encoded fluorescent heme sensors in combination with various modalities, such as confocal microscopy, flow cytometry, and microplate readers. Additionally, we describe a protocol for measuring total heme and its precursor protoporphyrin IX using a fluorometric assay that exploits porphyrin fluorescence. Together, the methods described enable the monitoring of total and bioavailable heme to study heme homeostatic mechanisms in virtually any cell type and organism.

Key words Heme, Iron, Fluorescent heme sensors, Bioavailable heme, Labile heme, Porphyrin

1 Introduction

1.1 General Background on Heme

Heme *b* (iron protoporphyrin IX), hereafter referred to as heme, is an essential metabolite that is required by virtually all life on Earth [1–5]. As a protein cofactor, it endows proteins with the ability to carry out a diverse array of functions that span electron transfer, chemical catalysis, and gas binding and transport. As a signaling molecule, the dynamic binding of heme to downstream proteins can regulate their activity and/or stability, leading to changes in metabolism and physiology. As a nutrient, heme is a major source of bioavailable iron. Ironically, the very same properties of heme that proteins exploit, including its redox activity, Lewis acidity, and

Oleh Khalimonchuk (ed.), *Iron Metabolism: Methods and Protocols*, Methods in Molecular Biology, vol. 2839, https://doi.org/10.1007/978-1-0716-4043-2_9,

hydrophobicity, can contribute toward heme cytotoxicity. Consequently, cells must tightly regulate the total concentration and bioavailability of heme such that it is sufficient to support metabolism but not too excessive to induce heme toxicity [1–5].

1.2 Total Heme

The total concentration of heme is governed by the relative rates of synthesis and degradation, which are relatively well-understood processes. Indeed, the structures and mechanisms of heme biosynthetic and degrading enzymes from diverse organisms are known. In contrast, the molecules and mechanisms that regulate heme bioavailability, including transporters, chaperones, and buffering factors, are largely unknown [1–5]. The ability to probe total and bioavailable heme across various cell types and organisms greatly enables fundamental research into heme homeostatic mechanisms and their dysregulation in disease. In this chapter, we describe protocols for measuring total and bioavailable heme *b*.

There are numerous methods to measure total heme, including UV/visible spectroscopy, fluorescence spectroscopy, reverse-phase liquid chromatography, and mass spectrometry [6]. All these methods require homogenization, digestion, and/or organic extractions of tissue or cell samples, followed by analysis of heme content using one of the aforementioned modalities. Herein, we describe a sensitive fluorescence assay for the quantification of heme that exploits the luminescence of protoporphyrin IX [7, 8]. Heme *b*, which is nonfluorescent due to the bound iron in the tetrapyrrole core, is measured by boiling tissue or cell extracts in oxalic acid, followed by measurement of porphyrin emission in a fluorescence microplate reader. Total heme is quantified by taking the difference between boiled (protoporphyrin IX + heme) and unboiled samples (protoporphyrin IX). Given the tight regulation of heme synthesis, in most cell and tissue samples, the unboiled samples will give little to no signal. However, in samples with defects in heme synthesis and/or iron availability, there may be elevated background porphyrin levels. Quantification is achieved by comparing the fluorescence signal in the unknown sample to a standard curve of known hemin concentrations. While most heme-requiring organisms have multiple heme types, *for example*, *b*, *c*, and *a*, in typical eukaryotes, heme *b* is the most abundant and therefore the largest contributor toward porphyrin emission upon its demetallation. For quantification of other heme types, it is necessary to rely on reverse-phase chromatographic and/or mass spectrometric methods to resolve the contributions from each heme type.

1.3 Bioavailable Heme

Total cellular heme can be thought to partition between exchange inert and labile heme-binding sites found in any number of heme interacting proteins [4, 9–11]. The former represents heme-binding proteins that do not readily exchange or dissociate heme on biologically meaningful timescales. The latter, however, constitutes hemoproteins that can exchange heme with other biomolecules and can equilibrate between bound and free states in the cell. We previously proposed that labile heme (LH) is relevant for understanding heme bioavailability and signaling in cells and organisms. Consequently, identifying the factors that regulate LH could reveal heme transporters, chaperones, and buffering factors that regulate heme distribution and availability in health and disease. Our approach toward measuring labile or bioavailable heme involves the use of the genetically encoded ratiometric fluorescent sensor, heme sensor 1 (HS1). While there are several previously reported genetically encoded fluorescence [10, 12–14] and activity-based sensors [15] and small molecule probes for heme [16, 17], to date, HS1 has been the most widely utilized [9, 18–27].

HS1 is an engineered tridomain protein consisting of cytochrome b_{562} (Cyt b_{562}) fused to enhanced green fluorescent protein (eGFP) and monomeric Katushka 2 (mKATE2) [9] (Fig. 1). Cyt b_{562} is a high-affinity hemoprotein from *Escherichia coli* that binds to the heme iron center using methionine 7 and histidine 102. Upon heme binding to Cyt b_{562}, eGFP fluorescence is quenched due to energy transfer from eGFP to heme. mKATE2 fluorescence, however, is relatively unaffected by heme binding to Cyt b_{562}.

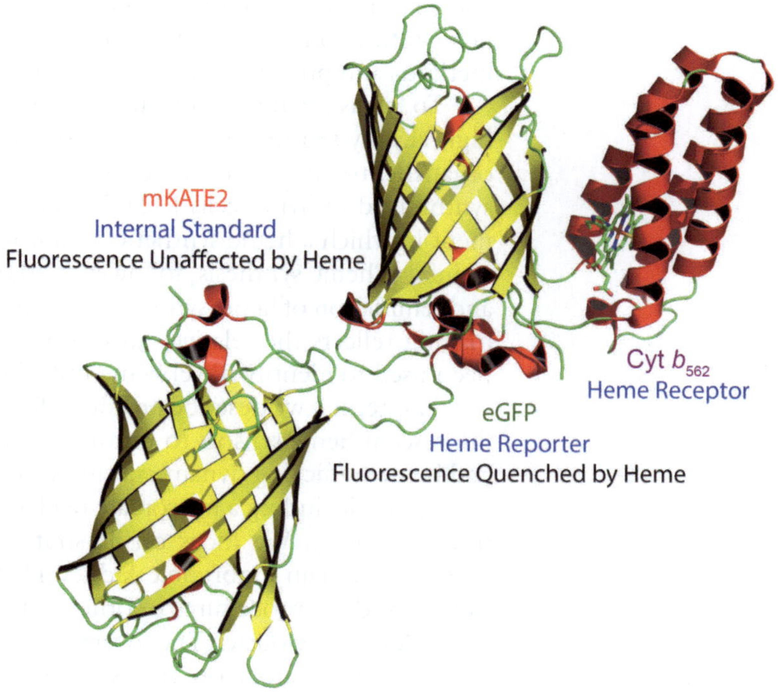

Fig. 1 Molecular model and design principles of the heme sensor 1, HS1. The model is derived from the X-ray structures of mKATE [Protein Data Bank (PDB) ID code 3BXB] and CG6 (PDB ID code 3U8P)

Thus, the eGFP/mKATE2 fluorescence ratio of HS1 provides a readout of heme bioavailability, with the ratio inversely correlating with heme bioavailability. The HS1 family of sensors can also easily be targeted to other organelles via the addition of well-known leader or retention sequences. Indeed, HS1 was previously targeted to the mitochondrial matrix and nucleus.

The prototype HS1 sensor binds both oxidation states of heme tightly, with ferric and ferrous heme dissociation constant K_D^{III} and K_D^{II} values of 3 nM and 1 pM, respectively [9, 24]. Since heme levels can vary between cells and organisms, we also engineered a weaker binding sensor variant, HS1-M7A, which has the heme coordinating methionine residue mutated to alanine. HS1-M7A exhibits K_D^{III} and K_D^{II} values of 1 μM and 25 nM, respectively [9, 24]. A third sensor variant that cannot bind heme due to the substitution of both heme coordinating ligands with alanine, HS1-M7A,H102A, can be used as a negative control to rule out heme-independent changes in sensor fluorescence ratio [9, 24]. Qualitative changes in sensor fluorescence ratio can be used to infer changes in heme bioavailability as long as HS1-M7A,H102A is invariant between experimental conditions compared to HS1 or HS1-M7A. For more quantitative measurements, the fluorescence ratios of the sensor under any given condition can be converted to a fractional heme occupancy if additional controls are conducted to record the sensor ratios associated with heme saturation or heme depletion, all of which are detailed in the methods and protocols described herein.

To measure nonsteady-state heme trafficking dynamics, the high-affinity sensor, HS1, can be used as a heme sink in different organelles to monitor the kinetics of heme distribution as it is being synthesized in living cells [22]. We describe a pulse-chase experiment in which a heme synthetic inhibitor, succinylacetone, is used to inhibit heme synthesis, followed by the removal of the inhibitor and reinitiation of heme synthesis. The kinetics of HS1 heme sensor loading reflects the relative rates of heme biodistribution and has been used to identify novel heme trafficking factors.

Altogether, we describe protocols for using genetically encoded fluorescent heme sensors to measure subcellular bioavailable heme and heme trafficking dynamics using various fluorescence detection modalities, including a microplate reader, confocal microscopy, and flow cytometry. We focus on illustrative applications on Baker's yeast and human embryonic kidney HEK293 cells, but the user can adapt these procedures to other cell lines and organisms as has been done previously. The overall intent of this chapter is to instruct investigators on robust quantitative methods to probe changes in total and bioavailable heme to better assess perturbations to heme homeostasis in different physiological contexts.

2 Materials

2.1 Yeast: Plasmids and Yeast Strain Transformations

1. yHS1; p415-GPD$_{pr}$-HS1; pDH013 (Addgene: plasmid 159150).
2. yHS1-M7A; p415-GPD$_{pr}$-HS1-M7A; pRH021 (Addgene: plasmid 159151).
3. yHS1-M7A,H102A; p415-GPD$_{pr}$-HS1-M7A,H102A; pJA010 (Addgene: plasmid 159152).
4. mito-yHS1; p415-GPD$_{pr}$-mito-yHS1; pMW016-2 (Addgene: plasmid 159153).
5. mito-yHS1-M7A; p415-GPD$_{pr}$-mito-yHS1-M7A; pAF001-3 (Addgene: plasmid 159164).
6. mito-yHS1-M7A,H102A; p415-GPD$_{pr}$-mito-yHS1-M7A, H102A; pAF003-4 (Addgene: plasmid 159165).
7. nuc-yHS1; p415-GPD$_{pr}$-nuc-yHS1; pDH046 (Addgene: plasmid 159166).
8. nuc-yHS1-M7A; p415-GPD$_{pr}$-nuc-yHS1-M7A; HS1-M7A-NLS (Addgene: plasmid 159167).
9. nuc-yHS1-M7A,H102A; p415-GPD$_{pr}$-nuc-yHS1-M7A, H102A; pCM009 (Addgene: plasmid 159168).
10. Empty vector plasmid; pRS415-GPD (Addgene: plasmid 183940).
11. Yeast strains of choice with leucine auxotrophy (*see* **Note 1**).
12. 50% (w/v) polyethylene glycol-3350 (PEG 3350) (Biotechnology grade) (*see* **Note 2**). To 25 mL of ultrapure water stirring in a 100-mL media bottle, slowly add 25 g of PEG 3350. Bring the final volume to 50 mL of ultrapure water. Autoclave for 15 min at 121 °C. Store at room temperature.
13. 1 M lithium acetate dihydrate (reagent grade). To 25 mL of ultrapure water stirring in a 100-mL media bottle, slowly add 3.3 g of lithium acetate (MW = 65.99 g/moL). Bring the final volume to 50 mL of ultrapure water. Autoclave for 15 min at 121 °C. Store at room temperature.
14. 0.1 M lithium acetate dihydrate (reagent grade). To 5 mL of 1 M lithium acetate in a 50-mL conical tube, add 45 mL of sterile ultrapure water. Invert the tube to mix and store at room temperature.
15. 10 mg/mL high-purity single-stranded salmon sperm DNA (ssDNA) (store at −20 °C). Upon receipt of ssDNA, aliquot 50 μL fractions into sterile microfuge tubes, heat to 100 °C for 10 min in a heat block, chill on ice for 5 min, and store at −20 °C. Avoid freeze-thawing the ssDNA and try to consume the entire aliquot during each use.

16. Water bath set to 30 °C.
17. Water bath set to 42 °C.
18. 30 °C standing incubator.
19. 30 °C shaking incubator.
20. −20 °C freezer.
21. −80 °C freezer.
22. Tabletop microcentrifuge.
23. Ultrapure water (Type 1 18.2 MΩ-cm). Autoclave for 15 min at 121 °C. Store at room temperature.
24. Sterilized microfuge tubes.
25. UV/visible spectrophotometer.
26. 1-cm pathlength cuvettes.
27. Vortex mixer.
28. 4.5-mm glass plating beads.

2.2 Yeast Cultivation

1. Ultrapure water (Type 1 18.2 MΩ-cm), both sterile and non-sterile. To sterilize, autoclave for 15 min at 121 °C. Store at room temperature.
2. Sterilized microfuge tubes.
3. Sterile 10-μL inoculating loops.
4. Sterile 1-μL inoculating loops.
5. 0.1% ergosterol solution: Weigh 15 mg of ergosterol into a 50-mL conical tube, and add 15 mL of 100% (200 proof) ethanol. Sonicate for 10–15 min and vortex as needed until completely dissolved.
6. 10% (v/v) Tween-80 (cell culture grade): To 45 mL of ultrapure water stirring gently in an Erlenmeyer flask, slowly add 5 mL of Tween-80. Allow to stir until fully dissolved.
7. 3% L-histidine solution: Dissolve 3 g of L-histidine monohydrochloride monohydrate in 80 mL of ultrapure water and make up to 100 mL. Store at room temperature.
8. 3% uracil solution: Dissolve 3 g of uracil in 80 mL of 0.3 M NaOH and make up to 100 mL. Store at room temperature.
9. 3% leucine solution: Dissolve 3 g of L-leucine in 80 mL of 0.15 M NaOH and make up to 100 mL. Store at room temperature.
10. 20% glucose solution: To 0.7 L of ultrapure water stirring in a graduated cylinder, slowly add 200 g of D-glucose until completely dissolved. Make up to 1 L with ultrapure water and autoclave for 15 min at 121 °C. Store at room temperature (RT).

11. Synthetic complete dropout powder lacking His, Leu, and Ura (SC-His-Leu-Ura); (Cat. #1327-030, Sunrise Science Products).
12. Yeast nitrogen base (YNB) without amino acids (AA) (YNB w/o AA); (Cat. #J386-500G, VWR).
13. 50-mL Luer Lock syringe.
14. 0.22-μm syringe filter.
15. Glycerol (high purity >99%; molecular biology, biotechnology, or cell culture grade). Autoclave for 15 min at 121 °C. Store at room temperature.
16. Yeast extract (bacteriological grade).
17. Peptone.
18. Agar (bacteriological grade).
19. 5-Aminolevulinic acid hydrochloride (5-ALA). Working stock solutions can be prepared by dissolving 50 mg of 5-ALA in 1 mL of sterile ultrapure water (298 mM 5-ALA) in a sterilized microfuge tube. Store at −20 °C or at room temperature.
20. 4,6-Dioxoheptanoic acid (succinylacetone (SA)). Working stock solutions can be prepared by dissolving 50 mg of SA in 1 mL of sterile ultrapure water (316 mM SA) in a sterilized microfuge tube. Store at −20 °C.
21. Synthetic complete media with ergosterol and Tween-80, without leucine (SCE-LEU): To 828.3 mL of ultrapure water in a 1 L bottle, add 1.66 g of SC-His-Leu-Ura powder (or as prescribed for 1 L of media on the instructions provided with the batch of dropout powder), 6.7 g of YNB, 3.8 mL of 3% L-histidine, and 2.9 mL of 3% uracil. Stir to mix and autoclave for 15 min at 121 °C. Once the bottle has cooled sufficiently that it can be touched without burning one's hand, add 100 mL of 20% D-glucose and 15 mL of 0.1% ergosterol. Finally, add the 50 mL of 10% Tween-80 by dispensing through the Luer Lock syringe fitted with the 0.22-μm syringe filter (*see* **Note 3**). Cap and gently swirl the bottle to mix. Store at 4 °C and use within 3 months.
22. Synthetic complete media with ergosterol and Tween-80 (SCE): To 822.5 mL of ultrapure water in a 1 L bottle, add 1.66 g of SC-His-Leu-Ura powder (or as prescribed for 1 L of media on the instructions provided with the batch of dropout powder), 6.7 g of YNB, 3.8 mL of 3% L-histidine, 2.9 mL of 3% uracil, and 5.8 mL of 3% leucine. Stir to mix and autoclave for 15 min at 121 °C. Once the bottle has cooled sufficiently that it can be touched without burning one's hand, add 100 mL of 20% D-glucose and 15 mL of 0.1% ergosterol. Finally, add the 50 mL of 10% Tween-80 by dispensing

through the Luer Lock syringe fitted with the 0.22-μm syringe filter (*see* **Note 3**). Cap and gently swirl the bottle to mix. Store at 4 °C and use within 3 months.

23. Synthetic complete media without leucine (SC-LEU): To 893.3 mL of ultrapure water in a 1 L bottle, add 1.66 g of SC-His-Leu-Ura powder (or as prescribed for 1 L of media on the instructions provided with the batch of dropout powder), 6.7 g of YNB, 3.8 mL of 3% L-histidine, and 2.9 mL of 3% uracil. Stir to mix and autoclave for 15 min at 121 °C. Once the bottle has cooled sufficiently that it can be touched without burning one's hand, add 100 mL of 20% D-glucose. Cap and gently swirl the bottle to mix. Store at room temperature and use within 12 months.
24. Synthetic complete media (SC): To 887.5 mL of ultrapure water in a 1 L bottle, add 1.66 g of SC-His-Leu-Ura powder (or as prescribed for 1 L of media on the instructions provided with the batch of dropout powder), 6.7 g of YNB, 3.8 mL of 3% L-histidine, 2.9 mL of 3% uracil, and 5.8 mL of 3% leucine. Stir to mix and autoclave for 15 min at 121 °C. Once the bottle has cooled sufficiently that it can be touched without burning one's hand, add 100 mL of 20% D-glucose. Cap and gently swirl the bottle to mix. Store at room temperature and use within 12 months.
25. Yeast extract–peptone–dextrose media (YPD): To 700 mL of ultrapure water, add 10 g of yeast extract and 20 g of peptone while stirring. Bring the final volume to 900 mL with ultrapure water. Autoclave for 15 min at 121 °C. Once the bottle has cooled sufficiently that it can be touched without burning one's hand, add 100 mL of 20% D-glucose. Cap and gently swirl the bottle to mix. Store at room temperature and use within 12 months.
26. Yeast extract–peptone–dextrose media with ergosterol and Tween-80 (YPDE): To 700 mL of ultrapure water, add 10 g of yeast extract and 20 g of peptone while stirring. Bring the final volume to 835 mL with ultrapure water. Autoclave for 15 min at 121 °C. Once the bottle has cooled sufficiently that it can be touched without burning one's hand, add 100 mL of 20% D-glucose and 15 mL of 0.1% ergosterol. Finally, add the 50 mL of 10% Tween-80 by dispensing through the Luer Lock syringe fitted with the 0.22-μm syringe filter (*see* **Note 3**). Cap and gently swirl the bottle to mix. Store at 4 °C and use within 3 months.
27. For solid media plates, prepare SCE-LEU, SCE, SC-LEU, SC, YPD, and YPDE as described above for 21–26, respectively, except add 20 g of agar prior to autoclaving. For SC and SCE media plates, also add 1 NaOH pellet along with the agar prior to autoclaving.

28. YPD or SC media plates containing 5-ALA (YPD + 5-ALA or SC + 5-ALA). Prepare the media plates as described in "27." After autoclaving the media and adding glucose, add 1 mL of a 50 mg/mL stock solution of 5-ALA in sterile ultrapure water.
29. Yeast freezing media: 35% (v/v) glycerol in YPD media.
30. 30 °C shaking incubator.
31. 30 °C standing incubator.
32. −20 °C freezer.
33. −80 °C freezer.
34. Petri dishes.
35. Optional: anaerobic chamber (https://coylab.com/products/anaerobic-chambers/).
36. Optional: anaerobic jar (https://www.bd.com/en-ca/offerings/capabilities/microbiology-solutions/environmental-systems/bd-gaspak-systems-and-accessories).

2.3 Mammalian Expression Plasmids

1. hHS1; pcDNA3.1-CMV$_{pr}$-hHS1; (Addgene: plasmid 159169).
2. hHS1-M7A; pcDNA3.1-CMV$_{pr}$-hHS1-M7A; (Addgene: plasmid 159170).
3. hHS1-M7A,H102A; pcDNA3.1-CMV$_{pr}$-hHS1-M7A,H102A (Addgene: plasmid 159171).
4. pEF5-FRT/V5-DEST-hHS1.
5. pEF5-FRT/V5-DEST-hHS1-M7A.
6. NLS-hHS1; pEF5-FRT/V5-DEST-NLS-hHS1.
7. NLS-hHS1-M7A; pEF5-FRT/V5-DEST-NLS-hHS1-M7A.
8. Mito-hHS1; pEF5-FRT/V5-DEST-NLS-Mito-hHS1.
9. Mito-hHS1-M7A; pEF5-FRT/V5-DEST-Mito-hHS1-M7A.

2.4 Cultivation and Transfection of HEK293 Cells

1. CO_2 incubator maintained at 37 °C and a 5% CO_2 atmosphere.
2. UV/visible spectrophotometer.
3. 1-cm pathlength cuvettes.
4. T75 cell culture flask (250 mL total volume, 75 cm^2 growth area, canted neck, and vented cap).
5. 6-well cell culture plate (2.5–3.0 mL total volume, flat bottom, 9.6 cm^2 growth area).
6. Ultrapure water (Type 1 18.2 MΩ-cm), both sterile and non-sterile. To sterilize, autoclave for 15 min at 121 °C. Store at room temperature.
7. Sterilized microfuge tubes.

8. Human embryonic kidney HEK293 cells (ATCC Cat. # CRL-1573).
9. Dulbecco's modified Eagle's medium (DMEM) with 4.5 g of glucose, without glutamine and without sodium pyruvate. Store at 4 °C.
10. Dulbecco's phosphate-buffered saline (DPBS).
11. Fetal bovine serum (FBS); heat inactivated. Store at −20 °C.
12. 5-Aminolevulinic acid hydrochloride (5-ALA) stock solution. See Subheading 2.2, **item 19** on its preparation and storage.
13. 4,6-Dioxoheptanoic acid (succinylacetone (SA)) stock solution. See Subheading 2.2, **item 20** on its preparation and storage.
14. L-ascorbic acid.
15. 2 kDa molecular weight cut off (MWCO) dialysis membrane.
16. Heme-depleted fetal bovine serum (HD-FBS): Weigh a quantity of ascorbic acid (MW 176.13 g/moL) to make a 10 mM solution in FBS. Add the weighed ascorbic acid powder directly to the FBS and stir to mix. Split the ascorbate–FBS solution into an appropriate number of 50-mL conical tubes (typically 10–20 mLs of FBS per 50-mL conical tube) and shake in an incubator at 37 °C. Heme dissociation from serum hemoproteins typically takes ~8 h and can be monitored by UV/visible spectroscopy (*see* **Note 4**). Transfer the ascorbate-treated FBS to a 2 kDa MWCO dialysis membrane and dialyze overnight (16–24 h) against 4 L of 1× DPBS, pH 7.4, slowly stirring at 4 °C. Repeat dialysis against DPBS two additional times for a total of three times. After dialysis, filter-sterilize the now HD-FBS into 50-mL conical tubes in a sterilized cell culture hood and store at −20 °C.
17. Regular heme (Reg) media: DMEM with 4.5 g of glucose, without glutamine and without sodium pyruvate, supplemented with 10% v/v FBS. Warm DMEM and FBS to 37 °C before use.
18. High heme (Reg + 5-ALA) media: Regular heme media with 1.2 mM 5-ALA. Add sufficient volume of the 50 mg/mL (298 mM) 5-ALA stock to regular media. Gently swirl to mix. This is typically prepared just prior to an experiment by adding 8 μL of the 5-ALA stock solution to 2 mL of a cell culture in a 6-well plate.
19. Low heme (HD + SA) media: DMEM with 4.5 g of glucose, without glutamine and without sodium pyruvate, supplemented with HD-FBS to 10% v/v, and 500 μM SA. Add 7.9 μL of 316 mM SA stock solution to 5 mL DMEM +10% HD-FBS media. Warm DMEM, HD-FBS, and 316 mM SA stock to 37 °C before use.

20. Opti-MEM™ (Gibco).
21. Lipofectamine™ LTX reagent with PLUS™ reagent (Invitrogen™).

2.5 Bioavailable Heme Measurements: General

1. Phosphate-buffered saline solution (PBS). To 800 mL of ultrapure water, add 8.0 g of NaCl, 0.2 g of KCl, 1.44 g of Na_2HPO_4, and 0.24 g of KH_2PO_4. Adjust the pH to 7.4 with microliter additions of concentrated HCl, and add ultrapure water to give a final volume of 1 L. The final concentrations of components in 1× PBS are 137 mM NaCl, 2.7 mM KCl, 10 mM Na_2HPO_4, and 1.8 mM KH_2PO_4. PBS can also be purchased from any number of commercial vendors.
2. Dulbecco's phosphate-buffered saline (DPBS) solution.
3. Dimethyl sulfoxide (DMSO) (cell culture grade).
4. Hemin chloride. Working stock solutions can be prepared by dissolving 10 mg of hemin chloride in 1 mL of DMSO, to make a 15.3 mM stock solution. Hemin stock solutions should be made fresh on the day of the experiment.
5. Digitonin. Stock solutions can be prepared by dissolving 10 mg of digitonin in 1 mL of DMSO to give a 13.1 mM or 1% stock solution. This can be aliquoted into 100 μL fractions and stored at −20 °C.
6. L-ascorbic acid. A stock solution can be prepared by dissolving 50 mg of L-ascorbic acid in 1 mL of PBS to give a 284 mM stock solution. This should be aliquoted and stored at −20 °C.
7. Cell permeabilization buffer. 131 μM digitonin, 1 mM ascorbate in PBS. To 9 mL of PBS in a conical tube, add 100 μL of the 13.1 mM digitonin stock solution, 35.2 μL of the 284 mM L-ascorbic acid stock solution, and 864.8 μL of PBS. This solution should be made fresh at the time of the experiment, and the total volume can be scaled based on the number of samples.
8. Heme sensor saturation buffer. 50 μM hemin chloride in "cell permeabilization buffer." To 5 mL of the "cell permeabilization buffer," add 16.3 μL of the 15.3 mM hemin stock solution. This solution should be made fresh at the time of the experiment, and the total volume can be scaled based on the number of samples.
9. Mammalian cell heme saturation buffer. 40 μM digitonin, 100 μM hemin chloride, 1 mM L-ascorbic acid in DMEM. Per 1 mL of prewarmed DMEM (prewarmed at 37 °C), add 3 μL of the 13.1 mM digitonin stock solution in PBS, 6.5 μL of the 15.3 mM hemin chloride stock solution in DMSO, and 3.5 μL of the 284 mM L-ascorbic acid stock solution in PBS.

This solution should be made fresh at the time of the experiment and the total volume can be scaled based on the number of samples.

10. Microfuge tubes.
11. Microcentrifuge.
12. Tabletop centrifuge with rotors capable of holding 50- and 15-mL conical tubes, for example, Eppendorf 5810R.
13. Water bath set to 30 °C.
14. Water bath set to 37 °C.
15. Cell counter.
16. UV/visible spectrophotometer.

2.6 Bioavailable Heme Measurements: Microplate Reader

1. 96-well plate; flat-bottom black wells.
2. Microplate reader capable of fluorescence kinetic measurements and that has the appropriate monochromator and/or filter sets for measuring the excitation and emission of eGFP (ex. 488 nm and em. 510 nm) and mKATE2 (ex. 588 nm and em. 620 nm) (*see* **Note 5**).

2.7 Bioavailable Heme Measurements: Microscopy

1. Confocal microscope equipped with a 60/63X objective lens, an argon ion laser to excite eGFP (488 nm), a helium–neon laser to excite mKATE2 (588 nm), and appropriate emission filters for eGFP (510 nm) and mKATE2 (620 nm) emission (*see* **Note 6**).
2. Microscope slides (75 × 25 × 1 mm).
3. Microscope cover glasses (18 × 18 mm).
4. Glass bottom 35-mm dish and #1.0 coverslip, 14-mm glass diameter, uncoated.

2.8 Bioavailable Heme Measurements: Flow Cytometry

1. Falcon® 5-mL polystyrene round-bottom tubes with 35-μm cell strainer cap.
2. Flow cytometer equipped with appropriate lasers and emission filters to monitor eGFP (ex. 488 nm and em. 510 nm) and mKATE2 (ex. 588 and em. 620 nm) fluorescence emission (*see* **Note 7**).

2.9 Total Heme Measurements

1. 2 M oxalic acid. To 250 mL of ultrapure water in a 1 L bottle, add 252.14 g oxalic acid dihydrate while stirring on a hot plate. Bring the final volume to 500 mL in ultrapure water. 2 M oxalic acid is not soluble in water at room temperature and must be heated for 20 min with a hot plate set to 170 °C for complete dissolution. This solution can be stored at room temperature, but it must be preheated with stirring prior to use for complete dissolution.

2. 20 mM oxalic acid. To 0.5 mL of 2 M oxalic acid in a 50-mL conical tube, add 49.5 mL of ultrapure water and mix by inversion. This solution is soluble at room temperature and can be stored at room temperature until needed.
3. Hemin chloride. Working stock solutions can be prepared by dissolving 3.2 mg of hemin chloride in 1 mL of 0.1 M NaOH, to make a 5 mM stock solution. The concentration of the hemin stock solution should be checked and verified by UV/visible spectroscopy by making a 250-fold dilution into ultrapure water and checking the absorbance at 612 nm and using the molar extinction coefficient of 4431 cm^{-1} M^{-1}.
4. Microfuge tubes.
5. Microtube cap locks.
6. Heat block that can accommodate microfuge tubes.
7. Microplate reader capable of fluorescence that has the appropriate monochromator and/or filter sets for measuring the excitation and emission of protoporphyrin IX (ex. 400 nm and em. 520 nm) (*see* **Note 5**).

2.10 Software

1. Microsoft Excel.
2. Graph Pad Prism v. 9.0 or more recent version (https://www.graphpad.com).
3. FlowJo v. 10.4.2 or more recent version (https://www.bdbiosciences.com/en-us/products/software/flowjo-v10-software).
4. ImageJ 1.54f or more recent version (https://imagej.net/software/fiji/downloads).

3 Methods

Carry out all procedures at room temperature unless otherwise specified. It is assumed that the user is well versed in best practices in sterile technique, microbial culturing, and maintenance of mammalian cell cultures.

3.1 Yeast Heme Sensor Transformation and Selection

Yeast transformations were accomplished using the lithium acetate method and are adapted from the method of Gietz and Schiestl [28].

1. From a frozen glycerol stock of the desired strain and mutants of *Saccharomyces cerevisiae*, patch cells onto a YPD or SC media plate (*see* **Note 8**) and incubate in a 30 °C incubator for 15–48 h until a thick lawn of cells appears.

2. For each plasmid transformation, inoculate a 10 mL overnight culture of the desired *Saccharomyces cerevisiae* strains and mutants and grow to mid-log phase in YPD or SC media (*see* **Note 8**) in 50-mL conical tubes, at 30 °C, and in a shaking incubator set at a shaker speed of 220 revolutions per minute (RPM) (*see* **Note 9**). Monitor cell growth by assessing solution turbidity by measuring the optical density at 600 nm (OD_{600}). An OD_{600} value of 1 corresponds to 2×10^7 cells/mL for haploid S288c yeast, including BY4741 and BY4742 strains. If strains overgrow, they can be rediluted to an OD_{600}/mL of 0.1–0.2 and cultured to an OD_{600}/mL of ~1, which takes 4–6 h.
3. For each plasmid transformation, harvest 10 OD_{600} of cells (~2×10^8 cells) by centrifugation at 3100 × *g* for 5 min at 4 °C in a tabletop centrifuge.
4. Decant the media, resuspend the cell pellet in 1 mL of sterile ultrapure water, and transfer to a sterile microfuge tube.
5. Pellet the cells by centrifugation at maximum speed (21,000 × *g*) for 2 min in a microcentrifuge.
6. Aspirate the water and resuspend the 10 OD_{600} cell pellet (~2×10^8 cells) in 100 μL of 0.1 M lithium acetate to give a final concentration of ~2×10^9 cells/mL.
7. Aliquot 50 μL of the 2×10^9 cells/mL cell suspension into two sterile microfuge tubes, one for the plasmid transformation and one for the no DNA negative control. Pellet the cells by centrifugation at maximum speed (21,000 × *g*) for 1 min in a microcentrifuge. Aspirate the 0.1 M lithium acetate.
8. To each cell pellet, now corresponding to 1×10^8 cells, add the following reagents in order:
   (a) 240 μL of 50% PEG 3350.
   (b) 36 μL of 1 M lithium acetate.
   (c) 10 μL of ssDNA (10 mg/mL).
   (d) 50 μL of plasmid DNA or sterile ultrapure water. For each empty vector or heme sensor plasmid transformation, use between 200 and 500 ng of plasmid DNA. Plasmid stock concentrations are typically maintained at ~500 ng/μL, requiring the addition of ~1 μL of plasmid DNA followed by the addition of 49 μL of sterile ultrapure water. The no DNA negative control will receive 50 μL of sterile ultrapure water.
9. Vortex the mixture for 10 min to generate a homogenous mixture. If necessary, triturate the mixture with a P1000 (1 mL) pipettor.
10. Incubate the cell suspension in a 30 °C water bath for 30 min.

11. Incubate the cell suspension in a 42 °C water bath for 20 min (*see* **Note 10**).
12. Pellet the cells by centrifugation at maximum speed (21,000 × *g*) for 2 min in a microcentrifuge. Aspirate the transformation media. Resuspend the cell pellet in 500 μL of sterile ultrapure water.
13. Plate 100 μL of the cell suspension on SC-LEU media plates (*see* **Note 8**). To spread the cells evenly, first add ~6–10 glass plating beads to the media plate followed by the addition of a 100 μL cell suspension. Spread the cells by orbitally shaking the plate in a circular motion. Be careful not to tilt the plate or shake so hard as to splatter the suspension onto the lid of the Petri dish.
14. After allowing the plate to dry, collect the used plating beads in a 1-L beaker (*see* **Note 11**) and place the plate in the 30 °C incubator. Single colonies of empty vector or sensor plasmid transformed yeast strains should be visible after ~2 days (*see* **Note 12**). No colonies should be observed on the media plate that was plated with cells that did not receive plasmid.
15. Once single colonies are large enough to be picked, usually after 2–3 days, 5–6 colonies should be scraped up using a 1 μL sterile inoculating loop and repatched onto a fresh SC-LEU plate (*see* **Note 8**) and grown for an additional 1–2 days at 30 ° C.
16. After 1–2 days of growth at 30 °C, transformants should be repatched into larger area patches corresponding to roughly 1/4th of the petri dish on a new media plate and incubated at 30 °C for 1–2 days.
17. After 1–2 days of growth at 30 °C, use a 10 μL inoculating loop to scrape up all the cells and resuspend in 0.5 to 1.0 mL of yeast freezing media. Vortex and place in the −80 °C freezer for long-term storage. For each isolated clone, make a separate freezer stock.

3.2 Yeast Cell Growth for Steady-State Bioavailable and Total Heme Measurements

1. From a frozen glycerol stock of the desired empty vector and sensor-expressing *Saccharomyces cerevisiae* strains and mutants, patch cells onto an SC-LEU media plate (*see* **Note 13**) and incubate in a 30 °C incubator for 15–48 h until a thick lawn of cells appear.
2. Inoculate triplicate 5 mL cultures of each desired empty vector and sensor-expressing *Saccharomyces cerevisiae* strains in SCE-LEU (*see* **Note 13**) media supplemented with or without 500 μM SA in 50 mL conical tubes (see **Note 14**), at 30 °C, and in a shaking incubator set at a shaker speed of 220 revolutions per minute (RPM) (*see* **Note 9**).

3. Grow to the desired cell density, growth state, or duration (*see* **Note 15**). Monitor cell growth by assessing solution turbidity by measuring the optical density at 600 nm (OD_{600}).
4. After the desired growth period, harvest 5 OD_{600} of cells, corresponding to 1×10^8 cells, from each culture by centrifugation at 3100 × *g* for 5 min at 4 °C in a tabletop centrifuge (*see* **Note 16**).
5. Decant or aspirate the spent media, resuspend the cell pellet in 1 mL of ultrapure water, and transfer it to a microfuge tube.
6. Pellet the cells by centrifugation at maximum speed (21,000 × *g*) for 2 min in a microcentrifuge.
7. Aspirate the water and resuspend the 5 OD_{600} cell pellet (~1×10^8 cells) in 1 mL of ultrapure water and pellet the cells by centrifugation at maximum speed (21,000 × *g*) for 2 min in a microcentrifuge (*see* **Note 17**).
8. After centrifugation, aspirate the water, leaving the "wet" cell pellet. If measuring total heme, the cell pellet can be stored at −80 °C indefinitely. If measuring steady-state heme bioavailability using the genetically encoded heme sensors, proceed to the next steps using a microplate reader (Subheadings 3.3 and 3.4), fluorescence microscopy (Subheading 3.7), or flow cytometry (Subheading 3.10, **Step 5**, Subheadings 3.11, and 3.12). If measuring total heme, proceed to Subheading 3.13.

3.3 Measuring Bioavailable Heme in Yeast Using a Microplate Reader

1. Resuspend the 5 OD_{600} cell pellets (~1×10^8 cells) in 1 mL of PBS to give a concentration of 5 OD_{600}/mL, corresponding to ~1×10^8 cells/mL (*see* **Note 18**).
2. For each strain being evaluated, pipette 150 μL (*see* **Note 19**) of the following cell suspensions into a black flat-bottom 96-well plate in triplicates:
   (a) 3 × 150 μL of the empty vector containing cells.
   (b) 3 × 150 μL of the desired heme sensor containing plasmid.
   (c) 3 × 150 μL of the succinylacetone-treated cells.
3. Using a microplate reader, record the eGFP (ex. 488 nm; em. 510 nm) and mKATE2 (ex. 588 nm; em. 620 nm) fluorescence intensities (*see* **Notes 20–23**).
4. Pellet the remaining 550 μL of the 5 OD_{600}/mL cell suspension left over from Subheading 3.3. **Step 1** by centrifugation at maximum speed (21,000 × *g*) for 5 min in a microcentrifuge, aspirate the media, and resuspend the cells in 550 μL of the "heme sensor saturation buffer" (*see* **Notes 24** and **25**). As discussed in **Note 25**, **Steps 4–10** may be omitted.
5. Incubate the cells for 30 min in a 30 °C water bath and mix by inversion every 10 min.

6. Pellet cells by centrifugation at maximum speed (21,000 × *g*) for 2 min in a microcentrifuge. Aspirate the media and resuspend the cells in 1 mL of PBS to wash the pellet of excess heme.
7. Pellet cells by centrifugation at maximum speed (21,000 × *g*) for 2 min in a microcentrifuge. Aspirate the media and resuspend the cells in 1 mL of PBS to again wash the pellet of excess heme.
8. Pellet cells by centrifugation at maximum speed (21,000 × *g*) for 2 min in a microcentrifuge. Aspirate the media and resuspend the cells in 550 μL of "cell permeabilization buffer."
9. For each strain being evaluated, pipette 150 μL (*see* **Note 19**) of the following cell suspensions into a black flat-bottom 96-well plate in triplicates:
   (a) 3 × 150 μL of the empty vector containing cells.
   (b) 3 × 150 μL of the desired heme sensor containing plasmid.
   (c) 3 × 150 μL of the succinylacetone-treated cells.
10. Using a microplate reader, record the eGFP (ex. 488 nm; em. 510 nm) and mKATE2 (ex. 588 nm; em. 620 nm) fluorescence intensities. Use the same instrument acquisition parameters, for example, excitation and emission wavelengths, gain settings, and slit widths, as from Subheading 3.3, **Step 3**.

3.4 Measuring Bioavailable Heme in Yeast Using a Microplate Reader: Analysis

1. Using software like Microsoft Excel or Graph Pad Prism, calculate the average intensity of eGFP and mKATE2 fluorescence emission values from empty vector cells. Subtract this intensity from the eGFP and mKATE2 channels associated with the heme sensor constructs.
2. For each sample replicate, calculate the background-subtracted eGFP/mKATE2 emission ratio. Qualitative changes in the sensor ratio can be used to infer perturbations in heme bioavailability across various test samples provided that the negative control sensor, HS1-M7A,H102A, exhibits a relatively invariant eGFP/mKATE2 ratio.
3. For quantitative heme imaging, the fractional heme occupancy of the sensor can be estimated using Eq. (1):

$$\%\text{Heme occupancy } (\alpha) = \frac{R - R_{\text{max}}}{R_{max} - R_{\text{min}}} \times 100 \qquad (1)$$

where R is the eGFP/mKATE2 fluorescence ratio of HS1 or HS1-M7A under any given experimental condition, R_{max} is the eGFP/mKATE2 fluorescence ratio associated with a heme saturated heme sensor, which is assessed in the calibration procedure in Subheadings 3.3, **Steps 4–10** or as assessed in **Notes 24** and **25**, and R_{min} is the eGFP/mKATE2 fluorescence ratio when the sensor is completely depleted of heme,

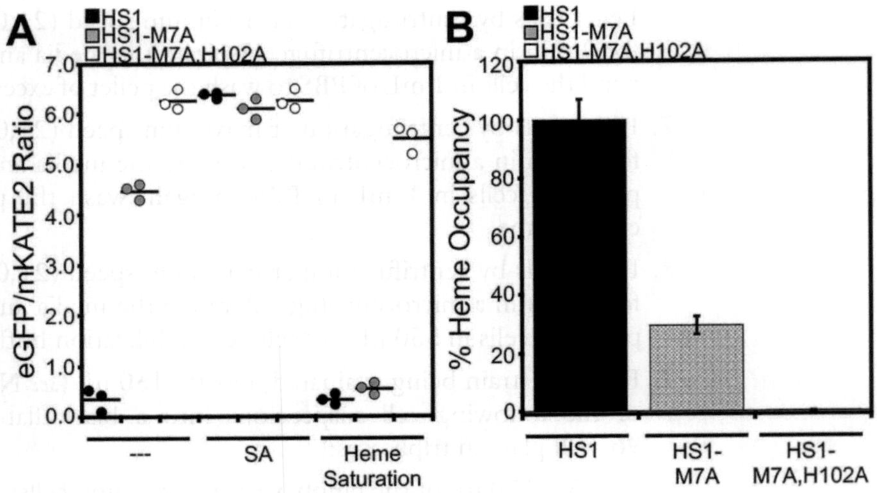

Fig. 2 Representative data from yeast demonstrating the heme dependence of the HS1 fluorescence ratios and their conversion to a fractional heme occupancy. (**a**) Heme sensor eGFP/mKATE2 ratios of HS1 (black), HS1-M7A (grey), and HS1-M7A,H102A (white) as a function of treatment with 500 μM succinylacetone (SA) or digitonin permeabilization and saturation with 50 μM hemin. (**b**) The eGFP/mKATE2 ratios for HS1, HS1-M7A, and HS1-M7A,H102A from panel A are converted to fractional heme occupancies based on Eq. (1), and the R_{min} derived from SA-treated samples and the R_{max} derived from cell permeabilization and hemin treatment

which is assessed from the succinylacetone-treated samples (*see* **Notes 26** and **27**). See Figs. 2a, b for a representative data set in yeast along with calibration controls.

4. If needed, the fractional heme occupancy can be converted to a buffered free heme concentration according to Eq. (2) (*see* **Note 28**):

$$[\text{Heme}] = \frac{\alpha \times K_D}{100 - \alpha} \quad (2)$$

where K_D is the heme sensor–heme dissociation constant and α is the heme occupancy of the sensor defined by Eq. (1). For reference, the HS1 ferric and ferrous heme dissociation constants K_D^{III} and K_D^{II} values are 3 nM and 1 pM, respectively. HS1-M7A exhibits K_D^{III} and K_D^{II} values of 1 μM and 25 nM.

3.5 Measuring Heme Trafficking Dynamics in Yeast Using a Microplate Reader

To measure heme trafficking dynamics, cells expressing the high-affinity heme sensor, HS1, in various compartments are heme-depleted using the heme synthetic inhibitor, succinylacetone. Next, following the removal of the inhibitor and reinitiation of heme synthesis, the kinetics of HS1 heme binding in different compartments is monitored by fluorescence, which reflects the relative rates of heme trafficking and biodistribution [22].

1. Inoculate triplicate 5-mL cultures of each desired *Saccharomyces cerevisiae* strain transformed with empty vector or yHS1, mito-yHS1, or nuc-yHS1 plasmids in SCE-LEU media supplemented with or without 500 μM SA in 50-mL conical tubes (*see* **Note 13**). The starting density should be OD_{600} ~ 0.01–0.03, corresponding to 2–6 × 10^5 cells/mL.
2. Grow cells at 30 °C in a shaking incubator set at a shaker speed of 220 RPM until the final density reaches an OD_{600}/mL ~ 1, corresponding to 2 × 10^7 cells/mL, which typically takes 14–16 h (*see* **Note 9**).
3. After culturing, harvest 1 OD_{600} of cells, corresponding to 2 × 10^7 cells, by pipetting the required volume into sterile microfuge tubes.
4. Pellet cells by centrifugation at maximum speed (21,000 × *g*) for 5 min in a microcentrifuge. Aspirate the media, wash the cell pellet by resuspending it in 1 mL of sterile ultrapure water, and pellet the cells again by centrifugation at maximum speed (21,000 × *g*) for 5 min in a microcentrifuge. Repeat this process once more for a total of two washes in ultrapure water (*see* **Note 17**).
5. Resuspend cells in 1 ml fresh SC-LEU to give a final concentration of 1 OD_{600}/mL. The cells that were precultured with succinylacetone (SA) should be split into two 500 μL fractions, one fraction should be treated with 0.8 μL of the SA stock solution (316 mM) to give a final concentration of 500 μM SA, and the other fraction should be left untreated. Thus, in total, there should be three types of sample conditions for each strain:
   (a) Cells that were precultured without SA and remain without SA (−/− SA).
   (b) Cells that were precultured with SA and remain with SA (+/+ SA).
   (c) Cells that were precultured with SA but are then diluted into media without SA (+/−SA).
6. Pipette duplicates of the 200 μL cell suspensions from each replicate culture into a black flat-bottom 96-well plate.
7. In a microplate reader preheated to 30 °C, record the eGFP (ex. 488 nm; em. 510 nm) and mKATE2 (ex. 588 nm; em. 620 nm) fluorescence emission values every 5 min for 4 h, with the plate being shaken at medium strength for 30 s prior to each read (*see* **Notes 20–22**).
8. If needed, conduct sensor calibration steps described in **Steps 4–10** in Subheading 3.3 and **Notes 24** and **25**. As discussed in **Note 25**, this step can be omitted.

3.6 Measuring Heme Trafficking Dynamics in Yeast Using a Microplate Reader: Analysis

1. Calculate HS1 fractional heme occupancy for each time point as described in **Steps 1–3** in Subheadings 3.4 and Eq. (1). R is derived from the background-subtracted eGFP/mKATE2 fluorescence ratio of yHS1, nuc-yHS1, or mito-yHS1 expressing cells (+/−SA) at every time point. $R_{\max}$ is derived from the background-subtracted eGFP/mKATE2 fluorescence ratio of yHS1, nuc-yHS1, or mito-yHS1 expressing cells that were continuously cultured without SA (−/−SA) at every time point. Alternatively, $R_{\max}$ can be estimated using the calibration procedure described in Subheadings 3.3, **Steps 4–10**, or as discussed in **Notes 24** and **25**. $R_{\min}$ is derived from the background-subtracted eGFP/mKATE2 fluorescence ratio of yHS1, nuc-yHS1, or mito-yHS1 expressing cells that were continuously cultured with SA (+/+SA) at every time point. Alternatively, $R_{\min}$ can be assessed as described in **Notes 26** and **27** from parallel cultures of cells expressing HS1-M7A, H102A or from *hem1*Δ cells expressing HS1. Figure 3 shows representative data from a heme trafficking dynamics experiment (*see* **Note 29**).
2. In Graph Pad Prism, the change in yHS1, mito-yHS1, or nuc-yHS1 heme occupancy as a function of time can be fit to the logistic function in Eq. (3) to extract trafficking rates to different compartments.

$$\left\{\%\mathrm{Bound}(\mathrm{t}) = [\mathrm{A}]/\left(1 + e^{-k(t - t1/2)}\right)\right. \tag{3}$$

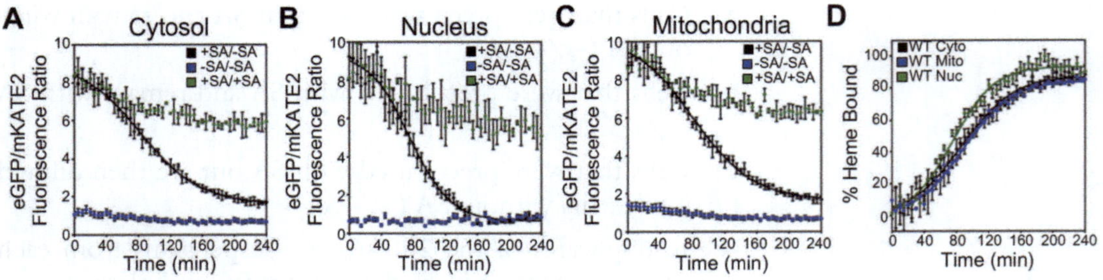

Fig. 3 Inter-compartmental heme trafficking dynamics as measured by the heme sensor HS1. WT cells expressing HS1 in the cytosol (**a**), nucleus (**b**), or mitochondria (**c**) were either depleted of heme by the addition of 500 μM succinylacetone (SA) before removal of SA to allow reinitiation of heme synthesis (+SA/−SA), or continually treated (+SA/+SA) or untreated (−SA/−SA) with SA. Upon reinitiation of heme synthesis in +SA/−SA samples, the rates of heme trafficking to the indicated subcellular locations were monitored by measuring the time-dependent change in sensor eGFP/mKATE2 fluorescence ratio. (**d**) The fractional heme occupancy of the sensor at each time point is calculated based on Eq. (1), with $R_{\min}$ derived from cultures that were continuously treated with SA (+SA/+SA) and $R_{\max}$ derived from cells continuously untreated with SA (−SA/−SA). HS1 is saturated with heme in heme-replete yeast (Fig. 1b), thereby enabling $R_{\max}$ to be determined by measuring the HS1 eGFP/mKATE2 ratio. The data in panel D are fit to the logistic function described by Eq. (3)

[A] is the maximal value of HS1 fractional saturation (amplitude), k is the first-order rate constant (min^{-1}), t is the time (min), and $t_{1/2}$ is the midpoint of the transition. The lag time can be defined as $t_{1/2}$–$2/k$ [29].

3.7 Measuring Steady-State Bioavailable Heme in Yeast Using Microscopy: Image Acquisition

1. Culture empty vector or desired heme sensor plasmid containing yeast as described in Subheading 3.2.
2. Resuspend the 5 OD_{600} cell pellets (~1×10^8 cells) in 1 mL of PBS to give a concentration of 5 OD_{600}/mL, corresponding to ~1×10^8 cells/mL (*see* **Note 30**).
3. Place 4 μL of the cell suspensions on a glass microscope slide and cover with a glass coverslip (*see* **Notes 31–32**).
4. Place the glass slide in the microscope for viewing. It is best to start an imaging experiment by first analyzing heme-deficient strains, for example, *hem1Δ* or succinylacetone-treated cells, expressing the desired heme sensor.
5. Optimize the image acquisition parameters, including the laser power, gain, and/or exposure time for optimal illumination of the eGFP and mKATE2 channels of sensor-expressing cells. Scan different fields of view to ensure acquisition parameters are being optimized for cells expressing median levels of fluorescence and not the highest or lowest luminescent cells. Once acquisition parameters are optimized for heme-depleted cells, check heme-replete cells, for example, non-succinylacetone-treated cells, to ensure eGFP emission intensities are dimmer than in heme-depleted cells and there is a sufficient signal from mKATE2 emitting cells that are not too high or too low. Lastly, image non-sensor-expressing cells to ensure that the background fluorescence on both channels is low. Ideal acquisition parameters will limit photobleaching, enable facile detection of the median or average mKATE2 emitting cells, and enable the greatest contrast in eGFP emission from heme-depleted and heme-replete cells. Optimization of acquisition parameters may be an iterative process that requires multiple cycles of checking heme-deficient and heme-replete cells before the selection of final parameters. Once parameters are selected, use the same settings for all the samples being analyzed (*see* **Note 33**).
6. Collect images from all samples, making sure to obtain ≥50 mKATE2-positive cells in total for each sample from multiple fields of view (*see* **Notes 33** and **34**).

3.8 Measuring Steady-State Bioavailable Heme in Yeast Using Microscopy: Image Processing and Analysis

1. Export/save confocal microscopy images for image processing and analysis in ImageJ/Fiji v. 1.54f (*see* **Notes 35–36**).
2. Determine background/autofluorescence for the eGFP and mKATE2 fluorescence channels in non-sensor-expressing cells (Fig. 4a). In ImageJ/Fiji, go to Analyze → Set Measurements → Check Min & Max Gray Value → and hit OK. Open an "eGFP" image from a non-sensor-expressing cell and use the

rectangle tool to draw a box around a cluster of one or more yeast cells that are individually resolved, making sure to also include a region of space that also does not include cells (Fig. 4a). Go to Analyze → Measure (or use the shortcuts "Ctrl+M" or just press "M") to display the minimum and maximum grayscale values for the intensities in the boxed region. Repeat this for different areas of the image and in at least three different image files representing different fields of view. Collect at least ~10 measurements in total for both the eGFP and mKATE2 channel images. Calculate the average of the "maximum" gray value intensities and record this value as the final background gray value for the eGFP and mKATE2 channels.

3. Subtract the background eGFP and mKATE2 values from the images corresponding to the sensor-expressing heme-depleted cells, for example, *hem1Δ* or succinylacetone-treated cells. Remove the background by going to Process → Math → Subtract and enter the value of the average background eGFP and mKATE2 intensities obtained from **Step 2** in Subheading 3.8. for the eGFP and mKATE2 channel images.

4. Colorize the micrographs and optimize the brightness and contrast for the eGFP and mKATE2 images. Apply color by going to Image → Lookup Tables. Use "green" for the eGFP channel images and "red" for the mKATE2 channel images. Then, go to Image → Adjust → Brightness/Contrast and adjust the scroll bar for "Maximum" until the desired brightness and contrast are achieved. "Minimum" is set to 0 (Fig. 4b). Scan through the other eGFP and mKATE2 channel images of the sensor-expressing heme-depleted cells to ensure that the selected "Minimum" and "Maximum" values give the appropriate brightness and contrast in all the other micrographs for this sample. This may be an iterative process requiring readjustments after multiple assessments of micrographs. Once the optimized "Maximum" and "Minimum" values are identified for the eGFP and mKATE2 channels, record these values and assign them to the other eGFP and mKATE2 micrographs collected for this sample. Note that if the exact same "Maximum" value cannot be assigned, select as close a value as possible. Once the new "Minimum" (usually 0) and "Maximum" values are assigned, select "Apply" to reset the dynamic range. The newly set dynamic range will become permanent and all data outside of that range will be deleted (*see* **Note 37**) (Fig. 4c).

5. Repeat **Steps 3** and **4** in Subheading 3.8. for all other sensor-expressing cells being analyzed to obtain background-corrected and colorized single-channel images. Use identical

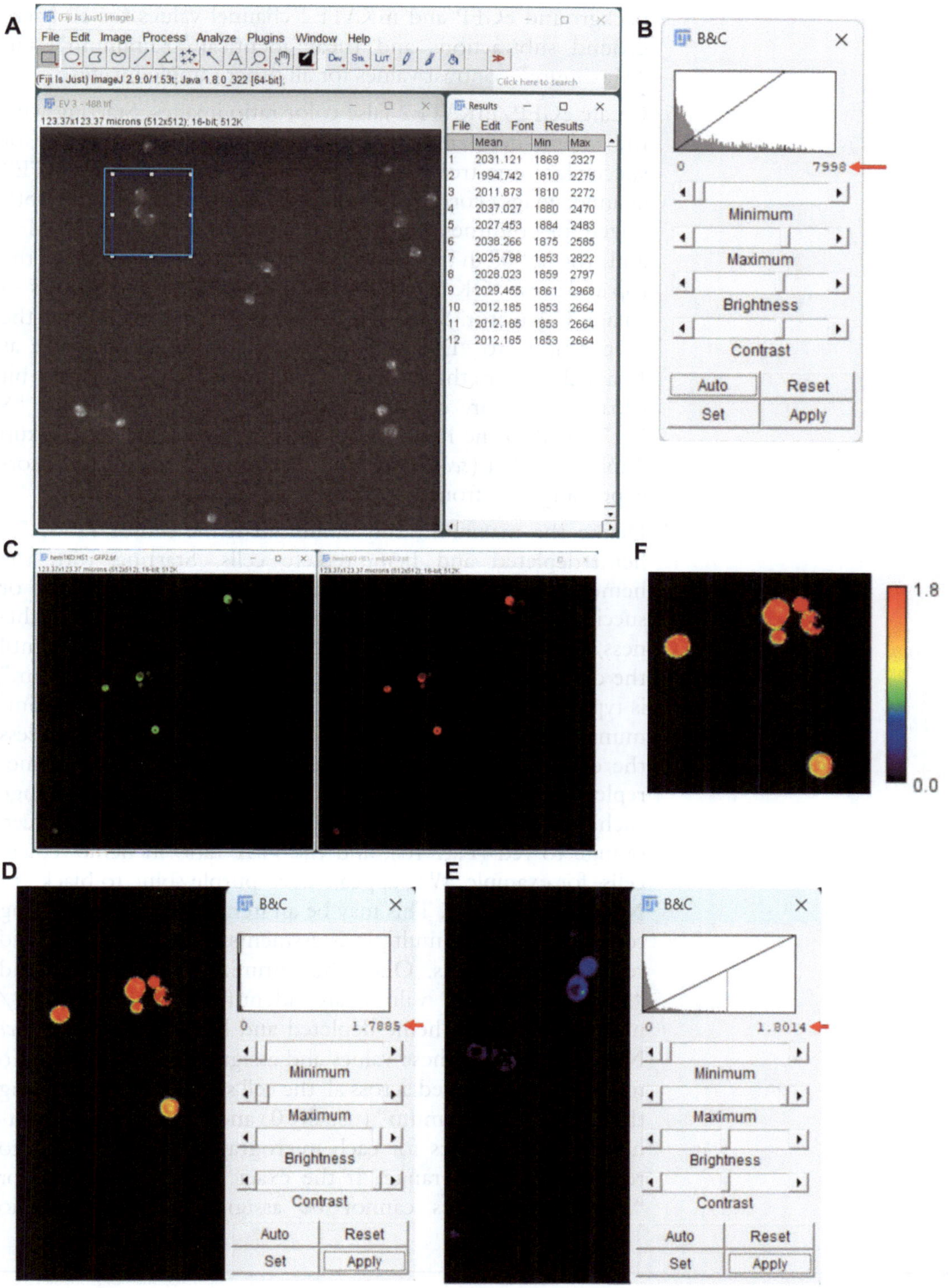

Fig. 4 Image analysis of the heme sensor fluorescence by confocal microscopy. (**a**) Determining the background fluorescence of yeast cells transformed with the empty vector plasmid. (**b**) Adjusting the brightness and contrast of heme-deficient sensor-expressing cells and noting the optimized "maximum" value (red arrow) that defines the boundaries of the pixel intensities for fluorescence emission. (**c**) Images from

background eGFP and mKATE2 channel values for all background subtractions and (near)-identical "Minimum" and "Maximum" contrast values for all other sample images.

6. Create eGFP/mKATE2 false color ratio images. Starting with the heme-deficient samples, for example, *hem1Δ* or succinylacetone-treated cells, open the eGFP and mKATE2 channel images corrected for background, colorized, and optimized for brightness/contrast. Go to Edit → Options → Misc and type in "0" in the "Divide by zero value" dialog box at the top and press "OK." To make the ratio image, go to Process → Image Calculator. Select the eGFP image as Image 1, set the "Operation" to "Divide," and select the mKATE2 image as Image 2. Ensure that both "Create new window" and "32-bit (float) result" are checked, then select "OK." Apply the "S Pet" LUT to the ratio image by selecting Image → Lookup Tables → S Pet (*see* **Note 38**). Repeat this process for sensor-expressing cells from all other sample conditions.
7. Define the maximum and minimum heme sensor ratios for heme-depleted and heme-replete cells. Starting with the heme-deficient samples, for example, *hem1Δ* or succinylacetone-treated cells, go to Image → Adjust → Brightness/Contrast and adjust the scroll bar for "Maximum" until the desired brightness and contrast are achieved. "Minimum" is typically set at or near 0. Once preliminary values of "Minimum" and "Maximum" brightness/contrast are noted, assess the effect of these values in the ratio micrograph for heme-replete HS1 expressing cells. Aim to define a dynamic range such that the sensor ratio in heme-deficient cells appears deep orange to red (Fig. 4d), and the HS1 ratio in heme-replete cells, for example, WT, appears deep purple/blue to black (*see* **Note 39**) (Fig. 4e). This may be an iterative process requiring readjustments after multiple assessments of heme-depleted and replete micrographs. Once the optimized "Maximum" and "Minimum" ratio values are identified for the eGFP/mKATE2 ratios in heme-depleted and heme-replete cells (*see* **Note 40**), record these values and assign them to all the ratio micrographs collected across all the cell samples. After entering the final new "Minimum" (usually 0) and "Maximum" brightness/contrast values for each micrograph, select "Apply" to reset the dynamic range. If the exact same "Minimum" or "Maximum" values cannot be assigned to various ratio

Fig. 4 (continued) the eGFP (green cells, left image) and mKATE2 (red cells, right image) channels that have been corrected for background fluorescence and optimized for brightness and contrast. (**d, e**) HS1 ratio images of *hem1Δ* (**d**) and wild-type (**e**) cells showing the adjusted dynamic range for the sensor ratios (red arrows). (**f**) HS1 ratio image of *hem1Δ* cells showing annotated lookup table (LUT) calibration bar

micrographs, select a value that is as close as possible. Finally, add a calibration bar to the image by selecting Analyze → Tools → Calibration bar. Set "Number of labels" to 2 and "Decimal places" to 1 (Fig. 4f).

3.9 HEK293 Cell Growth and Transfections for Steady-State and Total Bioavailable Heme Measurements

1. Grow HEK293 cells to confluency in "regular media" within a T75 flask incubated in a CO_2 incubator maintained at 37 °C and a 5% CO_2 atmosphere.
2. Seed ~1/24th of the HEK293 cells from a confluent T75 flask into 2-mL "regular media" cultures in each well of polystyrene-coated sterile 6-well plates. The total number of wells to inoculate is determined by the following formula:
   (a) Total number of wells = CG + {(R) × (CG) × (SV) × (3)}.
   (b) where "CG" is the number of comparison groups being tested, R is the number of replicates, and SV is the number of different sensor variant plasmids being transfected (*see* **Note 41**).
3. Grow cells to 30–50% confluency, which usually takes 1–2 days.
4. On the day of the transfection, for each replicate in a comparison group, for example, WT or mutant cell lines, change the media to 2-mL fresh regular media (2 wells) or 2 mL of HD + SA media (1 well) (*see* **Note 42**).
5. Incubate the cells in fresh media at 37 °C in a CO_2 incubator for 30–60 min.
6. During the cell incubation period in **Step 5**, prepare the "plasmid transfection master mix." The master mix is prepared by mixing, in order, 100 μL Opti-MEM, 2 μg of heme sensor plasmid (*see* **Note 43**), 2 μL of Lipofectamine Plus reagent, and 4 μL of the Lipofectamine LTX reagent. This is for one well and can be scaled up for the number of wells needed. Also, prepare "control transfection mix" that does not have the heme sensor plasmid to serve as an autofluorescence control; in this case, add a volume of sterile ultrapure water instead of the plasmid stock (*see* **Note 44**). Mix the "plasmid transfection master mix" and "control transfection mix" by gentle pipetting, and allow to equilibrate to room temperature for 5 min.
7. Pipette 100 μL of the "plasmid transfection master mix" or "control transfection mix" into each well. Allow the cells to grow for 48 h at 37 °C in a CO_2 incubator (*see* **Note 45**).
8. After 48 h, change the media to fresh regular or HD + SA media. For samples meant to serve as heme saturation controls, supplement the regular media with 1.25 mM 5-ALA (Reg + 5-ALA) (*see* **Note 46**).

9. Culture the cells for an additional 15–24 h before harvesting for flow cytometric analysis or imaging for microscopy (*see* **Note 47**).

10. For microscopy, repeat the procedure above, but at **Step 2**, seed the cells into an uncoated glass bottom 35-mm dish containing a 14-mm glass diameter and covered with a #1.0 coverslip. Proceed to Subheading 3.7 **Step 4** and Subheading 3.8. Prior to imaging, carefully aspirate the media from the culture dish. Wash each sample with 2 mL of DPBS prewarmed to 37 °C by applying the DPBS down the side of the well to avoid displacing the adhered cells, gently rock the dish orbitally several times, and aspirate the DPBS. Repeat the wash and add 2 mL of DPBS to each culture dish and place in a 37 °C incubator until ready for imaging.

3.10 Measuring Steady-State Bioavailable Heme Using Flow Cytometry: Sample Preparation

1. Harvest the cells. For HEK293 cells, start by aspirating the media from each well and then detach the cells from the plate by washing with 700 μL of DPBS (*see* **Note 48**). For yeast cells, follow the steps described in **Steps 3–4** in Subheading 3.5. Resuspend the cell pellet in 500 mL of PBS and proceed to **Step 5** in Subheading 3.10.

2. After cell detachment, transfer the entire 700 μL cell suspension to a 1.5 mL microcentrifuge tube. Rinse the well with 700 μL of DPBS, and transfer to the same microcentrifuge tube.

3. Pellet the cells by centrifugation for 4 min at 400 × *g* at 4 °C.

4. Carefully aspirate the supernatant and resuspend the pellet in 500 μL of DPBS.

5. Filter the cells into a 5 mL round-bottom polystyrene tube using a cell strainer cap (*see* **Note 49**).

3.11 Measuring Steady-State Bioavailable Heme Using Flow Cytometry: Data Collection

1. On the flow cytometer, set the instrument to collect logarithmic, area and height data for forward scatter (FSC), side scatter (SSC), fluorescein isothiocyanate (FITC) (for collection of eGFP emission), and PE-mCherry or PE-Texas Red (for collection of mKATE2 emission) channels.

2. In the worksheet area, create a dot (scatter) plot with FSC area (FSC-A) vs SSC area (SSC-A) axes, and two histogram plots—one each for FITC area (FITC-A) and PE-mCherry area (PE-mCherry-A). Set the histogram plot axes to display biexponential data (*see* **Note 50**). For yeast, the dot plot of FSC-A vs FSC-H is useful for delineating singlet cells which are defined as those that distribute linearly between both axes.

3. With the flow rate set to the lowest setting available (~1 μL/min), load the untransfected cells from the control group first,

and set the plots to display and record 5000–10,000 events (*see* **Note 51**). If analyzing yeast, load the "empty vector" transformed cells.

4. Adjust FSC and SSC gain settings to ensure that the majority of events are on scale, and mostly linear between both axes on the FSC-A vs SSC-A dot plot (*see* **Note 52**).
5. While viewing the FSC-A vs SSC-A dot plot, ensure that the FSC gain setting is sufficient to separate the actual cell population from cell debris (see "actual cells" gated population in Fig. 5a) (*see* **Note 53**). This step is not generally required for yeast cells.
6. Adjust the FITC and PE-mCherry channel laser power or gain settings to minimize the maximum values of the untransfected cells as assessed by histograms plotting a number of events against fluorescence intensity in the FITC-A or PE-mCherry-A channels.
7. Record the set number of events for the untransfected groups.
8. Duplicate the acquisition settings, including the laser powers and gain for the FSC, SSC, FITC, and mCherry channels from the control untransfected group, and use them to analyze the sensor-expressing cells from the control heme-deficient comparison group for optimizing FITC and PE-mCherry settings. Confirm that the FSC and SSC settings ported from the untransfected sample ensure that the majority of events are on the scale and mostly linear between both axes on the FSC-A vs SSC-A dot plots (**Step 4**) and while viewing the FSC-A vs SSC-A dot plot, that the FSC gain setting is sufficient to separate actual cells from debris (see "actual cells" gated population in Fig. 5a) (*see* **Note 53**).
9. Check that the current FITC and PE-mCherry laser power and gain settings are giving fluorescence emission intensities that are above the background values from the untransfected cells as assessed by the FITC-A and PE-mCherry-A histogram plots (*see* **Note 54**). If this is not the case, increase the laser power and gain settings as necessary and record the set number of events.
10. It may be necessary to toggle between the untransfected and sensor-expressing cells from the control heme-deficient groups to ensure emission intensity values are maximized and well differentiated from cells that do not express the sensor to optimize the acquisition parameters.
11. Once acquisition parameters are optimized, set the number of events to display to 10,000 and the stopping gate to record a minimum of 25,000 events (*see* **Note 55**). For yeast, record 50,000–100,000 events.

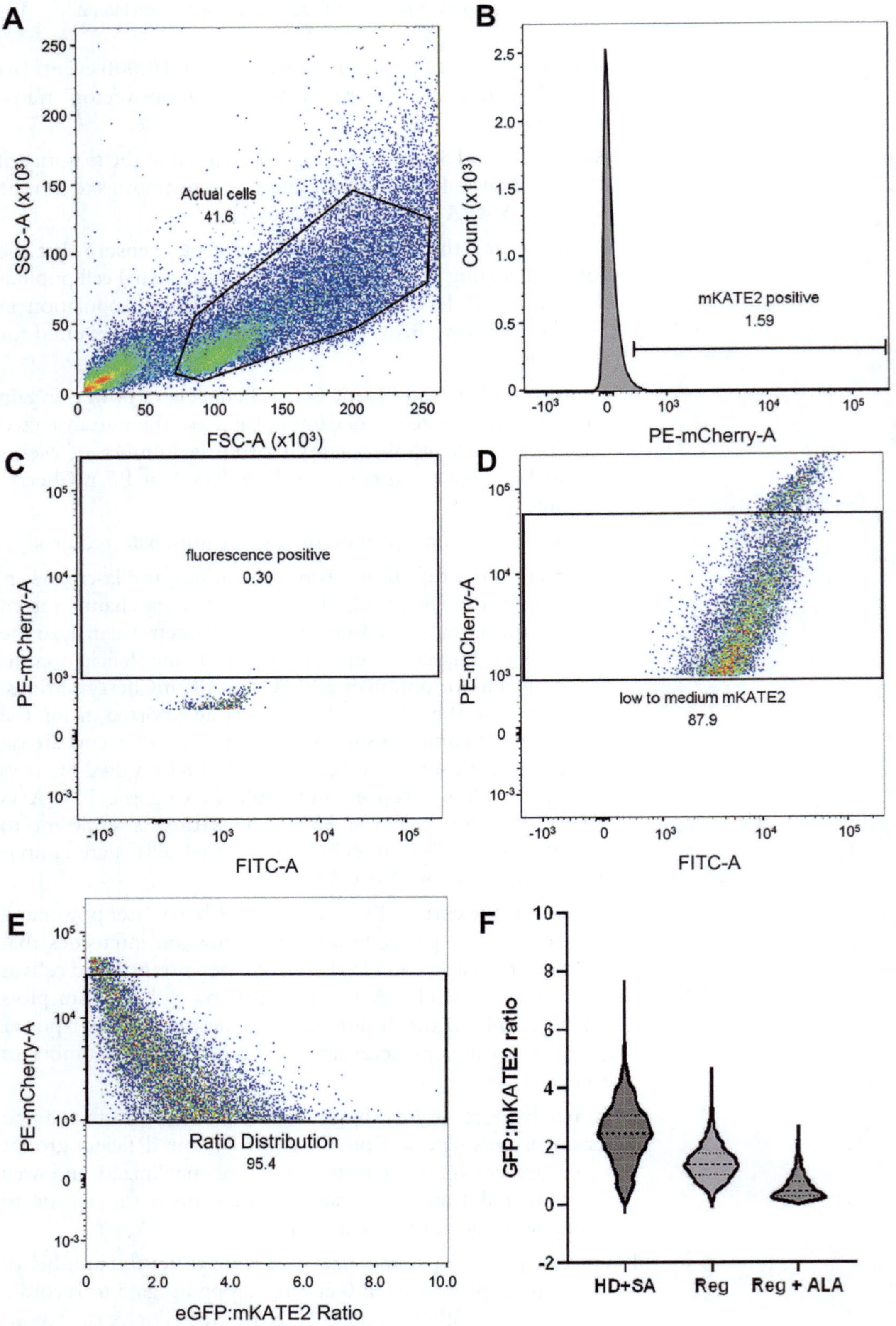

Fig. 5 Data analysis of heme sensor fluorescence by flow cytometry. (**a**) Differentiating intact cells from debris and other events during data acquisition. (**b**) Assigning the ranged gate for mKATE2-positive cells. (**c**) Gating for cells with eGFP and mKATE2 fluorescence above background. (**d**) Gating for eGFP-positive cells with low-

12. Duplicate the final acquisition settings for all parameters for all parameters and use them to analyze all samples, including the transfected and sensor-expressing heme-depleted cells that were used to initially optimize the acquisition settings.
13. Export and save data as FCS files.

3.12 Measuring Steady-State Bioavailable Heme Using Flow Cytometry: Data Analysis

In analyzing the data, cells should be gated to identify sensor-expressing cells as assessed by defining the population of cells in each sample that exhibit mKATE2 fluorescence above nonsensor-expressing cells. The eGFP/mKATE2 ratios are then assessed among the population of mKATE2 positive cells between the comparison groups.

1. Import the FCS data files into FlowJo.
2. Select sensor-expressing cells by identifying those that are mKATE2 positive. To do this, open the nonsensor-expressing sample and display the histogram plot for PE-mCherry-A. Using the ranged gate tool, select mKATE2 positive fluorescent cells by gating the region of the plot without fluorescence on the PE-mCherry axis and label it as "mKATE2 positive" (Fig. 5b). Apply this gate to all samples (*see* **Notes 56–58**).
3. Select actual intact cells. To do this, open the plot of the "mKATE2 positive cells" in the sensor-expressing control group, for example, HEK293 cells cultured in regular media, and set the axes to FSC area vs SSC area. Use a polygon gate to select the population of actual cells, and label them as "actual cells" (Fig. 5a). Apply this gate to all samples.
4. Select cells that exhibit eGFP and mKATE2 fluorescence above background levels. To do this, reopen the nonsensor-expressing group and open the plot at the "actual cells" gate level and set the axes to FITC-A vs PE-mCherry-A. Draw a gate that excludes 95–100% of cells by spanning the entire FITC-A axis and the region corresponding to intensities on the PE-mCherry-A axis that excludes most cells from the untransfected sample (Fig. 5c). Label this gate as "fluorescence positive," and apply it to all samples.
5. Exclude outlier cells that have very high/overblown sensor expression levels. Open the sensor-expressing control group, for example, cells cultured in regular media, and on the FITC-A vs PE-mCherry-A plot at the "fluorescence positive" gate level, draw a rectangle gate that spans the entire FITC-A axis

Fig. 5 (continued) to-medium mKATE2 fluorescence. (**e**) Final gating step to exclude high sensor-expressing cells. (**f**) Representative violin plots showing the eGFP/mKATE2 fluorescence ratios HS1-expressing HEK293 cells that were made to be heme-deficient (HD + SA), heme-replete (Reg), or heme-excessive (Reg + ALA) and includes cells with PE-mCherry-A values from ~1000 to between 50,000 and 100,000 (Fig. 5d) (*see* **Note 59**). Label as "low-to-medium mKATE2," and apply to all samples.

6. mKATE2 ratios for all sample/mKATE2 ratios for all sample groups. To do this, click on the *parent population* of any of the samples, including the heme-deficient and saturated samples, go to Tools → Cytometry → Derive Parameters and name the parameter as "eGFP:mKATE2" (or similar labeling). Under the "Formula" tab, click "Insert Reference," and select "FITC-A" (for eGFP) followed by the division sign (÷). Then click "Insert Reference" again and select PE-mCherry-A (for mKATE2).
7. Next, collapse the "Formula" tab and open the "Scale" tab. Change the maximum scale value to a number that more closely represents the spread of the majority of the cell population (usually between 10 and 20 for HEK293 cells and between 1 and 2 for yeast). Apply the changes and add the ratio to all samples at the parent level.
8. Assess the dependence of the sensor eGFP/mKATE2 ratio on sensor expression by plotting mKATE2 intensities against the eGFP/mKATE2 ratio in a population of cells. In a particular sample group, at the "low-to-medium mKATE2" gate level, set the dot plot axes to eGFP:mKATE2 vs PE-mCherry-A and draw a rectangle gate that excludes any cells with high PE-mCherry-A (mKATE2) values and a narrow spread of eGFP: mKATE2 ratios (Fig. 5e). Label as "ratio distribution," and apply to all samples. This gate further ensures the exclusion of ratios obtained from very high sensor expression (*see* **Note 60**).
9. To export the data to a spreadsheet, select the "ratio distribution" gate level for all samples (by holding Ctrl while clicking on each), and then, go to File → Documents → Export/ Concatenate → Export/Concatenate Populations. On the pop-up menu that is displayed, select "CSV—Scale values" under "Format" and set the destination folder and export the data. This step creates separate CSV files for each sample.
10. Use the FITC-A/PE-mCherry-A column values to make violin plots for each sample in a graphing package such as GraphPad Prism (*see* Fig. 5f for representative plots for each sample) (*see* **Notes 61** and **62**).
11. It is best to report the average median sensor eGFP/mKATE2 ratios from replicate flow cytometry experiments and show a representative histogram. Fractional heme occupancy of the sensor from each comparison group can be derived from Eq. (1) by inputting the median sensor fluorescence ratio under the "normal" test growth condition (R), $R_{\min}$ from heme-depleted cells and $R_{\max}$ from heme-saturated cells.

3.13 Measuring Total Heme Using the Porphyrin Fluorescence Assay: Data Collection

Quantification of cellular heme, which is nonfluorescent, involves de-metalating it in boiling oxalic acid to release the iron center, resulting in the formation of fluorescent protoporphyrin IX [7, 8]. Total heme is subsequently quantified by taking the difference between boiled (protoporphyrin IX + heme) and unboiled samples (protoporphyrin IX) and relating the difference in fluorescence to a calibration curve generated from hemin chloride standards.

1. Resuspend a 5 OD_{600} (1×10^8 cells) washed yeast cell pellet (see Subheading 3.2 on culturing) or a 2.5×10^5 to 5×10^5 washed cell pellet of HEK293 cells, or other mammalian cell type (see Subheading 3.9 on culturing) in 500 μL of 20 mM oxalic acid (*see* **Note 63**).
2. Incubate the samples at 4 °C overnight (12–15 h), protected from light in a covered box (*see* **Note 64**).
3. The next day, at the start of the experiment, prepare hemin chloride standards to generate a standard curve for heme and protoporphyrin IX quantification from cells. From the 5 mM hemin chloride stock solution in 0.1 M NaOH, make a 1 μM stock solution in ultrapure water. From this 1 μM hemin stock solution, prepare 1 mL of 0.5, 2, 5, 10, 25, 50, 75, 125, 250, and 500 nM hemin standards in ultrapure water. Aliquot the hemin standards into 2 × 500 μL fractions in new microcentrifuge tubes.
4. Add 500 μL of 2 M oxalic to each cell sample and hemin standards (*See* **Note 65**). Pipette to mix.
5. Split each sample into 2 × 500 μL fractions in microfuge tubes.
6. Transfer one fraction from each sample to a heat block pre-equilibrated to 100 °C and heat for 30 min. Place a microtube cap lock on these samples so that the microtube caps do not pop open during heating. The other fraction from each sample should be kept at room temperature (25 °C) for 30 min (*see* **Notes 66** and **67**).
7. Centrifuge all the samples for 3 min on a tabletop microcentrifuge at 21,000 × *g*.
8. Transfer the supernatants from the boiled and unboiled samples to new microfuge tubes and pipette 200 μL of each sample into a flat-bottom black fluorescence 96-well plate.
9. Measure the porphyrin fluorescence emission spectra for each sample by exciting at 400 nm and collecting emission between 550 and 700 nm. Alternatively, collect single wavelength emission by exciting at 400 nm and collecting emission at 605 nm and/or 665 nm.

3.14 Measuring Total Heme Using the Porphyrin Fluorescence Assay: Data Analysis

1. For each sample, including the hemin standards and the cell samples, subtract the unboiled fluorescence emission intensity (I) at 605 and/or 665 nm from the boiled fluorescence emission at the same wavelengths.
2. Make a hemin calibration curve by plotting the emission intensity at 605 and/or 665 nm against the hemin standard concentration. Fit the standard curve to a linear regression and calculate the slope (m), which reflects the change in fluorescence intensity per molar heme concentration: $\Delta I / \Delta\,[\text{Hemin}]$.
3. The moles of heme per cell are given by Eq. (4).

$$\text{Heme (mols) per cell} = \frac{\frac{[(I_{\text{boiled}} - I_{\text{unboiled}})]}{m} x\ V}{\#\text{of cells}} \tag{4}$$

I_{boiled} and I_{unboiled} reflect the emission intensities associated with boiled and unboiled samples and V is the volume, in liters, of the sample fractions (500×10^{-6} L in this protocol). The "number of cells" is half the cell number of the initial cell pellet since the cell pellet is split equally between boiled and unboiled samples.

The molar cellular concentration is given by Eq. (5).

$$\text{Heme (M) per cell} = \frac{\text{Heme (mols) per cell}}{\text{cell volume (L)}} \tag{5}$$

For haploid S288c *Saccharomyces cerevisiae*, cell volume is estimated to be 50–70 fL [9, 30], and for HEK293 cells, cell volume is estimated to be 1.2 pL [23, 31]. Figure 6 shows a representative protoporphyrin emission spectrum. The peak area between 550 and 700 nm can be used in place of intensities.

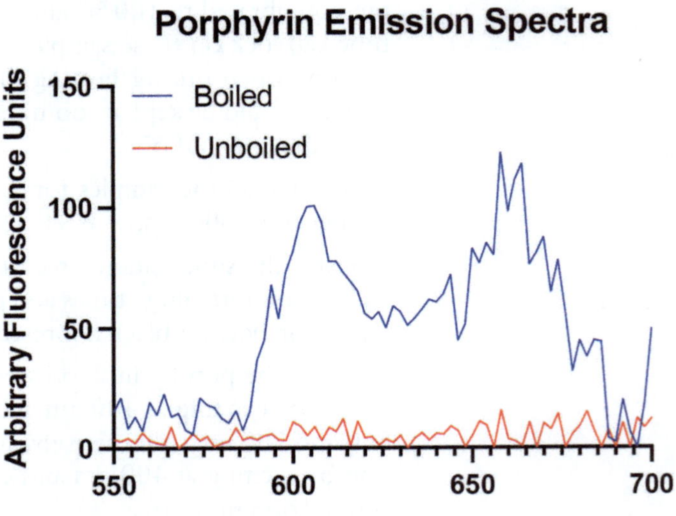

Fig. 6 Representative porphyrin emission spectra from HEK293 cells. 1×10^5 cells were either boiled or unboiled in oxalic acid, and porphyrin emission was measured upon excitation at 400 nm

4 Notes

1. Optional. The *hem1Δ* strain, which lacks the first heme biosynthetic enzyme, aminolevulinic acid synthase (ALAS), serves as a good negative control due to its zero-heme background. This strain must be generated or acquired from another laboratory.
2. Make fresh every 3 months.
3. Work next to a flame to maintain reagent sterility and follow laboratory safety guidelines. Caution: The ergosterol solution is in 100% ethanol which is flammable, so take extra precautions around the open flame when adding it to the growth media.
4. To monitor ascorbate-induced heme dissociation from serum hemoproteins in FBS, make a 1:4 dilution of the ascorbate-treated FBS in ultrapure water and measure the decrease in heme absorbance at 410 nm every 1–2 h for 6–8 h. It is best to take a UV/vis scan from 300 to 800 nm and monitor the decrease of the heme Soret peak at ~410 nm until there is no further change. This takes approximately 6–8 h to complete. An FBS sample that was not ascorbate-treated should be used as a control for each time point. The absorbance of the heme Soret peak should be relatively constant in non-ascorbate-treated FBS.
5. We use the BioTek Synergy Mx multimodal plate reader (Agilent Technologies), which employs a xenon flash light source and excitation and emission monochromators.
6. We use a Zeiss ELYRA LSM 780 super-resolution microscope equipped with a 63×, 1.4 numerical aperture oil objective. eGFP was excited with the 488 nm line of an argon ion laser, while mKATE2 was excited using the 594 nm line of a helium–neon laser. The 494–571-nm and 606–686-nm band-pass filters were used to filter emission for eGFP and mKATE2, respectively.
7. We perform flow cytometry experiments using a BD LSR Fortessa, BD LSR II, BD FACS Fusion, or BD FACS Aria III cytometers equipped with an argon laser (ex 488 nm) and yellow-green laser (ex 561 nm). eGFP is excited using the argon laser and is measured using a 530/30-nm band-pass filter, whereas mKATE2 is excited using the yellow-green laser and is measured using a 610/20-nm band-pass filter.
8. Modify the growth media with the appropriate dropout formulation or antibiotic supplementation required for selection as needed. If culturing *hem1Δ* strains, culture on media plates containing 5-ALA. If cells are oxidatively stressed, one can culture the strains in an anerobic chamber, for example, COY anerobic chambers (https://coylab.com/products/anaerobic-

chambers/), or an anerobic jar with a catalyst that consumes oxygen, for example, GAS PAK system (https://www.bd.com/en-ca/offerings/capabilities/microbiology-solutions/environmental-systems/bd-gaspak-systems-and-accessories). If strains are cultured anaerobically, use ergosterol and Tween-80 supplemented media, for example, YPDE or SCE media, with the appropriate dropout formulation or antibiotic supplementation required for selection as needed.

9. To ensure sufficient aeration, keep the caps of the conical tubes loose and tape the cap onto conical tube, so they do not fall off while the tubes are shaking.
10. The heat shock time can be varied for difficult-to-transform strains, including those that are less permeable to exogenous DNA, slow-growing, heat-sensitive, or oxidatively stressed strains. While 20 min is ideal for most strains, the heat shock time can be as little as 5 min or as long as 40 min, with a trade-off between compromised transformation efficiency or cell viability at low and high incubation times, respectively.
11. The plating beads can be rinsed in water, 70% ethanol, and dried using paper towels. Once dry, the plating beads can be autoclaved and reused.
12. If a lawn of colonies is observed such that individual single colonies are not isolable using a 1 μL inoculating loop, scrape up a loop full of yeast cells and steak to single colonies on a new media plate. After 1–2 days of growth at 30 °C, proceed to **Step 15** if isolable single colonies are well separated. If no colonies are observed from the initial transformation, try one or a combination of the following:
    (a) Vary the heat shock time of the transformation as discussed in **Note 10**.
    (b) Plate more cells after the transformation procedure (*see* **Step 12**).
    (c) Use more plasmid DNA (*see* **Step 7d**).
    (d) Use more cells in your transformation mix (*see* **Step 6**).
13. Tween-80, a source of oleic acid, and ergosterol supplementation are needed for culturing heme-deficient cells. Heme is essential for viability, and strains lacking the capacity to synthesize it are not viable. However, supplementation with ergosterol and oleic acid, two essential metabolites that are biosynthesized via heme-dependent proteins, allows for the partial rescue of yeast growth in strains that are unable to synthesize heme. Thus, cells that are supplemented with heme biosynthetic inhibitors, for example, succinylacetone, have defects in heme biosynthesis, for example, *hem1Δ* cells, or cultured in the absence of oxygen, which is required for heme synthesis, must be grown with ergosterol and oleic acid.

14. Optional. The *hem1Δ* strain, which lacks the first heme biosynthetic enzyme, aminolevulinic acid synthase (ALAS), serves as an additional negative control due to its zero-heme background. This strain must be generated or acquired from another laboratory. Culture empty vector or sensor-expressing *hem1Δ* cells on SCE-LEU media supplemented with or without 298 μM 5-ALA.
15. For measuring steady-state or total bioavailable heme in exponential phase yeast, inoculate a 5-mL SCE-LEU culture with an initial OD_{600} between 0.01 and 0.05. After 12–15 h of growth, cells will reach an OD_{600} ~ 1.0.
16. Harvest the same number of cells from each sample. Anywhere between 2 and 10 OD_{600} of cells, corresponding to 4×10^7 to 2×10^8 cells, can be harvested.
17. The wash steps are critical as residual ergosterol and Tween-80 from the media contribute to higher background levels in the eGFP fluorescence channel.
18. Depending on the number of cells harvested, the concentration can vary between 2 and 10 OD_{600}/mL of cells, corresponding to 4×10^7 to 2×10^8 cells/mL.
19. The volume of the cell suspension dispensed in the microplate can vary between 100 and 200 μL.
20. Appropriate gain and/or monochromator slit width settings should be selected to ensure that fluorescence intensities for all the samples remain within the dynamic range of the instrument. For instance, if the maximum fluorescence reading is 100,000 arbitrary fluorescence units (AFU), the gain and slit widths on the eGFP and mKATE2 channels should be adjusted to give counts on the order of ~500–10,000 AFU. This would ensure that a signal within the dynamic range of the instrument can still be read even if there is as much as a factor of 10 change in fluorescence intensity due to perturbations in heme and/or sensor expression levels. Gain and slit width settings in the eGFP channel can be optimized by ensuring that:
    (a) Relative to an empty vector strain, eGFP intensities are greatest in heme-depleted cells, for example, *hem1Δ* or SA-treated strains, since they exhibit the highest eGFP fluorescence emission due to the lack of heme binding to the sensor.
    (b) WT cells expressing HS1, the high-affinity heme sensor, are comparable or slightly elevated compared to empty vector cells since heme saturation of the sensor quenches eGFP fluorescence emission.

    Gain and slit width settings in the mKATE2 channel can be optimized by ensuring that:

(a) Relative to an empty vector WT strain, WT cells expressing HS1 should exhibit mKATE2 fluorescence emission intensities that are at least twofold to threefold higher and as much as ten-fold to 20-fold elevated.

21. Optional. The sedimentation of cells over time in the well of the microplate may contribute to time-dependent variations in the fluorescence intensity in the channels being measured. Pelleting cells in the microplate prior to measurement can help if this is an issue. In this case, a 96-well plate adapter and a holder are needed for the tabletop centrifuge.

22. Optional. Since cells sediment over time, better signal-to-noise ratios can be achieved using a 96-well microplate containing a flat optically transparent bottom and reading the samples from the bottom instead of the top. Further, the use of black plastic plates will mitigate background and stray light from adjacent sample wells.

23. Optional. If extended time course measurements are needed, for instance, to measure changes in heme availability due to acute exposures to drugs and other perturbates, the plate reader should be prewarmed to 30 °C if such a capability is available, and an extended kinetics run can be completed in which eGFP and mKATE2 fluorescence intensities are collected over a defined time interval and length. For measurements longer than 30 min, one can conduct experiments in cell suspensions of SC media rather than PBS to prevent nutrient deprivation. Since the microplates tend to dry out depending on the temperature and humidity, the time frame to collect continuous kinetic data is about 4–6 h.

24. This step is done to evaluate the sensor emission profile when it is completely saturated with heme. Toward this end, digitonin is used to permeabilize the cell membrane and excess exogenous heme is used to drive sensor heme binding. Since HS1 prefers reduced heme over oxidized heme, ascorbate is used to reduce heme to further drive heme binding to HS1.

    Since digitonin cannot permeabilize mitochondrial or nuclear membranes, an alternative strategy can be used to probe sensor fluorescence associated with heme saturation. This procedure involves digesting the yeast cell wall with Zymolyase, followed by incubating yeast spheroplasts with 0.1% Triton-X100, to permeabilize membranes, and incubating the cell suspension with 1 mM ascorbate and 50 μM hemin chloride. To accomplish this, resuspend washed sensor-expressing cells in "softening buffer" (100 mM Tris-HCl, pH = 9.4, 10 mM DTT) at a concentration of 10 OD_{600}/mL of cells. Incubate cells at room temperature, with occasional swirling, for 30–60 min. Wash and resuspend cells in "sorbitol buffer" containing 50 mM Tris-HCl, pH 7.4, 1.2 M Sorbitol. To this cell suspension in "sorbitol buffer," add 0.5 mg/mL of 100-T Zymolyase and incubate for 45 min at 30 °C with occasional inversion. Following this incubation period, wash the cells in "sorbitol buffer" and resuspend the cells in "sorbitol buffer" containing 1% Triton X-100, 1 mM ascorbate, and 50 μM hemin chloride at a final cell concentration of 5 OD_{600}/mL. Incubate this suspension at 30 °C for 30 min. After this incubation period, wash the cell pellet in sorbitol buffer and resuspend in sorbitol buffer containing 1 mM ascorbate at a final concentration of 5 OD_{600}/mL.

25. Alternatively, **Steps 4–10** in Subheading 3.3 can be omitted if one approximates the background fluorescence in the eGFP channel of empty vector cells to be equal to the eGFP intensity associated with a fully saturated heme sensor. This is because energy transfer between eGFP and heme is more than 99% efficient and eGFP fluorescence is almost completely quenched in the HS1 scaffold. Control experiments in our laboratory have previously validated this approximation in yeast for sensors expressed in the cytosol, nucleus, and mitochondria. We also have previously observed that in the WT S288c yeast strain, HS1 is quantitatively saturated with heme. Thus, the eGFP/mKATE2 ratios of HS1 expressed in WT cells can also be used to estimate the fluorescence emission associated with sensor saturation, thereby allowing for the omission of the calibration protocol described in **Steps 4–10** in Subheading 3.3.

26. Alternatively, R_{min} can be approximated from non-heme-binding HS1 variant, HS1-M7A,H102A. This is because the ratio of HS1-M7A, H102A is similar to the ratio observed for HS1 and HS1-M7A expressing cells treated with the heme biosynthetic inhibitor succinylacetone.

27. Alternatively, R_{min} can be approximated for HS1 or HS1-M7A expressed in *hem1Δ* cells. Indeed, the eGFP/mKATE2 ratios of HS1, HS1-M7A, and HS1-M7A,H102A are all similar in heme-deficient *hem1Δ* cells.

28. An estimate of buffered free heme assumes that the probe binds and equilibrates with the free heme pool in cells. Additional assumptions regarding the heme oxidation state must also be made given the current lack of oxidation-state selective heme sensors. Additionally, the heme sensors may not be sensing free cellular heme, but rather may only sense heme by exchanging heme with other labile heme-binding sites via associative mechanisms. Thus, it may be preferable to report sensor heme occupancy (Eq. 1) since it is unhindered by various assumptions that are difficult to validate at this time.

29. In some plate readers and plate types, there is nonuniform drying of the wells that results in heme-independent changes in sensor fluorescence ratios or otherwise noisy data. To circumvent this issue, the heme trafficking dynamics SA-pulse-chase assays can be conducted in 5- or 10-mL cultures in 50-mL conical tubes in a 30 °C shaking incubator. In this case, 200 μL aliquots of empty vector or HS1 expressing cells can be removed every 10–15 min and read in the plate reader as a single time point.
30. The concentration of yeast cells can vary between 1 and 5 OD_{600}/mL of cells, corresponding to 2×10^7 to 1×10^8 cells/mL. A higher cell density will lead to more cells per field of view during imaging, which may be advantageous for certain applications. However, too high a cell density may lead to the inability to clearly resolve individual cells or increase the likelihood of obtaining multiple layers of cells, which could affect Z-plane focusing.
31. Ensure the sample is well mixed by gently flicking the tube or pipetting up and down several times before pipetting onto the slide.
32. It may be necessary to lightly press down on the coverslip to remove any air pockets and ensure an even film of the cell suspension.
33. There may be gradual drying out of the sample if being illuminated for too long. Typical manifestation of this occurrence is the sudden movement of cells and clumping along a moving air–liquid interface. It may be necessary to make several slides for the same sample to collect sufficient data.
34. Collect images for the non-sensor-expressing negative control last.
35. Save in 16-bit TIFF format for best quality and perform all image processing and analysis on TIFF files.
36. Image processing and analysis may also be performed in other proprietary or open-source applications.
37. After each image processing step, save that new image as a different file, always in TIFF format (File → Save As → TIFF). Never save over your original image file. To make a high-quality color image for documents and presentations, go to File → Save As → PNG.
38. The S Pet LUT may not come bundled with some versions of ImageJ/Fiji. If this is the case, download S_Pet.lut from https://sourceforge.net/projects/bifijiplugins/ and place it in the LUT folder within the ImageJ/Fiji folder, and then launch or restart ImageJ/Fiji.

39. The dynamic range typically goes from 0 in heme-replete cells to some noninteger number in the heme-depleted cells. The dynamic range may vary between experimental trials.
40. To assess the heme sensor ratio associated with heme saturation, sensor calibration procedures like that described in Subheading 3.3. can be conducted prior to imaging.
41. It is recommended that there are at least two to three replicates (R). "Comparison group" (CG) refers to the number of test samples being compared. For example, if a WT and mutant cell line are being compared, the number of comparison groups is "2." "Sensor variant" (SV) refers to the type of sensor being transfected, for example, hHS1, mito-hHS1, nuc-hHS1, and hHS1-M7A. The coefficient 3 refers to the need for each comparison group to be manipulated to have low and high heme for sensor calibration, as well as the "normal" test growth condition, for example, regular media. The addition of the number of comparison groups, represented by "+SG," reflects the need to have one untransfected sample in each comparison group to measure cellular autofluorescence for background subtractions. In most instances, background autofluorescence is relatively invariant between sample groups and the "+SG" term can be reduced to "1." For example, if mitochondrial and cytosolic bioavailable heme were to be compared in duplicate WT and mutant cell lines, a total of 26 wells $\{(2 \times 2 \times 2 \times 3) + 2\}$ are needed. If the background autofluorescence of the WT and mutant cell lines are similar, a total of 25 wells $\{(2 \times 2 \times 2 \times 3) + 1\}$ are needed.
42. One regular media sample will be maintained in regular media to measure the "steady-state" eGFP/mKATE2 sensor ratios under the normal test condition. One regular media sample will be treated with 5-ALA or exogenous heme to increase heme saturation of the sensor to measure the eGFP/mKATE2 sensor ratio associated with heme saturation. The HD + SA sample is used to limit heme in cells to measure the eGFP/mKATE2 ratio associated with heme depletion. Some cell types can be depleted of heme with SA supplementation in regular media, without the need to use "HD" media. In this case, all three wells for each replicate can be resuspended in regular media.
43. In HEK293 cells, hHS1 is fractionally saturated with heme (~50% bound) and heme-responsive, whereas hHS1-M7A is 0% bound and relatively non-heme-responsive. Thus, in HEK293 cells, one should use the hHS1 sensor and corresponding organellar-targeted variants. A typical plasmid stock from a plasmid prep yields ~200–500 ng/μL of heme sensor plasmid, requiring 4–10 μL of plasmid in the transfection master mix per well.

44. A volume of Lipofectamine Plus reagent equal in μL to the μg of plasmid DNA is used. A volume of Lipofectamine LTX transfection reagent that is double the volume of the Lipofectamine Plus reagent is used. If transfections do not work well, plasmid DNA concentration or transfection strategy should be varied, including trying other lipid-based reagents or electroporation. Typical transfection efficiencies for transient transfection of the heme sensor plasmids in HEK293 cells are on the other of ~10%.

45. Culture time can span anywhere between 12 and 96 h depending on the purpose of the experiment.

46. As little as 350 μM 5-ALA can be used. Alternatively, supplementation of cells with 50 μM hemin chloride can also saturate hHS1 in HEK293 cells. When assessing different cell lines, which have varying capacities to synthesize or transport heme, try different concentrations of 5-ALA or hemin to determine the condition that works best to saturate the heme sensor. It is also possible to generate a heme-saturated sensor sample by applying a digitonin-based permeabilization procedure as described for yeast (Subheading 3.3, **Steps 4–10**). In this case, just prior to harvesting cells for flow cytometry (Subheading 3.10), resuspend cells in 1 mL of the "mammalian cell heme saturation buffer," transfer to a microfuge tube, and incubate in a 30 °C water bath for 30 min. After this, pellet cells at 400 × *g* for 4 min, aspirate the media, and wash the cell pellet with 1 ml DPBS at room temperature twice. Finally, resuspend the cell suspension in 500 μL of DPBS containing 1 mM ascorbate before being analyzed by flow cytometry.

47. In this protocol, the total time of cell growth from initial transfection to harvesting cells for analysis is 72 h and heme saturation is done by adding 5-ALA or hemin chloride ~24 h before analysis of fluorescence. The incubation times post-transfection and exposure to 5-ALA or hemin chloride can be varied as dictated by the objectives of the experiment.

48. Repeatedly stream the same 700 μL of DPBS over the cells. The cells will progressively slough off the base of the well with each successive application.

49. Place the cell suspension on the cell strainer cap, and gently tap the base of the tube against the bench surface to force the cells through the strainer mesh.

50. Dot plots of FSC area vs FSC height (FSC-H), SSC area vs SSC height (SSC-H), and FITC-A vs PE-mCherry-A may also be created to facilitate data collection.

51. Vortex the samples before loading if doing so manually, or add a shaking step if loading with an autosampler (tube or 96-well plate).

52. Restart the display of events after *each* change of settings to show only those events with the new settings. To check the number of events that are off-scale, place small rectangle gates on the right and top sides of the plot, and ensure that the percentage of events in those gates is <10% of events.
53. No actual gates will be applied in this step. The gate shown in Fig. 5a is for illustrative purposes.
54. The maximum values of fluorescence on various flow cytometers will differ.
55. The allowable number of events recorded will depend on cell viability for each experiment. The HD + SA samples are typically the least healthy as they are heme-depleted and are therefore the best for setting the total number of recorded events across all samples.
56. Ensure that the gate includes a small population (< 5%) of untransfected cells (Fig. 5b). They are needed for a subsequent gating step and will eventually be excluded.
57. The gates that follow this "mKATE2-positive" gate are hierarchical, with each subsequent gate being applied to the previously gated population.
58. The gate labels given at each step are suggestions and may be changed to any similar labeling that sufficiently identifies each population to the experimenter.
59. The choice of the upper limit will be determined by the sensor distribution in a population of cells in that particular sample or experimental trial, as well as the dynamic range of the instrument—higher upper limits may be used in instruments with maximum values in the 10^6–10^7 range.
60. This step may not be necessary as the cells with high sensor expression may already have been gated out at the previous gating step.
61. The relatively broad ratio distribution of the heme sensor reflects both heme-dependent and heme-independent perturbations to fluorescence emission intensities. In mammalian cells, unlike in yeast, one cannot culture cells with a zero-heme background. Thus, under heme-depleted conditions, there are still substantial levels of heme in mammalian cells. Moreover, heme depletion or saturation does not affect every cell equally within a population, further contributing to the broad ratio distributions.
62. It is best practice to evaluate how any conclusions derived from the experiment may change based on how one gates the samples. For instance, analyzing sensor ratios in the entire population of mKATE2 positive cells versus just the low-, medium-, low+medium-, or high-mKATE2-intensity groups will ensure

that one's conclusions are not dependent on how cells are gated. This must be weighed against the possibility that very high expression of the sensor, as assessed by high mKATE2 intensities, may itself affect bioavailable heme levels.

63. Prepare triplicate samples for each sample condition. Although not essential, it is best to also include control pellets generated from heme-deficient and/or excess cells. Since heme will be normalized per cell, it is important to accurately measure cell number. It is best to prepare samples such that an equal number of cells are being analyzed across the different comparison groups.
64. This step digests the cells.
65. 2 M oxalic acid is not soluble at room temperature. The solution must be heated with stirring for ~20 min with a hot plate set to 170 °C for complete dissolution prior to use. Heat the oxalic acid while the hemin chloride standards are being prepared.
66. The unboiled samples represent the background protoporphyrin IX levels in cells. The boiled samples represent the sum of the contributions from heme and protoporphyrin IX. Boiling in oxalic acid will induce the release of iron from nonfluorescent heme to generate fluorescent protoporphyrin IX.
67. Turn off the lights to prevent photodegradation of porphyrins. Protect samples from light by covering them in a box or with aluminum foil during sample processing. Alternatively, amber microtubes can be used.

Acknowledgments

This work was supported by the National Institutes of Health grants GM145350 and R01NS123168, a US National Science Foundation grant 1552791, and the Vasser-Woolley endowment.

References

1. Reddi AR, Hamza I (2016) Heme mobilization in animals: a metallolipid's journey. Acc Chem Res 49:1104–1110
2. Donegan RK, Moore CM, Hanna DA, Reddi AR (2019) Handling heme: the mechanisms underlying the movement of heme within and between cells. Free Radic Biol Med 133:88–100
3. Chambers IG, Willoughby MM, Hamza I, Reddi AR (2021) One ring to bring them all and in the darkness bind them: the trafficking of heme without deliverers. Biochim Biophys Acta, Mol Cell Res 1868:118881
4. Hanna DA, Martinez-Guzman O, Reddi AR (2017) Heme gazing: illuminating eukaryotic heme trafficking, dynamics, and signaling with fluorescent heme sensors. Biochemistry 56:1815–1823
5. Swenson SA, Moore CM, Marcero JR, Medlock AE, Reddi AR et al (2020) From synthesis to utilization: the ins and outs of mitochondrial heme. Cells 9

6. Hopp MT, Schmalohr BF, Kuhl T, Detzel MS, Wissbrock A et al (2020) Heme determination and quantification methods and their suitability for practical applications and every-day-use. Anal Chem
7. Michener JK, Nielsen J, Smolke CD (2012) Identification and treatment of heme depletion attributed to overexpression of a lineage of evolved P450 monooxygenases. Proc Natl Acad Sci USA 109:19504–19509
8. Sassa S (1976) Sequential induction of heme pathway enzymes during erythroid differentiation of mouse Friend leukemia virus-infected cells. J Exp Med 143:305–315
9. Hanna DA, Harvey RM, Martinez-Guzman O, Yuan X, Chandrasekharan B et al (2016) Heme dynamics and trafficking factors revealed by genetically encoded fluorescent heme sensors. Proc Natl Acad Sci USA 113:7539–7544
10. Leung GC, Fung SS, Gallio AE, Blore R, Alibhai D et al (2021) Unravelling the mechanisms controlling heme supply and demand. Proc Natl Acad Sci USA 118
11. Gallio AE, Fung SS, Cammack-Najera A, Hudson AJ, Raven EL (2021) Understanding the logistics for the distribution of heme in cells. JACS 1:1541–1555
12. Song Y, Yang M, Wegner SV, Zhao J, Zhu R et al (2015) A Genetically Encoded FRET Sensor for Intracellular Heme. ACS Chem Biol 10: 1610–1615
13. Dastpeyman S, Godin R, Cosa G, English AM (2020) Quantifying heme-protein maturation from ratiometric fluorescence lifetime measurements on the single fluorophore in its GFP fusion. J Phys Chem A 124:746–754
14. Abshire JR, Rowlands CJ, Ganesan SM, So PT, Niles JC (2017) Quantification of labile heme in live malaria parasites using a genetically encoded biosensor. Proc Natl Acad Sci USA 114:E2068–E2076
15. Yuan X, Rietzschel N, Kwon H, Walter Nuno AB, Hanna DA et al (2016) Regulation of intracellular heme trafficking revealed by subcellular reporters. Proc Natl Acad Sci USA 113: E5144–E5152
16. Kawai K, Hirayama T, Imai H, Murakami T, Inden M et al (2022) Molecular imaging of labile heme in living cells using a small molecule fluorescent probe. J Am Chem Soc 144: 3793–3803
17. Xu S, Liu HW, Chen L, Yuan J, Liu Y et al (2020) Learning from artemisinin: bioinspired design of a reaction-based fluorescent probe for the selective sensing of labile heme in complex biosystems. J Am Chem Soc 142:2129–2133
18. Weissman Z, Pinsky M, Donegan RK, Reddi AR, Kornitzer D (2021) Using genetically encoded heme sensors to probe the mechanisms of heme uptake and homeostasis in Candida albicans. Cell Microbiol 23:e13282
19. Sweeny EA, Singh AB, Chakravarti R, Martinez-Guzman O, Saini A et al (2018) Glyceraldehyde-3-phosphate dehydrogenase is a chaperone that allocates labile heme in cells. J Biol Chem 293:14557–14568
20. Sun F, Zhao Z, Willoughby MM, Zhou Y, Shao Y, Kang J, Chen Y, Chen M, Yuan X, Hamza I, Reddi AR, Chen C (2022) HRG-9 homologues regulate haem trafficking from haem-enriched compartments. Nature 610: 768–774
21. Mestre-Fos S, Ito C, Moore CM, Reddi AR, Williams LD (2020) Human ribosomal G-quadruplexes regulate heme bioavailability. J Biol Chem 295:14855–14865
22. Martinez-Guzman O, Willoughby MM, Saini A, Dietz JV, Bohovych I et al (2020) Mitochondrial-nuclear heme trafficking in budding yeast is regulated by GTPases that control mitochondrial dynamics and ER contact sites. J Cell Sci 133
23. Hanna DA, Moore CM, Liu L, Yuan X, Dominic IM et al (2022) Heme oxygenase-2 (HO-2) binds and buffers labile ferric heme in human embryonic kidney cells. J Biol Chem 298:101549
24. Hanna DA, Hu R, Kim H, Martinez-Guzman-O, Torres MP et al (2018) Heme bioavailability and signaling in response to stress in yeast cells. J Biol Chem 293:12378–12393
25. Donegan RK, Fu Y, Copeland J, Idga S, Brown G et al (2022) Exogenously scavenged and endogenously synthesized heme are differentially utilized by mycobacterium tuberculosis. Microbiol Spectr:e0360422
26. Dietz JV, Willoughby MM, Piel RB 3rd, Ross TA, Bohovych I et al (2021) Mitochondrial contact site and cristae organizing system (MICOS) machinery supports heme biosynthesis by enabling optimal performance of ferrochelatase. Redox Biol 46:102125
27. Li Y, Jiang Y, Paxman J, O'Laughlin R, Klepin S et al (2020) A programmable fate decision landscape underlies single-cell aging in yeast. Science 369:325–329
28. Gietz RD, Schiestl RH (1991) Applications of high efficiency lithium acetate trasformation of

intact yeast cells using single-stranded nucleic acids as carrier. Yeast 7:253–263

29. Arosio P, Knowles TP, Linse S (2015) On the lag phase in amyloid fibril formation. Phys Chem Chem Phys 17:7606–7618
30. Garber Morales J, Holmes-Hampton GP, Miao R, Guo Y, Munck E et al (2010) Biophysical characterization of iron in mitochondria isolated from respiring and fermenting yeast. Biochemistry 49:5436–5444
31. Chien HC, Zur AA, Maurer TS, Yee SW, Tolsma J et al (2016) Rapid method to determine intracellular drug concentrations in cellular uptake assays: application to metformin in organic cation transporter 1-transfected human embryonic kidney 293 cells. Drug Metab Dispos 44:356–364

Chapter 10

Recombinant Biogenesis and Analysis of Cytochrome *c* Species

Alicia N. Kreiman, Tania Yeasmin, and Molly C. Sutherland

Abstract

Recombinant expression and biogenesis of cytochrome *c* species is a simple and efficient method for the production of holocytochrome *c* species, thus presenting an avenue for the study of cytochrome *c* or the cytochrome *c* biogenesis pathways responsible for heme attachment. Here, we describe a method for recombinant *E. coli* production of holocytochrome *c* utilizing the System I (CcmABCDEFGH) bacterial cytochrome *c* biogenesis pathway, followed by analysis of cytochrome *c* species by cell lysis and heme stain.

Key words Cytochrome *c*, Cytochrome *c* biogenesis, Heme, Heme stain

1 Introduction

Cytochromes *c* are critical electron transport proteins found in nearly all organisms (reviewed in [1–6]). These proteins require the covalent attachment of heme for proper folding and function, a process called cytochrome *c* biogenesis (reviewed in [3–5, 7–11]). All known cytochromes *c* are matured by one of three pathways, termed System I, System II, and System III. Recombinant co-expression of a cytochrome *c* and a cytochrome *c* biogenesis pathway allows for the analysis of holocytochrome *c* species and the interrogation of biogenesis pathways. With hundreds of known cytochromes *c* playing important roles in many cellular processes [1, 5, 6, 12], the ability to identify biogenesis levels consistently and reliably is vital to the study of these proteins.

We utilize an optimized recombinant *E. coli* cytochrome *c* biogenesis assay, which allows for the study of a variety of cytochromes *c* and cytochrome *c* biogenesis pathways, including those encoded by difficult-to-culture organisms. Here, we describe this

---

Authors Alicia N. Kreiman and Tania Yeasmin have equally contributed to this chapter.

Oleh Khalimonchuk (ed.), *Iron Metabolism: Methods and Protocols*, Methods in Molecular Biology, vol. 2839,
https://doi.org/10.1007/978-1-0716-4043-2_10,

technique employing the well-developed bacterial *E. coli* System I pathway and *Bordetella pertussis* di-heme cytochrome c_4 [13–16]. Recombinant expression of these proteins in a strain lacking the endogenous cytochrome *c* biogenesis proteins allows for quantitative determination of heme attachment to cytochrome *c* via cell lysis, followed by SDS–PAGE protein separation, transfer to the nitrocellulose membrane, and heme staining [13, 17]. By exploiting the natural peroxidase activity of heme, heme in protein polypeptides (i.e., holocytochrome *c*) can be detected. Traditional in-gel heme stains utilizing 3,3′,5,5′-tetramethylbenzidine (TMBZ) have been replaced with the transfer of SDS–PAGE gels to nitrocellulose membranes and development utilizing an ultrasensitive enhanced chemiluminescent (ECL) substrate with femtogram protein-level detection and long stability coupled with imaging via a charge-coupled device (CCD) imaging system [17]. As described by Feissner et al, this method can detect the presence of heme at subpicomole levels, greatly increasing the sensitivity of heme detection [17]. We have consistently utilized this approach in recombinant cytochrome *c* biogenesis assays for quantification of the levels of cytochrome *c* maturation and interrogation of cytochrome *c* biogenesis proteins. Here, we describe the detailed protocol for cytochrome *c* biogenesis via recombinant expression, cell lysis, separation of cell lysates by SDS–PAGE, and visualization of cytochrome *c* species with heme staining.

2 Materials

Prepare all solutions using Milli-Q water and analytical-grade reagents. Prepare and store all reagents at room temperature (unless indicated otherwise).

2.1 Induction of Recombinant Cytochrome c Biogenesis Pathway Proteins and Cytochrome c Species

1. Recombinant *E. coli* lacking the endogenous cytochrome *c* biogenesis genes (e.g., RK103 (*E. coli* MG1655 Δ*ccm*::kan^R) [13] (*see* **Note 1**)) to construct experimental and control bacterial strains for assay.
2. Experimental strains include *E. coli* lacking the endogenous cytochrome *c* biogenesis genes transformed with an inducible plasmid for expression of the System I cytochrome *c* biogenesis pathway (e.g., pRGK385, pGEX GST:CcmABCDE(F:6XHis) GH [18] (*see* **Note 2**)) and a separate inducible plasmid with the cytochrome *c* of interest (e.g., pRGK332, pBAD cytochrome c_4:His [13] (*see* **Note 3** and **4**)).
3. Negative control strain includes *E. coli* lacking the endogenous cytochrome *c* biogenesis genes transformed with the pGEX vector plasmid and a separate inducible plasmid with the cytochrome *c* of interest (*see* **Note 5**).

4. Luria–Bertani (LB) broth: Dissolve 25 g of LB broth powder in 1 L of water. Sterilize by autoclaving.
5. 34 mg/mL Chloramphenicol (Cm): Dissolve 1.7 g of chloramphenicol powder in 50 mL 100% EtOH. No need to filter sterilize.
6. 50 mg/mL Carbenicillin (Carb): Dissolve 2.5 g of carbenicillin powder in 50 mL sterile water. Filter sterilize (*see* **Note 6**).
7. LB Agar with 50 μg/mL Carb and 20.4 μg/mL Cm: Add 40 g of LB agar powder to 1 L of water in an autoclavable bottle. Add a stir bar to the bottle. Sterilize by autoclaving. Cool the autoclaved bottles to approximately 45–50 °C. Use the sterile technique to add antibiotics: carbenicillin to a working concentration of 50 μg/mL and chloramphenicol to a working concentration of 20.4 μg/mL. Place the bottle on a stir plate and mix gently. Use the sterile technique to pour plates, solidify, and store at 4 °C.
8. 1 M isopropyl-B-D-thiogalactopyranoside (IPTG): Dissolve 11.9 g of IPTG powder in 50 mL sterile water. Filter sterilize.
9. 25% w/v arabinose (ara): Dissolve 125 g of arabinose powder in 500 mL sterile water. Filter sterilize.
10. Sterile culture tubes.
11. Inoculation loops.
12. 15-mL conical tubes.
13. Centrifuge capable of pelleting bacterial cultures in 15-mL conical tubes.
14. Incubator capable of growing *E. coli* cultures at 37 °C with appropriate agitation.

2.2 Lysis of Bacterial Cells

1. Bacterial lysis and protein solubilization reagent, the use of B-PER Bacterial Protein Extraction Reagent is described here (*see* **Note 7**).
2. 1 unit/μL DNase solution.
3. 1 M Mg^{2+} solution: Dissolve 203.3 mg of $MgCl_2$-$6H_2O$ in 80 mL of water. Transfer the contents to a graduated cylinder and adjust to a final volume of 100 mL. Autoclave to sterilize. Store at 4 °C.
4. 1.5-mL microcentrifuge tubes.
5. Microcentrifuge capable of pelleting bacterial cell debris.
6. NanoDrop with the capability to quantify total protein.

2.3 12.5% SDS–PAGE Separation of Bacterial Cell Lysates and Semi-dry Transfer

1. Separating SDS–PAGE buffer: 1.5 M Tris–HCl pH 8.8. Add about 400 mL water to a beaker with a stir bar. Weigh 90.8 g Tris base and add to the beaker while stirring. When dissolved adjust to pH 8.8 with HCl. Transfer the contents to a graduated cylinder and adjust to a final volume of 500 mL with water. Transfer to a glass bottle and autoclave.
2. Stacking SDS–PAGE buffer: 0.5 M Tris–HCl pH 6.8. Add about 400 mL water to a beaker with a stir bar. Weigh 30.0 g Tris base and add to the beaker while stirring. When dissolved adjust to pH 6.8 with HCl. Transfer the contents to a graduated cylinder and adjust to a final volume of 500 mL with water. Transfer to a glass bottle and autoclave.
3. 10% Sodium Dodecyl Sulfate (SDS) Solution: Add about 400 mL water to a beaker with a stir bar. Weigh 50.0 g of SDS and add to the beaker while stirring. When dissolved transfer the contents to a graduated cylinder and adjust to a final volume of 500 mL with water. Transfer to a glass bottle and store (*see* **Note 8**).
4. 30% bis/acrylamide (37:5:1).
5. 10% w/v ammonium persulfate (AP): Add 100 mg ammonium persulfate powder to 1 mL water (*see* **Note 9**).
6. N, N, N, N′—tetramethylethylenediamine (TEMED). Store at 4 °C.
7. Water-saturated butanol. Add 100 mL iso-butanol to 500 mL water in a bottle. Shake the bottle and allow phase separation to occur. Repeat shaking step 5×. The top layer will contain the water-saturated butanol.
8. SDS–PAGE running buffer: 25 mM Tris–HCl, 192 mM glycine, 0.1% SDS. Add about 3 L water to a large beaker with a stir bar. Weigh 60.0 g Tris base and add to the beaker while stirring. Weigh 288 g glycine and add to the beaker. Weigh 20.0 g SDS and add to the beaker (*see* **Note 10**). When contents are dissolved, transfer to 20 L carboy and fill to a final volume of 20 L using water (*see* **Note 11**).
9. 3× SDS–PAGE loading dye: 62.5 mM Tris–HCl pH 6.8, 2% SDS, 10% glycerol, bromophenol blue. Utilize a serological pipette to add 9.38 mL stacking SDS–PAGE buffer to a 50-mL conical tube. Weigh 1.5 g SDS and add to a conical tube, shake gently until dissolved. Place the conical tube on a scale and tare. Using a serological pipette, add 9.45 g glycerol to the conical tube. Weigh approximately 5 mg bromophenol blue and add to the conical tube. Bring the final volume to 25 mL in conical tube using water (*see* **Note 12**).

10. Broad range protein molecular weight (MW) ladder for SDS–PAGE.
11. SDS–PAGE gel running apparatus and power supply.
12. 0.2 micron nitrocellulose blotting membrane (*see* **Note 13**).
13. Thick chromatography paper.
14. Semi-Dry Transfer Buffer: 25 mM Tris–HCl, 192 mM glycine, 20% methanol. Add about 2 L water to a large beaker with a stir bar. Weigh 12.2 g Tris base and add to the beaker while stirring. Weigh 57.6 g glycine and add to the beaker while stirring. Use a graduated cylinder to measure 800 mL methanol and add to the beaker while stirring. When the solution is completely dissolved, add water to a final volume of 4 L. Store at 4 °C.
15. Semi-Dry SDS–PAGE Transfer Apparatus: Here, the use of the Bio-Rad Trans-Blot Turbo system is described (*see* **Note 14**).
16. Phosphate-Buffered Saline (PBS): 100 mM NaCl, 4 mM Na_2HPO_4 dibasic, 2 mM NaH_2PO_4 monobasic. Add about 3 L of water to a large beaker with a stir bar. Weigh 116.8 g NaCl and add to the beaker while stirring. Weigh 11.33 g Na_2HPO_4 dibasic and add to the beaker while stirring. Weigh 4.7 g NaH_2PO_4 monobasic and add to the beaker while stirring. When dissolved completely, transfer to 20 L carboy and adjust final volume to 20 L with water.
17. Forceps.
18. Plastic tray for rinsing nitrocellulose membrane.

2.4 Heme Staining

1. Ultrasensitive enhanced chemiluminescent (ECL) substrate with femtogram protein level detection and long stability for heme stain membrane development; here, the use of SuperSignal West Femto Maximum Sensitivity Substrate (Femto) is described (*see* **Note 15**).
2. 15-mL conical tube.
3. Plastic sleeves for membrane development (*see* **Note 16**).
4. Equipment for development of chemiluminescent substrates such as CCD camera/imager or X-ray film developer (*see* **Note 17**).

3 Methods

Carry out all procedures at room temperature, unless otherwise specified. Utilize the sterile technique to prevent contamination of bacterial cultures.

3.1 Induction of Recombinant Cytochrome c Biogenesis Pathway Proteins and Cytochrome c Species

1. Obtain isolated colonies utilizing LB agar plates supplemented with Carb and Cm from frozen stocks of recombinant *E. coli*. Experimental strains containing (1) pGEX plasmid encoding cytochrome *c* biogenesis pathway and (2) pBAD plasmid encoding cytochrome *c*. Negative control strains containing (1) pGEX vector and (2) pBAD plasmid containing cytochrome *c* (*see* **Notes 1–3**, and **18**).
2. Use sterile technique to prepare culture tubes for overnight growth. Aliquot 5 mL of LB broth to a sterile culture tube. Add the appropriate amount of antibiotics to the final concentration of 50 μg/mL carbenicillin and 20.4 μg/mL chloramphenicol (*see* **Notes 19** and **20**).
3. Prepare overnight cultures using sterile technique. Using a sterile loop, select a single, isolated colony obtained in **Step 1** of Subheading 3.1 and inoculate the 5 mL LB broth containing antibiotics from **Step 2** of Subheading 3.1 by swishing the sterile loop in the broth to dislodge the colony. Vortex the culture tube.
4. Grow cultures overnight at 37 °C in a tissue roller set to medium speed for approximately 16–18 h (*see* **Notes 21** and **22**).
5. Use sterile technique to prepare culture tubes for back dilution and induction. Each overnight culture prepared in **Step 3** of Subheading 3.1 will be back diluted into three culture tubes to obtain three technical replicates per strain. Aliquot 5 mL of LB broth to a sterile culture tube. Add the appropriate amount of antibiotics to the final concentration of 50 μg/mL carbenicillin and 20.4 μg/mL chloramphenicol (*see* **Note 23**).
6. Back-dilute overnight cultures with 1:6 dilution. In triplicate, aliquot 1 mL of the saturated overnight culture into the culture tubes prepared in **Step 5** of Subheading 3.1. For each strain of recombinant *E. coli*, there will be three technical replicates (*see* **Note 24**).
7. Grow back-diluted cultures at 37 °C in a tissue roller set to medium speed for 3 h (*see* **Note 21**).
8. Induce the expression of recombinant proteins. Remove culture tubes from the roller and utilize sterile technique to add induction reagents to final concentrations of 0.1 mM IPTG and 0.2% arabinose (*see* **Notes 25** and **26**).
9. Grow strains at 37 °C in a tissue roller set to medium speed for 3 h (*see* **Note 21**).
10. Collect cultures. Transfer the entire 6 mL culture from each tube to a 15-mL conical tube. Collect cell pellet by centrifugation at 5000 *g* for 10 min. Decant supernatant (*see* **Note 27**).
11. Freeze cell pellets overnight at −80 °C (*see* **Notes 28** and **29**).

3.2 Lysis of Bacterial Cells

1. Place 15-mL conical tubes from **Step 11** of Subheading 3.1 in a rack at room temperature.
2. Pipette 200 μL of B-PER reagent into each sample (*see* **Note 30**).
3. Incubate samples at room temperature for 20 min with no mixing.
4. After 20 min incubation, vortex samples to resuspend the cell pellet.
5. Incubate samples at room temperature for 20 min.
6. Prepare and label 1.5-mL microcentrifuge tubes. Three microcentrifuge tubes will be needed per sample.
7. Transfer the sample from 15-mL conical tube to 1.5-mL microcentrifuge tube by pipetting.
8. Pellet cell debris by centrifugation in a microfuge at max speed (approximately 15,000 rpm) for 10 min. A tight cell debris pellet should be evident. The holocytochrome *c* will be in the supernatant (*see* **Note 31**).
9. Transfer supernatant to a new, labeled 1.5-mL microcentrifuge tube by pipetting.
10. Pellet cell debris by centrifugation in a microfuge at max speed (approximately 15,000 rpm) for 10 min (*see* **Note 32**).
11. Transfer supernatant to a new, labeled 1.5-mL microcentrifuge tube by pipetting. This is the sample for analysis and will contain the matured holocytochrome *c*.
12. Determine the total concentration of protein in the sample via UV-vis spectrometric analysis, for example, NanoDrop protein A280 reading (*see* **Notes 33** and **34**).
13. To normalize samples for heme stain analysis, calculate the volume to obtain 50 μg of protein for each sample (*see* **Note 35**).
14. Freeze cell lysates at −80 °C until heme stain analysis is performed (*see* **Note 36**).

3.3 12.5% SDS–PAGE Separation of Bacterial Cell Lysates and Semi-dry Transfer

Gloves are required for all steps in this section.

1. Prepare a gel casting stand. Place a short glass plate flush with the raised sides of a spacer plate to create space in between for gel. Insert plates into the casting frame with a spacer plate to the back and lock closed with latches. Cover the bottom of clear stand holder with gray casting stand gasket and place the bottom of glass plates and green casting frame on top of gray casting stand gasket. Press to lift the handle on the clear stand, insert the glass plate below, and release to clip plates in place.

2. Prepare 12.5% separating gel. In a 15-mL conical tube, mix 2.35 mL 30% bis/acrylamide (37:5:1), 1.4 mL separating buffer, 55.5 μl 10% SDS, 1.95 mL sterile water. Once ready to cast, add 32 μL 10% AP and 2.7 μL TEMED. Mix gently and immediately use (*see* **Note 37**).
3. Cast separating gel. Using a 5-mL serological pipette, gently pipette ~5 mL 12.5% separating gel mixture between the two glass plates. Layer with water-saturated butanol by gently pipetting on top of the separating gel. Allow the gel to set for 35 min or until solidified (*see* **Notes 38** and **39**).
4. Rinse butanol off the separating gel with deionized water and ensure that the space above the gel in between the glass plates is completely dry (*see* **Note 40**).
5. Prepare 5% stacking gel. In a 15-mL conical tube, mix 510 μL 30% bis/acrylamide (37:5:1), 750 μL stacking buffer, 30 μL 10% SDS, 1.71 mL water. Once ready to cast, add 22.5 μL 10% AP and 1.5 μl TEMED. Mix gently and immediately use (*see* **Note 37**).
6. Cast 5% stacking gel. Using a pipette, layer the stacking gel solution on top of the solidified separating gel. Fill over the top of the glass plates and carefully insert well comb into the solution in the space between plates. Allow the gel to set for 35 min (*see* **Note 39**).
7. Take care not to trap bubbles in the wells. Sliding the comb in at a 45° angle can help prevent this.
8. Remove the well comb, place the solidified SDS–PAGE in the electrode assembly and place in the gel running box. Fill the gel running box with the SDS–PAGE running buffer.
9. Prepare B-PER lysate samples for SDS–PAGE separation (*see* **Note 41**). Add 50 μg of protein (calculated in **Step 13** of Subheading 3.2) to a 1.5-mL microcentrifuge tube. Add the appropriate amount of 3X loading dye (*see* **Note 42**).
10. Load MW protein ladder to the first well of SDS–PAGE gel.
11. Load samples prepared in **Step 9** of Subheading 3.3 to wells of the SDS–PAGE gel.
12. Attach the lid to the SDS–PAGE electrode and plug into the power supply.
13. Run SDS–PAGE gel at 85 V for 15 min.
14. Increase to 100 V and run until the dye front reaches the bottom of the gel. Turn off the power supply.
15. Prepare materials for transfer: 1 nitrocellulose membrane cut to the size of the SDS–PAGE gel. Six pieces of chromatography blotting paper cut slightly larger than the nitrocellulose

membrane. Incubate chromatography blotting paper and nitrocellulose membrane in separate containers of transfer buffer prior to beginning the next step (*see* **Note 43**).

16. Remove SDS–PAGE gel from electrode assembly. Separate the glass plates utilizing a spatula to gently pry apart the glass plates. The gel should remain on the glass plates. Remove the stacking gel.
17. Place three sheets of chromatography blotting paper wetted with semi-dry transfer buffer (from **Step 15** of Subheading 3.3) in the Bio-Rad Trans-Blot Turbo transfer cassette box.
18. Using forceps, gently place the nitrocellulose membrane wetted with semi-dry transfer buffer (from **Step 15** of Subheading 3.3) on top of the gel, ensuring the membrane is completely flush with the gel (*see* **Note 44**). Transfer the gel and membrane onto the blotting paper using the spatula to gently pry the gel off the glass plate. The membrane should rest directly on the blotting paper with the gel on top of the membrane.
19. Place three sheets of chromatography blotting paper wetted with semi-dry transfer buffer directly on top of the gel. Use a roller or glass pipette to gently remove any remaining air bubbles. Close the Bio-Rad Trans-Blot Turbo transfer cassette.
20. Place the transfer cassette in the Bio-Rad Trans-Blot Turbo machine (*see* **Note 45**). Turn the machine on. Select the StandardSD protocol (25 V, 1.0 A, 30 M) (*see* **Note 46**). Run program. After the program is complete, remove the transfer cassette and turn off the machine.
21. Remove the membrane from the transfer cassette and place in small plastic container with 15 mL PBS (*see* **Note 47**). Rinse the membrane with gentle shaking for 5 min. Discard PBS and replace with fresh 15 mL PBS and rinse the membrane with gentle shaking for 5 min (*see* **Notes 48** and **49**).

3.4 Heme Staining

1. Prepare CCD camera for chemiluminescent imaging.
2. Remove the membrane from PBS (see **Step 21** of Subheading 3.3) using forceps and gently blot off excess PBS on a paper towel.
3. Place the membrane in the plastic sleeve with the protein side facing upwards. Use forceps.
4. Mix 800 μL of Femto enhancer with 800 μL Femto peroxide in a 15-mL conical tube.
5. Pipette Femto solution from **Step 4** of Subheading 3.4 onto the membrane. Close the plastic sleeve over the membrane to ensure the membrane is evenly covered in Femto solution.

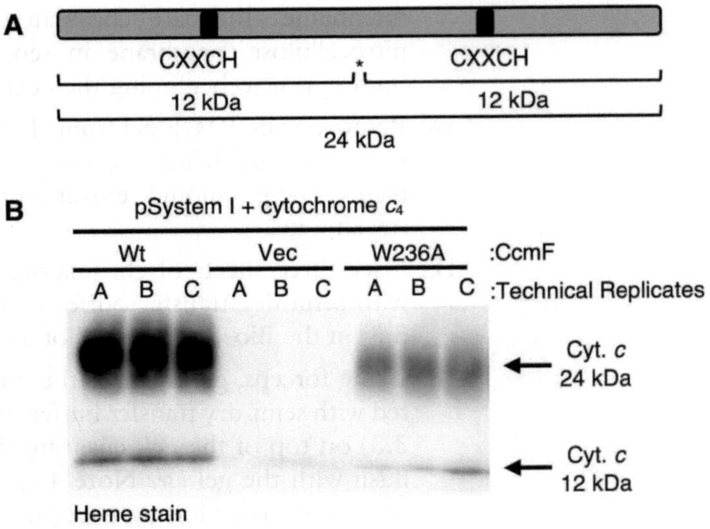

Fig. 1 Heme stain of holocytochrome *c* produced by recombinant *E. coli* assay. (**a**) Schematic of di-heme *B. pertussis* cytochrome c_4. 24 kDa protein contains two CXXCH heme attachment motifs (black boxes). In some instances, this cytochrome is proteolyzed resulting in 2, 12 kDa fragments [13]. (**b**) Heme stain of recombinant production of cytochrome *c*. The *E. coli* System I pathway (pRGK386, GST:CcmABCDE(F:His)GH) and *B. pertussis* cytochrome c_4:His (pRGK332) were recombinantly co-expressed in *E. coli* Δ*ccm*. Holocytochrome c_4 production was monitored by cell lysis and separation via SDS–PAGE; transfer to a nitrocellulose membrane and heme stain. Wt—the complete *E. coli* GST: CcmABCDE(F:His)GH pathway. Vec—pGEX vector. W236A—GST:CcmABCDE(F: His W236A)GH variant displays reduced cytochrome *c* biogenesis. Three technical replicates (A, B, C) per strain

6. Incubate the membrane in plastic sleeve covered in Femto solution for 4 min (*see* **Note 50**).
7. Using a paper towel or roller, remove excess Femto solution from the blot by pressing across the plastic sleeve and blot excess liquid onto a paper towel.
8. Place the membrane in CCD imager and develop according to protocol specific to the CCD imager (*see* **Note 51**). See Fig. 1 for an example of heme stain results.
9. If desired, heme stain can be quantified using the CCD imager software, ImageJ [19], or program of researchers' choice.

4 Notes

1. *E. coli* strain must lack the endogenous System I pathway (CcmABCDEFGH) to ensure that all cytochrome *c* biogenesis is due to recombinant expression from the plasmid. *E. coli* does not express endogenous cytochromes *c* under

aerobic conditions; thus, endogenous cytochromes *c* do not confound the analysis of these experiments. Here, we utilize RK103 (*E. coli* MG1655 Δ*ccm*::kan^R, deleted for all *ccm* genes) [13]. The researcher may substitute any bacterial strain that has the endogenous cytochrome *c* biogenesis genes deleted.

2. Here, we utilize recombinant expression of the *E. coli* System I pathway (pRGK386—pGEX GST:CcmABCDE(F:6XHis)GH [18]); however, this procedure has been successfully performed with recombinant expression of the System II [13, 20–23] and System III [24–27] pathways from different organisms. Thus, the researcher may construct a recombinant expression plasmid from their organism of interest for use in this assay.
3. Here, we utilize recombinant expression of an inducible *Bordetella pertussis* cytochrome c_4:6XHis (pRGK332—pBAD cytochrome c_4:His [13]); however, the substitution of any recombinantly expressed cytochrome *c* of interest is acceptable and many examples exist in the literature, for example [27–33].
4. Here, a pGEX (IPTG-inducible plasmid) and a pBAD (arabinose-inducible plasmid) are used for the recombinant expression of proteins. The use of two different inducing reagents allows for flexibility in induction conditions and timing. If needed, the amount of inducing reagent can be varied to optimize expression levels of holocytochrome *c*.
5. For the negative control strain, the empty vector for the cytochrome *c* biogenesis pathway plasmid (e.g., pGEX vector is used here) must be used to ensure that any cytochrome *c* biogenesis is due to recombinant expression from the plasmid.
6. Ampicillin can be used in place of carbenicillin for the selection of pGEX plasmids.
7. Here, B-PER Bacterial Protein Extraction Reagent is used. This is a commercially available bacterial lysis and protein solubilization solution for fast, efficient cell lysis and protein solubilization in this protocol. However, other single-step lysis reagents or small-scale mechanical cell lysis protocols (e.g., sonication) that efficiently lyse bacterial cells and solubilize proteins could be substituted.
8. SDS is a very fine powder, take proper precautions when weighing and do not aerosolize. Stir at low speed to prevent excess bubble formation. This will take a while to go into solution. If needed, slowly increase the amount of water in the beaker. Do not heat or autoclave.
9. This solution can be made in a larger volume (e.g. 5 mL), aliquoted into microcentrifuge tubes (e.g., 50 μL), and stored at −20 °C.

10. This may take some time to dissolve; if necessary, slowly add more water to the beaker. Stir at low speed to prevent excess bubbles.
11. SDS will make the solution bubbly. Pour slowly, and if necessary, allow solution to settle prior to adding water to a final volume of 20 L.
12. 3× SDS–PAGE loading dye does not contain dithiothreitol (DTT) or beta-mercaptoethanol (BME). These reducing agents will cause a loss of heme signal and the assay will not work.
13. PVDF can be used in place of nitrocellulose and will not affect the outcome of the analysis.
14. Here, we describe a semi-dry transfer to nitrocellulose utilizing the Bio-Rad Trans-Blot Turbo System. However, any standard transfer method and/or apparatus can be used (e.g., semi-dry, wet transfer) and will not affect the outcome.
15. An ultrasensitive enhanced chemiluminescent (ECL) substrate with femtogram protein-level detection and long stability is required for this assay. In this protocol, SuperSignal West Femto Maximum Sensitivity Substrate is used as an example; however, other comparable products with femtogram protein level detection and long stability may be substituted.
16. Any type of development envelope can be used. Examples include plastic wrap or plastic document sleeves cut to the size of the membrane.
17. A CCD camera/imager for chemiluminescent substrates with high sensitivity is recommended. If no CCD camera/imager is available, traditional development of chemiluminescent substrates utilizing X-ray film may be substituted; however, a reduction in sensitivity with film has been documented in the literature [17]. Data shown in Fig. 1 were imaged with the Azure Sapphire Biomolecular Imager.
18. Each experiment should include a negative control (pGEX vector and pBAD cytochrome *c*) to ensure that any heme attachment is due to recombinant cytochrome *c* biogenesis.
19. Sterile technique is essential at this step to prevent contamination. Utilize autoclaved sterilized culture tubes and media. Maintain the sterile field at all times by utilizing a flame.
20. Prepare the appropriate number of culture tubes for your experiment.
21. If a tissue roller is not available, culture tubes can be grown in a shaking incubator at 37 °C and 200 rpm.
22. Bacterial cultures should be grown to saturation prior to back dilution in **Step 6** of Subheading 3.1.

23. For best results, culture tubes prepared in **Step 5** of Subheading 3.1 can be incubated at 37 °C overnight or for a minimum of ~2 h to warm media to 37 °C. This minimizes shock to bacterial cultures upon back-dilution.
24. There is slight variability in the amount of holocytochrome *c* produced in each replicate; thus, the inclusion of technical replicates within each biological replicate is critical to obtain quantitative data.
25. It is recommended that the 1 M stock IPTG prepared in **Step 6** of Subheading 2.1 is diluted to 100 mM prior to induction to allow for accurate pipetting.
26. Increasing the concentration of arabinose *may* increase levels of cytochrome *c* biogenesis. Concentrations up to 0.8% have been reported in the literature (see [15]).
27. To completely decant supernatant, place 15-mL conical tube upside down on a paper towel to wick away residual media. Excess supernatant can freeze on top of cell pellet and is not ideal for cell lysis with B-PER reagent.
28. Freezing of pellets is not necessary, and researchers can proceed directly to Subheading 3.2 if time permits.
29. Cell pellets are stable for up to 1 month at −80 °C.
30. B-PER reagent should be added directly to frozen cell pellets. Cell pellets do not need to thaw prior to the addition of B-PER reagent.
31. In some instances, the lysate will be very viscous and will not form a tight cell pellet upon centrifugation. This can be resolved by the addition of 0.2 μL of 1 U/μL DNase and 1 μL 1 M Mg^{2+} at a final concentration of 1 μM to the lysate. Incubate tubes containing DNase and Mg^{2+} on ice with shaking for 30 min. Repeat **Step 8** of Subheading 3.2, and a tight cell debris pellet should be produced.
32. Multiple clearing centrifugations are required to ensure all cell debris is removed from samples prior to analysis. The cell debris pellet should be barely visible at this step. If a large cell debris pellet is present, repeat **Steps 9–10** of Subheading 3.2.
33. NanoDrop A280 reading can calculate protein concentration based directly on the sample absorbance at 280 or 205 nM and the protein-specific extinction coefficient. This application is appropriate for purified proteins that contain residues of tryptophan (Trp) or tyrosine (Tyr) cysteine disulfide bonds (Cys-Cys) and exhibit absorbance at 280 nM. Contaminants or buffers that absorb ~280 nM might affect protein concentration.

34. When measuring protein mixtures, authors recommend using 1 Abs = 1 mg/mL "sample type."
35. Depending on the expression of the cytochrome *c* of interest more or less protein may be required. This amount can be adjusted after heme staining is performed in Subheading 3.4.
36. Minimize freeze/thaw cycles of cell lysates. Reduction in heme signal will occur upon multiple freeze/thaw cycles. If needed, prepare and store 50 μg aliquots at −80 °C for analysis.
37. Upon the addition of AP and TEMED, gel will immediately begin to solidify. Must work quickly to cast gel before it solidifies.
38. Water-saturated butanol layer ensures the separating gel solidifies evenly. This ensures the interface between the separating and stacking gels is even, leading to clearly resolved protein bands.
39. The authors have found that gels typically solidify in 35 min. To ensure the gel has solidified, a small amount of the separating gel solution, prepared in **Step 2** of Subheading 3.3, can be left in the 15-mL conical tube. This remaining amount will solidify at approximately the same rate as the cast gel and can be used as proxy for the cast gel.
40. Gently place a paper towel between the glass plates and tilt plates to allow the paper towel to wick away water. Repeat if necessary, until all water is removed from between plates.
41. Three technical replicates should be loaded for each recombinant *E. coli* strain, and these were prepared during the back-dilution **Step 6** of Subheading 3.1. If multiple gels are required, ensure that a wild-type control is included on each gel for normalization.
42. Do not boil or heat B-PER lysate samples. Heating will cause loss of heme signal. Samples should be loaded onto SDS–PAGE gel immediately after preparation.
43. Wear gloves at all times when preparing reagents for transfer to prevent contamination from skin proteins.
44. If needed, use the forceps to gently lift and place the membrane on the gel multiple times until there are no visible bubbles. Air bubbles will inhibit transfer in that area of the membrane and negatively impact the results.
45. An alternative transfer method may be substituted based on standard laboratory protocols. For example, wet transfer or semi-dry transfer.
46. If using the Bio-Rad Trans-Blot Turbo machine, an alternative program may be required depending on the molecular weight of the cytochrome *c* being analyzed. Refer to manual.

47. Membrane should be completely covered in PBS. Adjust the volume of PBS if needed.
48. Rinsing of the membrane in PBS removes residual transfer buffer from the membrane and reduces background during membrane development.
49. Membrane may be stored in PBS at 4 °C overnight in a sealed or covered plastic container.
50. Incubation time can be adjusted based on your protein amount and signal. Heme has a natural peroxidase activity; thus, more heme bound will lead to a higher signal intensity during development.
51. Imaging time will vary based up CCD imager. Expect to see heme stain polypeptides in your experimental sample and not in your vector control samples (*see* Fig. 1). If needed, the membrane may be washed in PBS for 2 × 5 min with shaking and re-imaged (follow **Steps 1–8** of Subheading 3.4) to obtain a better exposure. Unlike standard chemiluminescent protocols, heme stains can be repeated and imaged up to three times with minimal changes to signal intensity.

Acknowledgments

We thank Amber L. Grunow for the generation of heme stain in Fig. 1. We thank Robert G. Kranz for the generous use of RK103, pRGK332, and pRGK386. This publication was supported by the National Institute of General Medical Sciences of the National Institutes of Health under Award Numbers R35GM142496 to M.C.S. and T32GM133395 to A.N.K. The content is solely the responsibility of the authors and does not necessarily represent the official views of the National Institutes of Health.

References

1. Bertini I, Cavallaro G, Rosato A (2006) Cytochrome *c* : occurrence and functions. Chem Rev 106:90–115. https://doi.org/10.1021/cr050241v
2. Ow Y-LP, Green DR, Hao Z, Mak TW (2008) Cytochrome c: functions beyond respiration. Nat Rev Mol Cell Biol 9:532–542. https://doi.org/10.1038/nrm2434
3. Kranz RG, Richard-Fogal C, Taylor J-S, Frawley ER (2009) Cytochrome c biogenesis: mechanisms for covalent modifications and trafficking of heme and for heme-iron redox control. Microbiol Mol Biol Rev 73:510–528. https://doi.org/10.1128/MMBR.00001-09
4. Babbitt SE, Sutherland MC, Francisco BS et al (2015) Mitochondrial cytochrome c biogenesis: no longer an enigma. Trends Biochem Sci 40:446–455. https://doi.org/10.1016/j.tibs.2015.05.006
5. Alvarez-Paggi D, Hannibal L, Castro MA et al (2017) Multifunctional cytochrome *c* : learning new tricks from an old dog. Chem Rev 117:13382–13460. https://doi.org/10.1021/acs.chemrev.7b00257
6. Santucci R, Sinibaldi F, Cozza P et al (2019) Cytochrome c: an extreme multifunctional protein with a key role in cell fate. Int J Biol Macromol 136:1237–1246. https://doi.org/10.1016/j.ijbiomac.2019.06.180

7. Sanders C, Turkarslan S, Lee D-W, Daldal F (2010) Cytochrome c biogenesis: the Ccm system. Trends Microbiol 18:266–274. https://doi.org/10.1016/j.tim.2010.03.006
8. Barker PD, Ferguson SJ (1999) Still a puzzle: why is haem covalently attached in c-type cytochromes? Structure 7:R281–R290
9. Allen JWA, Barker PD, Daltrop O et al (2005) Why isn't "standard" heme good enough for c-type and d1-type cytochromes? Dalton Trans:3410–3418
10. Verissimo AF, Daldal F (2014) Cytochrome c biogenesis system I: an intricate process catalyzed by a maturase supercomplex? Biochim Biophys Acta 1837:989–998. https://doi.org/10.1016/j.bbabio.2014.03.003
11. Mavridou DAI, Ferguson SJ, Stevens JM (2013) Cytochrome c assembly. IUBMB Life 65:209–216. https://doi.org/10.1002/iub.1123
12. Hüttemann M, Pecina P, Rainbolt M et al (2011) The multiple functions of cytochrome c and their regulation in life and death decisions of the mammalian cell: From respiration to apoptosis. Mitochondrion 11:369–381. https://doi.org/10.1016/j.mito.2011.01.010
13. Feissner RE, Richard-Fogal CL, Frawley ER et al (2006) Recombinant cytochromes c biogenesis systems I and II and analysis of haem delivery pathways in Escherichia coli. Mol Microbiol 60:563–577. https://doi.org/10.1111/j.1365-2958.2006.05132.x
14. Richard-Fogal CL, Frawley ER, Feissner RE, Kranz RG (2007) Heme concentration dependence and metalloporphyrin inhibition of the system I and II cytochrome c assembly pathways. J Bacteriol 189:455–463. https://doi.org/10.1128/JB.01388-06
15. San Francisco B, Sutherland MC, Kranz RG (2014) The CcmFH complex is the system I holocytochrome c synthetase: engineering cytochrome c maturation independent of CcmABCDE. Mol Microbiol 91:996–1008. https://doi.org/10.1111/mmi.12510
16. Sutherland MC, Jarodsky JM, Ovchinnikov S et al (2018) Structurally mapping endogenous heme in the CcmCDE membrane complex for cytochrome c biogenesis. J Mol Biol 430:1065–1080. https://doi.org/10.1016/j.jmb.2018.01.022
17. Feissner R, Xiang Y, Kranz RG (2003) Chemiluminescent-based methods to detect subpicomole levels of c-type cytochromes. Anal Biochem 315:90–94
18. Richard-Fogal CL, Frawley ER, Bonner ER et al (2009) A conserved haem redox and trafficking pathway for cofactor attachment. EMBO J 28:2349–2359. https://doi.org/10.1038/emboj.2009.189
19. Rasband WS (1997) ImageJ. US National Institutes of Health, Bethesda. http://imagej.nih.gov/ij/
20. Frawley ER, Kranz RG (2009) CcsBA is a cytochrome c synthetase that also functions in heme transport. Proc Natl Acad Sci 106:10201–10206. https://doi.org/10.1073/pnas.0903132106
21. Kern M, Scheithauer J, Kranz RG, Simon J (2010) Essential histidine pairs indicate conserved haem binding in epsilonproteobacterial cytochrome c haem lyases. Microbiology 156:3773–3781. https://doi.org/10.1099/mic.0.042838-0
22. Kern M, Eisel F, Scheithauer J et al (2010) Substrate specificity of three cytochrome c haem lyase isoenzymes from Wolinella succinogenes: unconventional haem c binding motifs are not sufficient for haem c attachment by NrfI and CcsA1. Mol Microbiol 75:122–137. https://doi.org/10.1111/j.1365-2958.2009.06965.x
23. Sutherland MC, Tran NL, Tillman DE et al (2018) Structure-function analysis of the bifunctional CcsBA heme exporter and cytochrome *c* synthetase. mBio 9:e02134–e02118. https://doi.org/10.1128/mBio.02134-18
24. San Francisco B, Bretsnyder EC, Kranz RG (2013) Human mitochondrial holocytochrome c synthase's heme binding, maturation determinants, and complex formation with cytochrome c. Proc Natl Acad Sci 110:E788–E797. https://doi.org/10.1073/pnas.1213897109
25. Babbitt SE, Hsu J, Kranz RG (2016) Molecular basis behind inability of mitochondrial holocytochrome *c* synthase to mature bacterial cytochromes: defining a critical role for cytochrome c α Helix-1. J Biol Chem 291:17523–17534. https://doi.org/10.1074/jbc.M116.741231
26. Mendez DL, Babbitt SE, King JD et al (2017) Engineered holocytochrome *c* synthases that biosynthesize new cytochromes *c*. Proc Natl Acad Sci 114:2235–2240. https://doi.org/10.1073/pnas.1615929114
27. Espino-Sanchez TJ, Wienkers H, Marvin RG et al (2023) Direct tests of cytochrome *c* and c_1 functions in the electron transport chain of malaria parasites. Proc Natl Acad Sci USA 120:e2301047120. https://doi.org/10.1073/pnas.2301047120

28. Pollock WBR, Rosell FI, Twitchett MB et al (1998) Bacterial expression of a mitochondrial cytochrome *c*. Trimethylation of Lys72 in yeast *iso* -1-cytochrome *c* and the alkaline conformational transition. Biochemistry 37:6124–6131. https://doi.org/10.1021/bi972188d
29. Jensen HM, Albers AE, Malley KR et al (2010) Engineering of a synthetic electron conduit in living cells. Proc Natl Acad Sci USA 107: 19213–19218. https://doi.org/10.1073/pnas.1009645107
30. Richard-Fogal CL, San Francisco B, Frawley ER, Kranz RG (2012) Thiol redox requirements and substrate specificities of recombinant cytochrome c assembly systems II and III. Biochim Biophys Acta 1817:911–919. https://doi.org/10.1016/j.bbabio.2011.09.008
31. Gupta D, Sutherland MC, Rengasamy K et al (2019) Photoferrotrophs produce a PioAB electron conduit for extracellular electron uptake. mBio 10:10.1128/mBio.02668-19
32. Su L, Fukushima T, Ajo-Franklin CM (2020) A hybrid cyt c maturation system enhances the bioelectrical performance of engineered Escherichia coli by improving the rate-limiting step. Biosens Bioelectron 165:112312. https://doi.org/10.1016/j.bios.2020.112312
33. Keffer JL, McAllister SM, Garber AI et al (2021) Iron oxidation by a fused cytochrome-porin common to diverse iron-oxidizing bacteria. mBio 12:e01074–e01021. https://doi.org/10.1128/mBio.01074-21

Chapter 11

Ultra-Performance Liquid Chromatography (UPLC) Analysis of Heme Biosynthesis Intermediates

Hector A. Bergonia and John D. Phillips

Abstract

The utilization of ultra-performance liquid chromatography (UPLC) to analyze the various intermediates in the heme biosynthetic pathway is presented. The first product, ALA, was derivatized to a highly fluorescent pyrrolizine; PBG, the second intermediate, was enzymatically converted to uroporphyrinogen, and all the porphyrinogen intermediates were oxidized in acid to form fluorescent porphyrins. Heme was measured as hemin. The stable porphyrin forms of the intermediates, are then resolved and quantified by UPLC. Further details about the various methods are discussed to promote successful UPLC analyses. Method variations that may be preferable in certain situations are also presented.

Key words UPLC (ultra-performance liquid chromatography), ALA (∂-aminolevulinic acid), PBG (porphobilinogen), Porphyrins, Heme, Porphyria

1 Introduction

Much of the work done on heme biosynthesis was in the early twentieth century, Dr. Hans Fisher received the Nobel Prize in 1930 for his work defining the chemical pathway of heme biosynthesis [1]. There has been a great deal of work over the last century on the metabolic regulation of heme biosynthesis since the pathway intermediates have cellular toxicity when they accumulate in high amounts. Ultra-performance liquid chromatography (UPLC) is the premier analytical technique used to separate and quantify chemical substances of interest in complex mixtures, specifically in the case of porphyrins due to the specificity, sensitivity, and speed of analysis. This has been made possible through a combination of several advancements in chromatography technology from the traditional HPLC (high-performance liquid chromatography). UPLC uses smaller columns with smaller particle sizes (≤2 μm versus ≥15 μm), but with a much greater surface area per unit column volume for analyte separations, thus allowing greater efficiency and

Oleh Khalimonchuk (ed.), *Iron Metabolism: Methods and Protocols*, Methods in Molecular Biology, vol. 2839,
https://doi.org/10.1007/978-1-0716-4043-2_11,

higher resolution. The solvent delivery system has been engineered to handle the much higher backpressures that arise from the more densely packed columns. Column heaters are often employed to reduce the viscosities of the eluting solvents so that higher flow rates can be achieved. Column materials have also evolved; the support matrices used in UPLC columns are specifically designed to be more chemically inert and to withstand the elevated pressure, temperature, and greater pH range of solvents. All of the above improvements in the mechanics and chemistry of solute separation combined with advances in the various detectors' sensitivity and range have been major advancements. UPLC systems thus provide greater sensitivity, resolution, and speed of compound quantitation compared to traditional systems [2].

The UPLC system in combination with a PDA (photodiode array detector) and a fluorescence detector has been extremely useful in studying the heme biosynthetic pathway. Since heme is an essential requirement for life [3, 4], the ability to synthesize or acquire it is, therefore, paramount to the fitness of an organism. The mammalian heme biosynthetic pathway consists of eight dedicated enzyme-catalyzed steps starting with the production of ALA (∂-aminolevulinic acid) [5–7]. Like its predecessor HPLC, UPLC is now routinely used to quantify the various intermediates of the heme biosynthetic pathway, up to the final product heme, as listed in Table 1. In this table, the steps in the heme biosynthesis pathway are named after the enzyme that catalyzes the chemical transformation of the substrate to the product. The initial two steps of the pathway form pyrroles that are then combined to form a tetrapyrrole, the basic building block of heme. The next five steps of the pathway involve intermediates that are porphyrinogens, the reduced forms of the tetrapyrrole. Hydroxymethylbilane is used by uroporphyrinogen III synthase to generate the III-isomer, uroporphyrinogen III. Hydroxymethylbilane can also circularize nonenzymatically to form uroporphyrinogen I, in a rapid reaction. The final two steps are the oxidation of the tetrapyrrole and the insertion of iron into the macrocycle. While it is not listed in Table 1, under normal conditions, a small amount of zinc (II) protoporphyrin IX (ZnPPIX) can be formed instead of heme (iron (II) protoporphyrin IX), where zinc replaces iron in the final step [8]. Each of the pathway intermediates has toxicities associated with it, and as such, evolutionarily there is minimal accumulation of any heme precursor intermediate. In humans, any genetic or biosynthetic process abnormality that occurs along the pathway usually leads to a type of porphyria. The phenotype of the porphyria depends on which pathway intermediate is present in excess and the extent of the excess metabolite that has accumulated. Hence, the ability to quantify any intermediate along the heme biosynthetic pathway is essential in studying the direct and indirect genetic and environmental factors that can affect heme production.

Table 1
UPLC detection methods of intermediates in heme biosynthesis

Heme pathway step (Enzyme)	Substrate	Intermediate product	UPLC detection of product
1. Aminolevulinic acid synthase (ALAS)	Glycine, succinyl CoA	∂-Aminolevulinic acid (ALA)	Derivative fluorescence
2. ALA dehydratase (ALAD)	ALA	Porphobilinogen (PBG)	Porphyrin fluorescence
3. Porphobilinogen deaminase (PBGD)	PBG	Hydroxymethylbilane, (HMB)	Porphyrin fluorescence
4. Uroporphyrinogen III synthase (UROS)	HMB	Uroporphyrinogen III	Porphyrin fluorescence
5. Uroporphyrinogen decarboxylase (UroD)	Uroporphyrinogen	7-, 6-, 5-, and 4-carboxylporphyrinogen	Porphyrin fluorescence
6. Coproporphyrinogen oxidase (COPOX)	Coproporphyrinogen III	Protoporphyrinogen IX (PPIX)	Porphyrin fluorescence
7. Protoporphyrinogen oxidase (PPOX)	Protoporphyrinogen IX	PPIX	Porphyrin fluorescence
8. Ferrochelatase (FECH)	PPIX	Heme	Absorbance at 398 nm

For practicable UPLC quantification, heme biosynthesis intermediates need to be modified to stable and more easily detectable chemical forms. To do this, ALA is derivatized to a fluorescent product, 2,6-diacetyl-1,5-dimethyl-7-(2-carboxyethyl)-3- pyrrolizine [9]. Porphobilinogen (PBG) is enzymatically converted to hydroxymethylbilane using porphobilinogen deaminase (PBGD). Hydroxymethylbilane nonenzymatically cyclizes to uroporphyrinogen I [10–12], which is then oxidized to uroporphyrin I. The reason the porphyrinogens are oxidized is because they are difficult to measure by absorbance; however, once the macrocycle is oxidized, it becomes fluorescent and its fluorescence is easily detected and highly sensitive (Fig. 1). The porphyrinogen intermediates are easily oxidized upon exposure to oxygen and UV light to give rise to the corresponding porphyrins. The process is made faster under acidic conditions. The porphyrin intermediates of the next few heme biosynthesis steps are very specific and are easily quantified in a fluorescence detector. Heme is nonfluorescent due to the paramagnetic (having unpaired electrons) property of iron (II) or iron (III) quenching of the fluorescence [13]. However, ZnPPIX is fluorescent because the full d electron shell of zinc (II) that makes it diamagnetic is not able to quench the fluorescence of the oxidized macrocycle [14, 15]. Heme readily oxidizes upon exposure to air

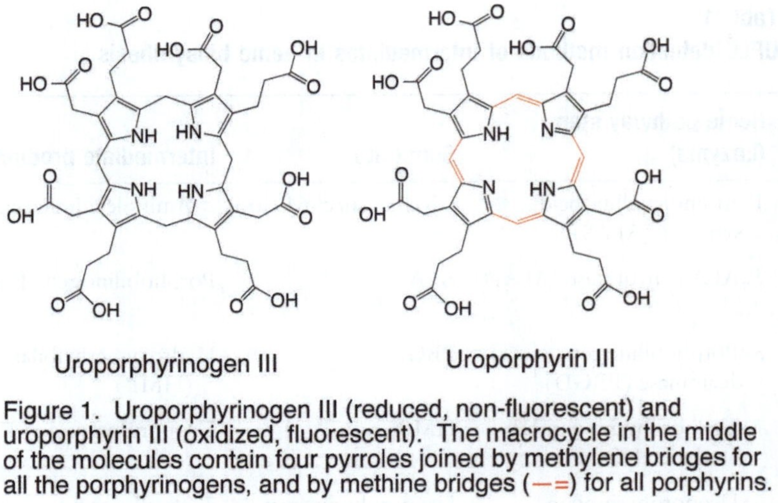

Fig. 1 Uroporphyrinogen III (reduced, nonfluorescent) and uroporphyrin III (oxidized, fluorescent). The macrocycle in the middle of the molecules contains four pyrroles joined by methylene bridges for all of the porphyrinogens, and by methine bridges (– =) for all porphyrins after 6-electron oxidation

during sample processing but remains nonfluorescent. It is thus analyzed as iron (III) protoporphyrin IX and is detected using absorbance readings from the PDA.

2 Materials

All reagents are analytical grade. Solvents are HPLC grade. Water is ultrapure or 18 MΩ-cm.

1. 50 mM potassium phosphate, pH 7.4 (*see* **Note 1**).
2. BCA protein assay kit.
3. ALA derivatization agent (DA): Mix water, 37% formaldehyde, ethanol, and acetylacetone in a ratio of 107:5:15:23 by volume (*see* **Note 2**).
4. 100 mM Tris–HCl, pH 7.6.
5. PBGD in 100 mM Tris–HCl, pH 7.6 (*see* **Note 3**).
6. 3 M HCl aqueous (*see* **Note 4**).
7. EA extraction solvent: Mix four volumes of ethyl acetate to one volume of glacial acetic acid.
8. NES 2x extraction solvent: 0.2 N NaOH, 6 mM EDTA, 4% m/v SDS. Completely dissolve EDTA in NaOH before adding SDS to ensure complete dissolution.
9. 0.2% v/v formic acid in water. Add 2 mL concentrated formic acid to 1 L water.

10. 0.2% v/v formic acid in methanol. Add 2 mL concentrated formic acid to 1 L methanol.
11. 1 M ammonium acetate, pH 5.16. Dissolve 77 g ammonium acetate into about 800 mL water. Adjust the pH to 5.16 with glacial acetic acid. Add more water to the final 1 L volume. Put the solution through a 0.2-micron glass fiber filter. Always use a freshly prepared solution.
12. 100 mM ammonium acetate.
13. Acetonitrile.
14. Methanol.

3 Methods

3.1 Sample Homogenization: Homogenize All Nonfluid Samples

1. Cultured mammalian cells. Resuspend about 100 μL pellets of washed cells in about four volumes of 50 mM potassium phosphate, pH 7.4. Homogenize the suspension by microtip sonication with five cycles of 3 s *on* then 3 s *off* at 20% amplitude setting of a 500-watt sonicator while in an ice bath (*see* **Note 5**).
2. Mammalian tissues. Put about 100 mg of tissue into a 2-mL bead mill tube preloaded with 1.4 mM ceramic beads. Add 400 μL of 50 mM potassium phosphate, pH 7.4. Agitate the mixture in a bead mill shaker for 30 s (*see* **Note 6**).

3.2 BCA Protein Determination

1. Mix 50 volumes of BCA Protein Reagent A with one volume of BCA Protein Reagent B.
2. Dispense 200 μL aliquots onto a clear bottom 96-well plate.
3. Dilute the homogenate sample with water to three different concentrations, like 1/20, 1/40, and 1/60.
4. Add 10 μL of the diluted homogenate to a well. Analyze each dilution in duplicate.
5. Make a standard curve series in 0.1 mg/mL increments from 0.0 to 0.8 mg/mL BSA in water and analyze simultaneously, preferably on the same microplate.
6. Incubate at 37 °C for 30 min, let cool, and stabilize to room temperature for 15 min.
7. Measure the absorbance at 562 nM in a microplate reader (*see* **Note 7**).

3.3 ALA Derivatization

1. Dilute the homogenate to about 2 mg/mL with water.
2. Mix 50 μL of the resulting mixture with 150 μL DA.
3. Incubate at about 100–103 °C for 5.0 min.
4. Cool immediately in an ice bath for at least 1 h in the dark.

5. Centrifuge for 10 min at 16,000 g in a microfuge at 4 °C.
6. Take the supernatant containing the fluorescent ALA derivative for UPLC analysis. Spike extra homogenate with known amounts of ALA to make the standard curve (*see* **Note 8**).

3.4 Enzymatic Deamination of Porphobilinogen

1. To 40 μL homogenate, add 200 μL of reaction mixture containing 100 mM Tris–HCl (pH 7.6), 120 μM succinylacetone, and about 50 μg/mL purified PBGD (*see* **Note 9**).
2. Incubate at 37 °C for 0.5 h. Add 80 μL 6 M HCl to stop the reaction and precipitate the proteins.
3. Mix well and let stand under UV light for 30 min to ensure total oxidation of all the product uroporphyrinogen to uroporphyrin.
4. Centrifuge at 16,000 g for 10 min at room temperature.
5. Analyze porphyrins using UPLC. For background porphyrin that may be present in the PBGD stock, analyze the enzyme that has been deactivated by heating in a boiling water bath for 10 min.

3.5 Porphyrins and Porphyrin Isomers

1. Add the same volume of 3 M HCl to the homogenate.
2. Incubate at 37 °C for 0.5 h or at room temperature for 2 h (*see* **Note 10**).
3. Centrifuge at 16,000 g for 10 min at room temperature.
4. Analyze the supernatant using UPLC.

3.6 Protoporphyrin

1. Add 200 μL of EA to 50 μL of homogenate.
2. Vortex for 60 s. Spin at 16,000 g for 1 min at room temperature.
3. Analyze the supernatant using UPLC.

3.7 Heme

1. Add the same volume of NES 2x to homogenate (*see* **Note 11**).
2. Vortex the mixture for 60 s.
3. Add 4 volumes of acetonitrile and mix for another 60 s.
4. Centrifuge at 16,000 g for 10 min at room temperature.
5. Analyze the supernatant using UPLC.

3.8 UPLC Instrument Settings

All the instrument settings for the UPLC analyses described here are performed using a Waters Acquity UPLC system, which includes a binary solvent manager, sample manager, a fluorescence detector, a PDA detector, a column heater, and an Acquity UPLC BEH C18, 1.7 μM, 2.1 Å 100 mM column (*see* **Note 12**). The usual volume of prepared sample injected into the system is 10 μL.

1. ALA: Set the fluorescence detector at 370 nM excitation and 460 nM emission. Keep the sample chamber dark and at 4 °C. Keep the flow rate constant at 0.3 mL/min and the column at 50 °C for a total run time of 11 min. Designate 0.2% formic acid in water as Solvent A and 0.2% formic acid in methanol as Solvent B, the total being 100%. Use the following elution gradient schedule with the percent Solvent A at each successive step is as follows: 0 min, 80%; 5 min, 55%; 5.5 min, 1%; 7.5 min, 1%; and 8 min, 80%. Set the gradient for solvent composition from 0 to 5 min at Waters Gradient 5 (convex, with a higher slope at 0 min compared to that at 5 min), and that from 5.0 to 5.5 min at Waters Gradient 7, concave. Keep all other composition gradients linear. Inject 10 μL of supernatant containing the derivatized ALA into the UPLC.
2. Porphyrins: Set the fluorescence detector at 404 nM excitation and 618 nM emission. Keep the samples at room temperature. Set the flow rate at 0.4 mL/min and the column at 60 °C for the total run time of 8 min. The 0.2% formic acid in water is Solvent A while 0.2% formic acid in methanol is Solvent B. The gradient schedule with the percent Solvent A at each step is as follows: 0 min, 60%; 4 min, 20%; 4.1 min, 1%; 5.5 min, 1%; 5 min, 60%. Set all gradient steps as linear except for the one at 0–4 min which is Waters Gradient 5 (convex).
3. Porphyrin isomers: Set the fluorescence detector at 404 nM excitation and 630 nM emission. Keep the samples at room temperature. Set the flow rate at 0.3 mL/min and the column at 60 °C for the total run time of 12 min. Here Solvent A is 1 M ammonium acetate pH 5.16, while methanol is Solvent B. The linear gradient schedule with the percent Solvent A at each step is as follows: 0 min, 75%; 6 min, 35%; 7 min, 1%; 8.5 min, 1%; 9 min, 80% (*see* **Note 13**).
4. 4. Protoporphyrin: Set the fluorescence detector at 404 nM excitation and 630 nM emission. Keep the samples at room temperature. Set the flow rate at 0.4 mL/min and the column at 60 °C for the total run time of 8 min. Again 0.2% formic acid in water is Solvent A while 0.2% formic acid in methanol is Solvent B. The linear gradient schedule with the percent Solvent A at each step is as follows: 0 min, 80%; 2.5 min, 1%; 4.5 min, 1% and 5 min, 20%.
5. Heme: Set the PDA detector to collect absorbance readings at 398 nM throughout the chromatographic run. Keep the samples at room temperature. Use 100 mM ammonium acetate as Solvent A and methanol as Solvent B. Use the same solvent gradient progression as for protoporphyrin (*see* **Note 14**).

4 Notes

1. Different buffers close to pH 7 and up to 100 mM may be used for homogenization without significantly affecting the methods used to quantify most of the intermediates. The exception is ALA, where buffer salt concentration needs to be below 5 mM to avoid interference with color development during derivatization [9]. Buffer solutions can become more important if the sample homogenate were to be measured for something else, such as enzyme activity. Water is also a convenient homogenization medium since the appropriate buffer can be added later for a different kind of measurement.
2. The mixture needs to be warmed slightly, as in hand warmth, to render the components soluble.
3. PBGD has been purified from R. spheroids [16]. It may also be cloned for protein expression.
4. The final concentration of HCl in the acidified sample must be around 1.5 M for efficient separation from components that may bind to porphyrins and to allow maximal extraction and fluorescence [17]. One volume of 6 M HCl may be added to three volumes of dilute and low protein samples to reduce dilution of the analyte porphyrins.
5. For other pellet volumes, approximate the volume of the packed cell pellet by visually comparing it with a known volume of water in an identical container; this is almost always a 1.5 mL microfuge tube. The amount of buffer added can range from three to four volumes of the cell pellet. An ice bath during sonication may be necessary if an enzyme analysis is also performed. Use the highest power setting that will not splatter the sample suspension. Homogenization can be checked by smearing about 5 μL on a slide and putting this under a microscope to look for intact cells. More sonication cycles may be performed as needed. For bacterial cells, 12 or more sonication cycles may be needed, and an ice bath is needed to keep the sample from overheating and denaturing the protein.
6. Soft tissues may be sonicated at a higher setting than that for cells. Tissues may also be broken down using a Potter Elvehjem homogenizer with a glass tube and a PTFE pestle. The pestle is turned using a hand drill set at maximum speed. The 2-mL size homogenizer can go down to 500 μL total volume of tissue plus buffer. Six 8-second cycles of up and down strokes are sufficient. Another alternative is the stainless steel Polytron or Tekmar homogenizer, which can process a total volume as low as 1 mL. About 60 s at maximum speed is sufficient to break up tough tissue material to uniformity.

7. The contents of the microplate wells may not have been mixed thoroughly enough before incubation; hence, the color development in the wells may not have been uniform enough. Therefore, ensure that the microplate reading cycle has an active automix function that activates for a few seconds before actually making measurements. Repeat reading at least three times using the automixer until the individual coefficients of variation between duplicates stop changing significantly. The dilutions may be changed to suit the range of concentrations encountered. In this method, the linearity of the standard curve starts to degrade above 0.8 mg/mL. This protein quantification allows normalization of the results of the analyses of the intermediates to per milligram of sample protein. Also, assays of the pathway intermediates tend to give directly proportional results when done at different protein concentrations below 10 mg/mL.
8. Standard curves can be prepared to be applicable to a wide range of values from close to 5 nM to over 50 μM, depending on the sample.
9. The amount of PBGD can vary depending on the batch preparation. A separate experiment may be done by incubating 10 μM PBG (the intended maximum value on the standard curve) with increasing amounts of PBGD for 30 min until the amount of uroporphyrinogen formed does not increase with additional enzyme present. The product in the PBG assay is analyzed by UPLC the same way as for porphyrins. However, it is not quantified as uroporphyrin and calculated back to PBG although four PBG molecules are used up to make one uroporphyrin. Instead, the total porphyrin produced is correlated to the corresponding known PBG concentration on the standard curve.
10. All porphyrins are usually oxidized and hence fluorescent after sample processing. If in doubt, expose to UV light for 0.5 h or to direct fluorescent lighting for 2 h at room temperature.
11. The NES 2x extraction [18] ensured solubilization of the heme samples where it is difficult to extract by EA [19]. The addition of acetonitrile enhanced the precipitation of proteins, but this step may even be omitted for more dilute samples. Hemin may be heated with oxalic acid to release the central iron atom and produce fluorescent PPIX [20], but we discovered that the treatment also modified the vinyl side chains such that more fluorescent peaks besides that of PPIX were detected by UPLC.
12. UPLC systems are rarely identical among different laboratories. Thus, a method adapted from another laboratory often necessitates modifications to achieve the desired elution behavior of the analyte under consideration. The methods

enumerated here began as modifications of published methods and then evolved to become even very different versions as further experimentation, and optimizations were accumulated and more published biochemical considerations and modifications are included, together with technological changes and advancements of equipment used. These current methods started as HPLC methods for porphyrins [10]. These included 8-, 7-, 6-, 5-, and 4-carboxylporphyrin (uroporphyrin, heptacarboxylporphyrin, hexacarboxylporphyrin, pentacarboxylporphyrin, and coproporphyrin, respectively). The same thing happened to the methods for porphyrin isomers [21], protoporphyrin IX, and heme [19]. The PDA module in the UPLC system is usually set to measure absorbance along with the fluorescence detector. The PDA data collected can be used to verify the identity of the fluorescent analyte peak, if present in amounts high enough to be measured for absorbance.

13. Porphyrin isomers refer to isomers I and III for the 8-, 7-, 6-, 5-, and 4-carboxylporphyrins. Both isomers are converted to coproporphyrinogen by uroporphyrinogen decarboxylase, but only isomer III is acted upon by coproporphyrinogen III synthase and is able to proceed down the rest of the heme synthesis pathway. Since isomer III standards for a few of these porphyrins are difficult to obtain commercially and the fluorescence behaviors of both isomers for uroporphyrin and coproporphyrin are almost the same, the III isomers are quantified based in the corresponding I isomer fluorescence, albeit at a different elution time.
14. PPIX fluorescence can also be measured simultaneously with hemin absorbance. The same thing can be done with zinc protoporphyrin IX (ZnPPPIX), but with fluorescence detection set at 406 nM excitation and 586 nM emission. Although the fluorescence detector can be programmed to alternate between PPIX and ZnPPIX fluorescence settings, this drastically reduces the number of readings from the preferred 20 per second to just one per second. It is, therefore, not practical to analyze all three at the same time in the UPLC run at low analyte concentrations.

Acknowledgments

This work is done at the Iron and Heme Core facility at the University of Utah Center for Iron and Heme Disorder and was supported by a grant from the NIH National Institute of Diabetes and Digestive and Kidney Diseases, Grant number U54DK110858.

References

1. Fischer H (1966) On haemin and the the relationships between haemin and chlorophyll. Elsevier. Available via Hans Fischer—Nobel Lecture. NobelPrize.org. Nobel Prize Outreach AB 2023. https://www.nobelprize.org/uploads/2018/06/fischer-lecture-3.pdf
2. Kumar A, Saini G, Nair A, Sharma R (2012) UPLC: a preeminent technique in pharmaceutical analysis. Acta Pol Pharm 69(3):371–380
3. Dailey HA, Medlock AE (2022) A primer on heme biosynthesis. Biol Chem 403(11–12):985–1003. https://doi.org/10.1515/hsz-2022-0205
4. Celis AI, DuBois JL (2019) Making and breaking heme. Curr Opin Struct Biol 59:19–28. https://doi.org/10.1016/j.sbi.2019.01.006
5. Phillips JD (2019) Heme biosynthesis and the porphyrias. Mol Genet Metab 128(3):164–177. https://doi.org/10.1016/j.ymgme.2019.04.008
6. Deacon AC, Elder GH (2001) ACP Best Practice No 165: front line tests for the investigation of suspected porphyria. J Clin Pathol 54(7):500–507. https://doi.org/10.1136/jcp.54.7.500
7. Bissell DM, Anderson KE, Bonkovsky HL (2017) Porphyria. N Engl J Med 377(9):862–872. https://doi.org/10.1056/NEJMra1608634
8. Labbe RF, Vreman HJ, Stevenson DK (1999) Zinc protoporphyrin: a metabolite with a mission. Clin Chem 45(12):2060–2072
9. Bergonia HA, Franklin MR, Kushner JP, Phillips JD (2015) A method for determining delta-aminolevulinic acid synthase activity in homogenized cells and tissues. Clin Biochem 48(12):788–795. https://doi.org/10.1016/j.clinbiochem.2015.04.023
10. Phillips JD, Kushner JP (1999) Measurement of uroporphyrinogen decarboxylase activity. In: Maines MD, Costa LG, Reed DJ, Sassa S, Sipes IG (eds) Current Protocols in Toxicology. Wiley, New York, pp 8.4.1–8.4.3
11. Straka JG, Kushner JP, Pryor MA (1982) Uroporphyrinogen decarboxylase. A method for measuring enzyme activity. Enzyme 28(2–3):170–185
12. Bishop DF, McBride L, Desnick RJ (1982) Fluorometric coupled-enzyme assay for delta-aminolevulinate synthase. Enzyme 28(2–3):94–108. https://doi.org/10.1159/000459094
13. Volchkov VV, Ivanov VL, Uzhinov BM (2010) Induced intersystem crossing at the fluorescence quenching of laser dye 7-amino-1,3-naphthalenedisulfonic acid by paramagnetic metal ions. J Fluoresc 20(1):299–303. https://doi.org/10.1007/s10895-009-0555-y
14. Hopf FR, Whitten DG (1975) Photochemistry of porphyrins and metalloporphyrins. In: Smith KM (ed) Porphyrins and metalloporphyrins. Elsevier, Amsterdam, pp 667–700
15. Yang W, Chen X, Su H, Fang W, Zhang Y (2015) The fluorescence regulation mechanism of the paramagnetic metal in a biological HNO sensor. Chem Commun (Camb) 51(47):9616–9619. https://doi.org/10.1039/c5cc00787a
16. Jordan PM, Shemin D (1973) Purification and properties of uroporphyrinogen I synthetase from Rhodopseudomonas spheroides. J Biol Chem 248(3):1019–1024
17. Perkins SL, Johnson PM (1989) Loss of porphyrins from solution during analysis: effect of sample pH and matrix on porphyrin quantification in urine by "igh-performance" liquid chromatography. Clin Chem 35(7):1508–1512
18. Deroost K, Lays N, Noppen S, Martens E, Opdenakker G, Van den Steen PE (2012) Improved methods for haemozoin quantification in tissues yield organ-and parasite-specific information in malaria-infected mice. Malar J 11:166. https://doi.org/10.1186/1475-2875-11-166
19. Peter F, Growcock G, Strunc G (1978) Fluorometric determination of erythrocyte protoporphyrin in blood, a comparison between direct (hematofluorometric) and indirect (extraction) methods. Clin Chem 24(9):1515–1517
20. Morrison GR (1965) Fluorometric microdetermination of heme protein. Anal Chem 37:1124–1126. https://doi.org/10.1021/ac60228a014
21. Lim CK, Li FM, Peters TJ (1988) High-performance liquid chromatography of porphyrins. J Chromatogr 429:123–153. https://doi.org/10.1016/s0378-4347(00)83869-4

Chapter 12

Radiolabeling Heme and Porphyrin with ^{14}C-Glycine or ^{14}C δ-Aminolevulinic Acid

Diane M. Ward, Samantha Gillis, and Yvette Y. Yien

Abstract

Radiolabeling enables the quantitation of newly synthesized heme and porphyrin, allowing us to distinguish heme synthesis rates from total cellular heme. Here, we describe a protocol for labeling heme with ^{14}C-glycine or ALA and the sequential extraction of heme and porphyrin from the same samples for quantitation by liquid scintillation.

Key words Heme, Porphyrin, Radioactivity, Glycine, Metabolism

1 Introduction

Radiolabeling techniques to quantitate de novo heme and porphyrin synthesis have played an essential role in advancing studies of iron metabolism. Even with current advances in mass spectrometry-based metabolomics techniques, the extraction and detection of porphyrins in a mix of cellular metabolites remains challenging for several reasons. First, porphyrins are light-sensitive, requiring handling under limited lighting conditions. Second, heme and porphyrin are miscible in lipid and are solubilized under similar conditions as lipids, which are vastly more abundant in the cell; the low concentration of heme/porphyrin in comparison to other metabolites causes numerous challenges in mass spectrometric detection. Non-isotopic methods such as HPLC provide a static snapshot of total cellular heme/porphyrin rather than a dynamic view of heme metabolism. In combination with non-isotopic quantitation methods and enzymatic assays, radiolabeling techniques provide insight into the dynamics of substrate utilization in regulation of heme synthesis.

Newly synthesized heme is typically labeled with ^{55}Fe [1, 2], ^{14}C-glycine, or ^{14}C-aminolevulinate (ALA) [3–5]. The advantage of ^{14}C glycine and ALA labeling is the ability to simultaneously

Oleh Khalimonchuk (ed.), *Iron Metabolism: Methods and Protocols*, Methods in Molecular Biology, vol. 2839,
https://doi.org/10.1007/978-1-0716-4043-2_12,

quantify free protoporphyrin within the same experiment. In particular, labeling with ^{14}C-ALA standardizes the amount of ALA entering the labeled heme population, thus allowing one to isolate changes in the heme synthesis pathway that occur downstream of ALAS.

This protocol is a modification of a method initially described by Ponka and Schulman [5]. Briefly, cultured cells are glycine-starved in minimal essential media (MEM) for an hour prior to labeling. ^{14}C-glycine or ^{14}C-ALA labeling is carried out in MEM supplemented with dialyzed FBS using holo-transferrin as an iron source. One may also use Fe-PIH or other iron chelates as a source of iron during labeling as previously described [5, 6]. Cells are lysed in RIPA buffer containing Triton X-100, and radiolabeled heme and porphyrins are extracted in ethylacetate. Subsequently, heme is isolated by sequentially washing the ethylacetate extract with sodium acetate and HCl. Protoporphyrin is obtained by washing the porphyrin fraction with saturated sodium acetate and HCl. Labeled heme and protoporphyrin are quantitated by liquid scintillation.

2 Materials

2.1 Cell Culture

1. Murine erythroleukemia and 3 T3 cells should be cultured in complete DMEM media.
   - DMEM (Gibco 11,995–065) supplemented with.
     (a) 1% penicillin/streptomycin (Gibco 15140-122).
     (b) 1% L-glutamine (Gibco 25031-081).
     (c) 10% fetal bovine serum (FBS) (R&D Biosystems S11550).
2. Differentiation is carried out in total DMEM media supplemented with 2% DMSO (Corning MT25950CQC).
3. G1E-ER4 cells should be cultured in the following media as described [7].
   - IMDM (Gibco 12440-046) supplemented with.
     - 15% fetal bovine serum (FBS) (GeminiBio 100-525).
     - 1.25% penicillin/streptomycin (Gibco 15140-122).
     - 125 uM monothioglycerol (MTG) (MP Biomedicals 155723).
     - 50 ng/mL stem cell factor (SCF) (455-MC/CF).
     - 2 U/mL erythropoietin (EPO) (PHC9631).
4. Differentiation is carried out by adding 1×10^{-8} mM estradiol (Sigma E2758-1g).

2.2 Labeling Reagents

- Minimum essential media (MEM), Gibco #12360038.
- Labeling media–MEM media supplemented with dialyzed FBS, 10% final concentration (Sigma F0392).
- Holo-transferrin (10 mg/mL). Holo-transferrin powder can be obtained from Sigma (T4132). Dissolve all the powder in the vial with sterile water (cell culture grade) to a concentration of 10 mg/mL. Store in 200 uL aliquots at −20 °C.
- ^{14}C-glycine (Perkin Elmer or American Radiochemicals), stock concentration 0.1 uCi/uL or ^{14}C-ALA (American Radiochemicals), stock concentration 0.1 uCi/uL.

2.3 Heme and Porphyrin Extraction

Prepare all reagents using reverse osmosis water.

- RIPA buffer (store at 4 °C).

  150 mM NaCl.

  50 mM Tris pH 8.0.

  1% NP40.

  1% Triton X-100.

  0.5% sodium deoxycholate.
- Protein assay dye reagent concentrate (Biorad #5000006).
- Bovine serum albumin stock (10 mg/mL) for protein quantification.
- Ethylacetate:acetic acid (3:1).
- 3% sodium acetate.
- 1.5 N HCl.
- Saturated sodium acetate.

  Note: Dissolve at 50 °C and then cool to room temperature before using to ensure true saturation.
- 0.1 N HCl.
- Scintillation fluid (Ultima Gold, Perkin Elmer #6013329).

3 Methods

3.1 Cell Culture

(a) MEL cells are seeded at a density of approximately 50 k cells/mL and grown in 50 mL complete DMEM media in 50 mL until the concentration is about 1.5 million cells/mL, as determined with a hemacytometer. To differentiate MEL cells, seed undifferentiated MEL cell samples at a cell concentration of 3×10^4cells/mL in 50 mL complete DMEM media (1.5 million cells) for 72 h to obtain actively growing cultures. To differentiate, seed these cells at a concentration of 3×10^5cells/ml (15 million cells) in 50 mL media with 2%

DMSO for 72 h. Cell density and DMSO concentration (ranging from 1.5% to 2%) can vary between labs, so these are conditions that should be optimized.

(b) G1E-ER4 cells were grown in complete IMDM media in 100 mL media until the concentration reaches 1–1.5 million cells/mL. To differentiate, undifferentiated G1E-ER4 cell samples were set up with 7.5 million cells in 50 mL media and left to expand for 48 h. G1E-ER4 cells were differentiated at a concentration of 7 × 10^5cells/mL (35 million cells) in 50 mL media with 1 × 10^{-8} mM estradiol for 48 h.

(c) For both cell lines, 40 million cells should be used for each sample on the day of labeling.

(d) Adherent cells (we have used 3 T3 cells; adapt growth conditions to cell type of choice) should be actively growing and about 80% confluent on the day of labeling. 6 wells of cells grown in a 6-well plate should be used for each sample. 3 T3 cells are grown in complete DMEM media.

3.2 Labeling and Heme/Porphyrin Extraction

1. 1 h before labeling, remove growth media from cells and glycine-starve cells with pre-warmed MEM media, which contains no glycine. 1 well of 3 T3 cells in 6-well plates are starved in 1 mL MEM. For MEL/G1E-ER4 cells, spin down 40 million cells and resuspend in 10 mL MEM. Leave cells in a CO_2 incubator at 37 °C.
2. 3 T3 cells: Remove MEM media. Replace with 2 mL labeling media per well. Add 15 uL of holotransferrin (10 mg/mL) into each well. Incubate for 30 min at 37 °C. A full 6-well plate is 2 technical replicates, with 3 wells per technical replicate. MEL/G1E-ER4 cells: Spin down and aspirate MEM media. Resuspend 40 million cells in 4 mL of labeling media and dispense into 2 wells of a 12-well plate. Add 15 uL of holotransferrin (10 mg/mL) into each well. Incubate for 30 min at 37 °C.
3. Add 1.5 uCi (15 uL) of ^{14}C-glycine or ^{14}C-ALA to each well. When choosing which label to use, see Note 1– ALA is generally less efficiently taken up by the cells resulting in lower signal.
4. Incubate for 2 h at 37 °C.
5. Washing: Adherent cells: aspirate labeling media and wash once with 5 mL ice cold PBS. Erythroid suspension cells: spin down cells in each well in a 2 mL Eppendorf tube at 2000 g for 5 min. Wash once in 2 mL ice cold PBS. Resuspend well, gently vortex. Spin down at 2000 g for 5 min and remove the PBS.
6. Cell lysis: Adherent cells: Add 500 uL ice cold RIPA buffer in *one* well. Resuspend cells, pipet into second and third well. 3 wells of cells are used for 1 technical replicate. You will end up with 2 tubes of 500 uL lysates from a 6-well plate. Erythroid

suspension cells: Resuspend each cell pellet in 500 uL of ice cold RIPA buffer.

Note: At this point, you can freeze the cell lysates at −80 °C and continue subsequent steps later. We do not recommend freezing the cell pellet and resuspending at a later stage—when we did this, we found that the results were less consistent even across technical replicates.

7. Protein quantitation: Keep 5 uL of cell lysate for protein quantitation with the Bradford assay. Make a standard curve with dilutions of 10 mg/mL BSA stock for comparison.
8. Extraction of total heme/porphyrin:
   (a) Add 500 uL ethylacetate: acetic acid (3:1) to the lysate and vortex for 1 full minute, then centrifuge at maximum speed for 5 min.

   Notes: Sufficient vortexing is extremely important to reliably extract heme and porphyrin from the lysate. Insufficient vortexing will result in an inconsistent and weak signal. Some protocols call for centrifuging in a 4 °C centrifuge. In our experience, this is unnecessary.

   (b) Transfer 350–400 uL of the top layer into a 1.5 uL centrifuge tube. This is the heme/porphyrin fraction.

   Note: When carrying out the extraction in differentiated erythroid cells, the top layer should be pink and the bottom layer should be colorless. This is an indication that the heme has been correctly extracted.

   (c) Repeat extraction from the cell lysate—500 uL ethylacetate:acetic acid (3:1)—to the lysate and vortex for 1 full minute and then centrifuge at maximum speed for 5 min.
   (d) Pool 300–350 uL of the top layer with the tube from (b).
9. Wash the pooled ethylacetate layers with 3% sodium acetate. This is done by adding 750 uL of 3% sodium acetate to the pooled ethylacetate layers, vortexing for 30 s, and centrifuging at maximum speed for 5 min.

   Notes: The interface between layers may be blurred when centrifugation is carried out at room temperature due to heat generation, making it difficult to distinguish between the layers. Simply remove the tubes from the centrifuge and let them cool. The interface will become clear when the tubes have cooled to room temperature.

10. Separation of heme and porphyrin:
    (a) Transfer 300 uL of the top layer to a new tube.
    (b) Add 600 uL of 1.5 N HCl.
    (c) Vortex for 1 min and centrifuge at maximum speed for 5 min.

Notes: If centrifugation is done at room temperature, let the tubes cool after centrifugation to see the interface clearly. The top ethylacetate layer from this step is heme; the bottom layer contains porphyrins.

11. Quantitation of heme: Transfer 200 uL of the top layer to a scintillation vial. If using differentiated erythroid cells, the top layer may be pink from heme (this can be an internal quality check for the extraction) if the cells are well-differentiated. Add 10 mL of scintillation fluid and count in a scintillation counter using the ^{14}C program.
12. Porphyrin extraction:
    (a) Split the bottom layer into 2 tubes (i.e., 1 sample is split into 2 tubes). This is approximately 350 ul in each tube.
    (b) To each tube, add 500 uL of saturated sodium acetate and 500 uL of ethylacetate:acetic acid (3:1).
    (c) Vortex for 1 min.
    (d) Centrifuge at maximum speed for 5 min.
    (e) The top layer contains coproporphyrin and protoporphyrin.
13. Protoporphyrin extraction:
    (a) Transfer 300 ul of the top layer to a new tube. Pool the top layers from the same sample together (total 600 ul)
    (b) Wash with 600 ul 0.1 N HCl and vortex for 30 s.
    Note: This step eliminates the coproporphyrins.
    (c) Centrifuge at max speed for 5 min.
    (d) Transfer 600 uL of the top layer to a scintillation vial.
    Note: This extract contains protoporphyrins, which consist of protoporphyrin IX and auto-oxidized protoporphyrinogen IX.
    (e) Add 10 mL of scintillation fluid and count using the 14C program.
    (f) For analysis, normalize heme and porphyrin radioactive counts (CPM or DPM) to protein concentration. (Refer to Note 2 on normalization controls and Note 3 on omitting the decolorization steps described in the original protocol (Ref 5).

4 Notes

1. We obtained lower counts with ^{14}C-ALA as compared to ^{14}C-glycine used as a label, despite the ALA being committed to heme synthesis, while glycine is used in many other processes such as protein synthesis. We speculate that this discrepancy reflects the relative uptake of the labeled molecules into the cell,

as glycine (although a nonessential amino acid) is frequently obtained from the diet or extracellular environment and has specific cellular transporters [5, 8], while most ALA is synthesized from glycine and succinyl CoA and transported within the cell, not from the extracellular environment.

2. While it is common to normalize the data to protein concentration, it is sometimes advisable to normalize to cell number. This is particularly true when there is a concern that treatments (such as erythroid differentiation) or genetic modifications (such as defects in erythroid differentiation that cause hemoglobin deficiencies—a major component of protein content in red cells) have altered total protein content within the cells.
3. The original protocol described by Ponka and Schulman called for decolorizing heme extracts with H_2O_2 as the color of heme can quench the signal in liquid scintillation. For the numbers of erythroid cells used in this protocol, it has not been necessary, but it may be required when labeling larger numbers of cells or reticulocytes as previously described.

Acknowledgments

The work was supported by R35 GM133560 to Y.Y.Y.

References

1. Roy CN, Penny DM, Feder JN, Enns CA (1999) The hereditary hemochromatosis protein, HFE, specifically regulates transferrin-mediated iron uptake in HeLa cells. J Biol Chem 274:9022–9028
2. Shaw GC, Cope JJ, Li L, Corson K, Hersey C, Ackermann GE, Gwynn B, Lambert AJ, Wingert RA, Traver D, Trede NS, Barut BA, Zhou Y, Minet E, Donovan A, Brownlie A, Balzan R, Weiss MJ, Peters LL, Kaplan J, Zon LI, Paw BH (2006) Mitoferrin is essential for erythroid iron assimilation. Nature 440:96–100
3. Seguin A, Takahashi-Makise N, Yien YY, Huston NC, Whitman JC, Musso G, Wallace JA, Bradley T, Bergonia HA, Kafina MD, Matsumoto M, Igarashi K, Phillips JD, Paw BH, Kaplan J, Ward DM (2017) Reductions in the mitochondrial ABC transporter Abcb10 affect the transcriptional profile of heme biosynthesis genes. J Biol Chem 292:16284–16299
4. Bonkowsky HL, Healey JF, Sinclair PR, Sinclair JF, Shedlofsky SI, Elder GH (1983) Iron and the liver: acute effects of iron-loading on hepatic heme synthesis of rats. Role of decreased activity of 5-aminolevulinate dehydrase. J Clin Invest 71:1175–1182
5. Ponka P, Schulman HM (1985) Regulation of heme synthesis in erythroid cells: hemin inhibits transferrin iron utilization but not protoporphyrin synthesis. Blood 65:850–857
6. Grillo AS, SantaMaria AM, Kafina MD, Cioffi AG, Huston NC, Han M, Seo YA, Yien YY, Nardone C, Menon AV, Fan J, Svoboda DC, Anderson JB, Hong JD, Nicolau BG, Subedi K, Gewirth AA, Wessling-Resnick M, Kim J, Paw BH, Burke MD (2017) Restored iron transport by a small molecule promotes absorption and hemoglobinization in animals. Science 356:608–616
7. Weiss MJ, Yu C, Orkin SH (1997) Erythroid-cell-specific properties of transcription factor GATA-1 revealed by phenotypic rescue of a gene-targeted cell line. Mol Cell Biol 17:1642–1651
8. Dufay JN, Fernández-Murray JP, McMaster CR (2017) SLC25 Family member genetic interactions identify a role for HEM25 in yeast electron transport chain stability. G3 (Bethesda) 7:1861–1873

Chapter 13

Heterologous Expression and Purification of Eukaryotic ALA Synthase from *E. coli*

Breann L. Brown

Abstract

This chapter presents a method for the heterologous expression and purification of human ALA synthase from *Escherichia coli*. Mature ALAS is produced with an N-terminal hexahistidine affinity tag followed by a SUMO fusion tag for solubility and ease of purification. The plasmid is introduced into competent *E. coli* cells, and robust protein expression is induced with IPTG. The ALAS cofactor, pyridoxal 5′-phosphate, is inserted during protein production to yield an active enzyme upon purification. After cell lysis, the tagged ALAS protein is isolated via a multistep purification that involves an initial nickel-affinity step, affinity tag cleavage and removal, and a final size exclusion chromatography polishing step. Importantly, this protocol is amenable to various ALAS truncations and mutations, opening the door to understanding ALAS biology and its intersections with iron utilization across several organisms.

Key words Heme, Mitochondria, Aminolevulinic acid, Aminolevulinic acid synthase, Pyridoxal 5′--phosphate, Protein expression and purification, Iron

1 Introduction

For decades, researchers have employed heterologous protein expression and purification from *Escherichia coli* to uncover fundamental mechanisms that drive diverse biological processes [1]. The high rates of use are largely attributed to the fast growth, cost-effectiveness, and ease of manipulation of this bacterial expression platform compared to insect or mammalian cells [2]. With this methodology, a target gene of interest can be cloned into an expression vector harboring various affinity and solubility tags that allow for robust overexpression and protein solubility in *E. coli* [3–7]. Along with a diversity of expression plasmids, multiple expression strains are engineered to produce challenging proteins of interest, including labile, toxic, and membrane proteins [8].

Oleh Khalimonchuk (ed.), *Iron Metabolism: Methods and Protocols*, Methods in Molecular Biology, vol. 2839,
https://doi.org/10.1007/978-1-0716-4043-2_13,

The first step of heme biosynthesis in α-proteobacteria and non-plant eukaryotes is catalyzed by the enzyme 5-aminolevulinic acid synthase (ALAS) [9, 10]. Eukaryotic ALAS localizes to the mitochondrial matrix where it is activated via binding to the pyridoxal 5′-phosphate (PLP) cofactor to produce aminolevulinic acid (ALA) from the condensation of glycine and succinyl-CoA [11, 12]. ALA is eventually converted to heme, an iron-containing porphyrin ring, after additional enzymatic steps that occur in the cytosol and mitochondrion. In humans, ALAS exists as two isoforms, and aberrations in the expression and activity of either form are linked to several disorders [13–16]. One such disease, X-linked sideroblastic anemia, results in severe iron overload due to decreased porphyrin synthesis stemming from deficiencies in the erythroid-specific enzyme ALAS2 [14, 17]. Synthesis of this enzyme is, in turn, controlled by iron availability. The *ALAS2* mRNA contains a 5′ iron-responsive element (IRE) that is occupied by iron-responsive proteins (IRPs) in low iron conditions, inhibiting *ALAS2* translation [18–20]. Thus, ALAS bridges iron availability to the regulation of heme synthesis to maintain healthy red blood cell production.

This protocol describes a method to express and purify active, cofactor-bound eukaryotic ALAS protein. ALAS is produced in a pET28b-derivative vector that expresses an N-terminal hexahistidine (His_6) affinity tag followed by a SUMO (small ubiquitin-related modifier) fusion tag [21]. The SUMO tag promotes protein solubility and can be recognized and cleaved by the yeast Ulp1 protease (SUMO protease 1) revealing the native N-terminus [22, 23]. Under the indicated expression conditions, ALAS is covalently bound to the PLP cofactor. The protein is initially purified via immobilized metal affinity chromatography (IMAC) using Ni-NTA agarose resin. The His_6-SUMO tag is then digested and removed by a second IMAC step. A final preparative gel filtration step produces active protein with greater than 95% purity. The purified sample is amenable to multiple downstream methods, such as enzymatic assays and structural biology approaches, that can lend greater insight into the role ALAS homologs play in heme biosynthesis and iron utilization.

2 Materials

2.1 Sodium Dodecyl Sulfate-Polyacrylamide Gel Electrophoresis (SDS-PAGE)

1. Pre-cast 4–12% Bis–Tris polyacrylamide gels.
2. 4× Laemmli sample buffer: prepared with reducing agent.
3. MES running buffer: 50 mM MES (2-[N-morpholino]ethanesulfonic acid), 50 mM Tris base, 1 mM EDTA, 0.1% (w/v) SDS.

4. Coomassie R250 stain: 0.3% w/v Coomassie R250, 10% acetic acid, 45% methanol, prepared in water.
5. Coomassie de-stain: 45% methanol, 10% acetic acid, prepared in water.

2.2 Protein Expression (See Note 1)

1. pET28b-His_6SUMO-humanALAS$2_{54–587}$ (*see* **Note 2**).
2. BL21-CodonPlus (DE3)-RIL chemically competent *E. coli* cells (*see* **Note 3**).
3. Glucose: 20% (w/v) prepared in water and sterile filtered.
4. SOB medium: 5 g/L yeast extract, 20 g/L tryptone, 0.584 g/L NaCl, 0.186 g/L KCl, 2.4 g/L $MgSO_4$, autoclaved.
5. SOC medium: 100 μL 20% glucose per 1 mL SOB medium, prepared fresh.
6. LB agar plates: Prepared with 1X kanamycin and chloramphenicol.
7. LB medium: 10 g/L tryptone, 5 g/L yeast extract, 10 g/L NaCl, autoclaved.
8. Kanamycin: 50 mg/mL (1000×) prepared in water and sterile filtered.
9. Chloramphenicol: 20 mg/mL (1000×) prepared in 100% ethanol and sterile filtered.
10. Isopropyl β-D-1-thiogalactopyranoside (IPTG): 1 M solution prepared in water and sterile filtered.

2.3 Protein Purification

1. Pyridoxal 5′-phosphate (PLP): 4 mM prepared in water and stored at −20 °C.
2. Ni-NTA lysis buffer: 25 mM Hepes-KOH, pH 8.0, 100 mM KCl, 400 mM NaCl, 2 mM $MgCl_2$, 20 mM imidazole, 10% glycerol, 1 mM DTT, 20 μM PLP.
3. Ni-NTA elution buffer: 25 mM Hepes-KOH, pH 8.0, 100 mM KCl, 400 mM NaCl, 2 mM $MgCl_2$, 250 mM imidazole, 10% glycerol, 1 mM DTT, 20 μM PLP.
4. Dialysis buffer: 25 mM Hepes-KOH, pH 7.0, 150 mM KCl, 10% glycerol, 1 mM DTT, 20 μM PLP.
5. SEC buffer: 25 mM Hepes-KOH, pH 7.0, 150 mM KCl, 10% glycerol, 0.5 mM TCEP.
6. EDTA-free protease inhibitor tablets.
7. High-pressure homogenizer.
8. Ni-NTA agarose resin.
9. 1.5 × 15 cm glass chromatography column.
10. Ulp1 (SUMO protease 1) (*see* **Note 4**).
11. Dialysis tubing: 3.5–10 kDa molecular weight cut off (MWCO).

12. Dialysis tubing clips.
13. Dialysis float buoys.
14. Centrifugal filters: 30 kDa MWCO.
15. Fast protein liquid chromatography (FPLC) system.
16. HiLoad Superdex 200 pg 16/600 gel filtration column.

3 Methods

3.1 Protein Expression

3.1.1 Bacterial Transformation

1. Add 1 μL plasmid DNA (≥ 50 ng/μL) to 20 μL BL21-CodonPlus (DE3)-RIL competent cells in a sterile 1.5 mL tube.
2. Incubate on ice for 30 min.
3. During incubation, prepare SOC media in a sterile 1.5 mL tube (500 μL per reaction) and incubate at 37 °C.
4. Heat shock cells at 42 °C for 45 s then cool on ice for 5–10 min.
5. Add 450 μL SOC media to cells and grow at 37 °C with shaking (~400 rpm) for 60–75 min.
6. Plate 25–50 μL of the transformation onto an LB agar plate containing 1× kanamycin and chloramphenicol and incubate overnight at 37 °C.

3.1.2 Test Expression (See **Note 5**)

1. Remove the LB agar plate from the incubator and choose three individual colonies.
2. Inoculate three separate 14 mL sterile tubes each containing 3 mL LB media, 3 μL kanamycin, 3 μL chloramphenicol, and one bacterial colony.
3. Let cultures grow at 37 °C with shaking at 220 rpm for 4 h.
4. Remove 1 mL of the culture from the original tube and transfer to a fresh sterile tube.
5. To the second tube, add 1 μL 1 M IPTG and return all tubes to the 37 °C shaking incubator for 2.5 h.
6. Remove 20 μL of each sample and mix with 5 μL 4× Laemmli sample buffer containing reducing agent in a 1.5 mL tube. Boil samples at 100 °C for 5 min and spin briefly.
7. Load samples onto an acrylamide gel and perform electrophoresis according to the manufacturer's specifications.
8. After briefly staining with Coomassie and de-staining, identify which culture produces the most robust protein overexpression by comparing the uninduced to induced samples across the three colonies.

9. Prepare an overnight starter culture containing 50 mL LB media, 50 μL kanamycin, 50 μL chloramphenicol, and 50 μL of the uninduced culture that displayed the strongest overexpression (*see* **Note 6**).
10. Grow culture overnight at 37 °C and 220 rpm.

3.1.3 Large Expression

1. Prewarm the desired amount of LB media at 37 °C. Cultures are typically grown in 1 L portions in 2.8 L Fernbach-style baffled flasks.
2. Inoculate each liter of media with 1 mL kanamycin, 1 mL chloramphenicol, and 10 mL of the overnight starter culture.
3. Let cultures grow with shaking at 37 °C and 220 rpm to an $OD_{600} = 0.6–0.8$.
4. Chill cultures at 4 °C for 15 min.
5. Add 0.5 mL 1 M IPTG to each liter and induce for 4 h at 22 °C and 220 rpm.
6. Harvest cells via centrifugation for 10 min at 6000× *g* (*see* **Note 7**).
7. Transfer cell pellets to 50 mL conical tubes and store at −80 °C.

3.2 Protein Purification (See Note 8)

3.2.1 Ni-NTA IMAC Purification

1. Resuspend the cell pellet in five- to tenfold excess (w/v) Ni-NTA Lysis Buffer and one EDTA-free protease inhibitor tablet per 50 mL buffer.
2. Lyse the cell suspension via three passes through a high-pressure homogenizer at 15,000 psi.
3. Transfer the lysate to 50 mL Oak Ridge centrifuge tubes and spin at 30,000× *g* for 20 min at 4 °C to separate soluble protein supernatant from the pellet.
4. While lysate is spinning, equilibrate 5 mL of Ni-NTA agarose resin (10 mL of slurry in 20% ethanol) into water by batch equilibration. Transfer the resin to a 50 mL conical tube, add ~40 mL of ultrapure water, and then spin at 1000× *g* for 2 min in a swinging bucket rotor. Carefully pour off the supernatant (pelleted resin will be extremely loose), add another 40 mL of water, and resuspend the resin by inverting the tube. Spin again.
5. Repeat wash twice more with 30 mL Ni-NTA lysis buffer.
6. Add the soluble protein supernatant to Ni-NTA resin, using additional 50 mL conical tubes as needed to contain the entire supernatant.
7. Equilibrate the protein with Ni-NTA resin while rotating at 4 °C for 1 h.

8. Spin the resin-protein mixture at 1000× *g*, 2 min, and collect supernatant (Flow Through).
9. Add 25 mL lysis buffer to pelleted resin and resuspend by briefly inverting the conical tube. Spin once more at 1000× *g* for 2 min and collect buffer wash.
10. Repeat once more with an additional 25 mL lysis buffer.
11. Resuspend the Ni-NTA resin in 10 mL lysis buffer and transfer to a glass column.
12. Continue washing the resin with lysis buffer until no more protein elutes (*see* **Note 9**).
13. Elute His_6-tagged protein with Ni-NTA elution buffer until no more protein elutes, typically ~5 column volumes (CV) Elution Buffer (*see* **Note 10**).
14. Add Ulp1 protease to eluate, transfer the solution to dialysis tubing, and dialyze overnight at 4 °C against ≥1 L Dialysis Buffer (*see* **Notes 11** and **12**).

3.2.2 Subtraction IMAC and Gel Filtration Purification

1. Equilibrate 5 mL Ni-NTA agarose resin in dialysis buffer (*see* **Note 13**).
2. Remove the protein from overnight dialysis and filter with 0.2 μm PES or PVDF membrane.
3. To remove the cleaved His_6-SUMO tag, add the filtered protein to the equilibrated Ni-NTA resin in a gravity column and collect Flow Through, which contains the untagged protein.
4. Continue washing the resin with dialysis buffer until no more protein elutes.
5. Combine the Flow Through and buffer wash fractions and concentrate with centrifugal filters to 5 mL then filter with a 0.2 μm membrane (*see* **Note 14**).
6. Inject up to 5 mL protein onto a Superdex 200 16/600 column equilibrated in SEC Buffer (*see* **Note 15**).
7. Pool the appropriate fractions as determined via SDS-PAGE analysis and concentrate with a centrifugal filter to the minimum possible volume that can be attained without precipitation.
8. Filter the purified protein with a 0.2 μm membrane (*see* **Note 16**).
9. Aliquot into small portions (10–50 μL) and snap freeze in liquid nitrogen. Store at −80 °C for future use.

4 Notes

1. This protocol was optimized using various ALAS constructs cloned with an N-terminal hexahistidine (His_6) tag, followed by a SUMO fusion to aid in protein expression, solubility, and purification [23]. Although human ALAS2 is noted here, the protocol is also amenable to other eukaryotic homologs such as yeast ALAS (Hem1) [24–27]. However, the final buffer pH should be adjusted so that it is greater than one pH unit away from the protein isoelectric point (pI).
2. The mitochondrial ALAS homologs are normally expressed as their mature forms. This plasmid was previously reported and received from T. A. Baker [24]. The pET28b vector harbors a kanamycin resistance selection marker.
3. The BL21-CodonPlus (DE3)-RIL cell line contains the rare *argU*, *ileY*, and *leuW* tRNA genes to aid in protein overexpression. These cells are chloramphenicol resistant. If using a different competent cell line, remove the chloramphenicol from subsequent steps.
4. Yeast Ulp1 (SUMO protease 1) is commercially available but also can be expressed and purified from *E. coli* with a His_6 affinity tag [28].
5. The test expression step is optional and does not need to be performed after each transformation once a single optimal DNA clone has been identified and purified via miniprep.
6. A 50 mL overnight culture is sufficient for 4 L of large growth. Scale up as necessary.
7. No more than 2 L of large cultures should be combined into a single pellet to prevent complications from aggregation during purification.
8. All steps should be performed at 4 °C or with buffers chilled to 4 °C as allowed. A diagnostic SDS-PAGE analysis should be performed with samples obtained after IMAC purification, affinity tag removal, and gel filtration to confirm ≥95% purity.
9. It is recommended to qualitatively test for the presence of protein using a colorimetric protein dye, such as Bradford Reagent. Approximately 190 μL dye (1X) is mixed with 10 μL of sample in a clear-bottom 96-well plate. Fractions containing protein will result in a color change. The appropriate buffer should be used as a negative control.
10. The presence of ALAS is easily observed by a strong yellow color due to the covalently bound PLP. However, protein dye can be used if purifying ALAS variants that do not bind PLP tightly.

11. Approximately 100 μL of Ulp1 (30–50 μM) for every 30 mL of eluate is sufficient for greater than 90% cleavage. The cleavage efficiency should also be confirmed via SDS-PAGE.
12. The dialysate should be protected from light to prevent the degradation of the PLP [29].
13. The same Ni-NTA resin can be used for both the initial and subtraction IMAC purification steps. After eluting the His_6-tagged protein, strip resin with 10 CV 2 M imidazole, pH 8.0, wash with 10 CV water, and store at 4 °C.
14. Excessive concentration can cause increased aggregation and precipitation. Therefore, it may be necessary to perform multiple gel filtration runs depending on the maximum allowable sample volume for the column in use.
15. The protein should elute in a position commensurate with a homodimer. Protein standards should be run using the indicated column in SEC Buffer to confirm the correct sample elution volume.
16. Conversely, to prevent loss of sample from filtering, the protein can be centrifuged at 12,000× *g* for 10 min at 4 °C. Carefully transfer the supernatant to a fresh 1.5 mL tube before aliquoting.

References

1. Gileadi O (2017) Recombinant protein expression in E. coli: a historical perspective. Methods Mol Biol 1586:3–10. https://doi.org/10.1007/978-1-4939-6887-9_1
2. Langlais C, Korn B (2006) Recombinant protein expression in bacteria. In: Encyclopedic reference of genomics and proteomics in molecular medicine. Springer, Berlin/Heidelberg, pp 1609–1616. https://doi.org/10.1007/3-540-29623-9_4800
3. Terpe K (2003) Overview of tag protein fusions: from molecular and biochemical fundamentals to commercial systems. Appl Microbiol Biotechnol 60(5):523–533. https://doi.org/10.1007/s00253-002-1158-6
4. Smith DB, Johnson KS (1988) Single-step purification of polypeptides expressed in *Escherichia coli* as fusions with glutathione S-transferase. Gene 67(1):31–40. https://doi.org/10.1016/0378-1119(88)90005-4
5. Butt TR, Edavettal SC, Hall JP, Mattern MR (2005) SUMO fusion technology for difficult-to-express proteins. Protein Expr Purif 43(1):1–9. https://doi.org/10.1016/j.pep.2005.03.016
6. Kapust RB, Waugh DS (1999) *Escherichia coli* maltose-binding protein is uncommonly effective at promoting the solubility of polypeptides to which it is fused. Protein Sci 8(8):1668–1674. https://doi.org/10.1110/ps.8.8.1668
7. LaVallie ER, DiBlasio EA, Kovacic S, Grant KL, Schendel PF, McCoy JM (1993) A thioredoxin gene fusion expression system that circumvents inclusion body formation in the *E. coli* cytoplasm. Biotechnology (NY) 11(2):187–193. https://doi.org/10.1038/nbt0293-187
8. Rosano GL, Morales ES, Ceccarelli EA (2019) New tools for recombinant protein production in *Escherichia coli*: a 5-year update. Protein Sci 28(8):1412–1422. https://doi.org/10.1002/pro.3668
9. Kikuchi G, Kumar A, Talmage P, Shemin D (1958) The enzymatic synthesis of delta-aminolevulinic acid. J Biol Chem 233(5):1214–1219
10. Gibson KD, Laver WG, Neuberger A (1958) Initial stages in the biosynthesis of porphyrins. 2. The formation of delta-aminolaevulic acid from glycine and succinyl-coenzyme A by particles from chicken erythrocytes. Biochem J 70(1):71–81. https://doi.org/10.1042/bj0700071
11. Shemin D, Rittenberg D (1946) The biological utilization of glycine for the synthesis of the

protoporphyrin of hemoglobin. J Biol Chem 166(2):621–625
12. Shemin D, Kumin S (1952) The mechanism of porphyrin formation; the formation of a succinyl intermediate from succinate. J Biol Chem 198(2):827–837
13. Bishop DF, Henderson AS, Astrin KH (1990) Human delta-aminolevulinate synthase: assignment of the housekeeping gene to 3p21 and the erythroid-specific gene to the X chromosome. Genomics 7(2):207–214. https://doi.org/10.1016/0888-7543(90)90542-3
14. Bottomley SS, May BK, Cox TC, Cotter PD, Bishop DF (1995) Molecular defects of erythroid 5-aminolevulinate synthase in X-linked sideroblastic anemia. J Bioenerg Biomembr 27(2):161–168. https://doi.org/10.1007/BF02110031
15. Granick S (1963) Induction of the synthesis of delta-aminolevulinic acid synthetase in liver parenchyma cells in culture by chemical that induce acute porphyria. J Biol Chem 238: 2247–2249
16. Whatley SD, Ducamp S, Gouya L, Grandchamp B, Beaumont C, Badminton MN, Elder GH, Holme SA, Anstey AV, Parker M, Corrigall AV, Meissner PN, Hift RJ, Marsden JT, Ma Y, Mieli-Vergani G, Deybach JC, Puy H (2008) C-terminal deletions in the ALAS2 gene lead to gain of function and cause X-linked dominant protoporphyria without anemia or iron overload. Am J Hum Genet 83(3):408–414. https://doi.org/10.1016/j.ajhg.2008.08.003
17. Cotter PD, Baumann M, Bishop DF (1992) Enzymatic defect in "X-linked" sideroblastic anemia: molecular evidence for erythroid delta-aminolevulinate synthase deficiency. Proc Natl Acad Sci USA 89(9):4028–4032. https://doi.org/10.1073/pnas.89.9.4028
18. Cox TC, Bawden MJ, Martin A, May BK (1991) Human erythroid 5-aminolevulinate synthase: promoter analysis and identification of an iron-responsive element in the mRNA. EMBO J 10(7):1891–1902
19. Melefors O, Goossen B, Johansson HE, Stripecke R, Gray NK, Hentze MW (1993) Translational control of 5-aminolevulinate synthase mRNA by iron-responsive elements in erythroid cells. J Biol Chem 268(8): 5974–5978
20. Bhasker CR, Burgiel G, Neupert B, Emery-Goodman A, Kuhn LC, May BK (1993) The putative iron-responsive element in the human erythroid 5-aminolevulinate synthase mRNA mediates translational control. J Biol Chem 268(17):12699–12705
21. Wang KH, Sauer RT, Baker TA (2007) ClpS modulates but is not essential for bacterial N-end rule degradation. Genes Dev 21(4): 403–408. https://doi.org/10.1101/gad.1511907
22. Bird LE (2011) High throughput construction and small scale expression screening of multi-tag vectors in *Escherichia coli*. Methods 55(1): 29–37. https://doi.org/10.1016/j.ymeth.2011.08.002
23. Malakhov MP, Mattern MR, Malakhova OA, Drinker M, Weeks SD, Butt TR (2004) SUMO fusions and SUMO-specific protease for efficient expression and purification of proteins. J Struct Funct Genom 5(1–2):75–86. https://doi.org/10.1023/B:JSFG.0000029237.70316.52
24. Kardon JR, Yien YY, Huston NC, Branco DS, Hildick-Smith GJ, Rhee KY, Paw BH, Baker TA (2015) Mitochondrial ClpX activates a key enzyme for heme biosynthesis and erythropoiesis. Cell 161(4):858–867. https://doi.org/10.1016/j.cell.2015.04.017
25. Brown BL, Kardon JR, Sauer RT, Baker TA (2018) Structure of the mitochondrial aminolevulinic acid synthase, a key heme biosynthetic enzyme. Structure 26(4):580–589 e584. https://doi.org/10.1016/j.str.2018.02.012
26. Kardon JR, Moroco JA, Engen JR, Baker TA (2020) Mitochondrial ClpX activates an essential biosynthetic enzyme through partial unfolding. elife 9. https://doi.org/10.7554/eLife.54387
27. Tran JU, Brown BL (2023) The yeast ALA synthase C-terminus positively controls enzyme structure and function. Protein Sci 32(4):e4600. https://doi.org/10.1002/pro.4600
28. Wang X, Liu H, Liu Y, Li Y, Yan L, Yuan X, Zhang Y, Wu Y, Liu J, Zhang C, Chu Y (2016) A novel strategy for the preparation of codon-optimized truncated Ulp1 and its simplified application to cleavage the SUMO fusion protein. Protein J 35(2):115–123. https://doi.org/10.1007/s10930-016-9654-1
29. Ang CY (1979) Stability of three forms of vitamin B6 to laboratory light conditions. J Assoc Off Anal Chem 62(5):1170–1173

Chapter 14

Direct Spectroscopic Ferrochelatase Assay

Harry A. Dailey and Amy E. Medlock

Abstract

Ferrochelatases (E.C. 4.99.1.1) catalyze the insertion of ferrous iron into either protoporphyrin IX to make protoheme IX or coproporphyrin III to make coproheme III. Ferrochelatase activity in extracts or purified protein can be measured via several assays. Here, we describe a rapid real-time direct spectroscopic ferrochelatase assay for both protoporphyrin and coproporphyrin ferrochelatases.

Key words Ferrochelatase, Heme biosynthesis, Iron, Porphyrin, Chelation

1 Introduction

Ferrochelatases (E.C. 4.99.1.1) are enzymes that catalyze the insertion of ferrous iron into porphyrin macrocycles. While ferrous iron is the physiological metal substrate for all ferrochelatases, there are different types of ferrochelatases that utilize different porphyrins as substrates, either protoporphyrin IX or coproporphyrin III [1]. Protoporphyrin ferrochelatases are part of the protoporphyrin-dependent (PPD) heme synthesis pathway found in proteobacteria and eukaryotes. Protoporphyrin ferrochelatases catalyze the terminal step in the protoporphyrin-dependent heme synthesis pathway to produce protoheme IX and two protons. Coproporphyrin ferrochelatases are part of the coproporphyrin-dependent (CPD) pathway, which is found in most diderm bacteria in the Firmicutes and Actinobacteria Phyla. Coproporphyrin ferrochelatases catalyze the penultimate step of the CDP pathway to produce coproheme III, which is then decarboxylated to protoheme IX. The CDP pathway was not discovered until 2015 [2]; thus, it is important to note that all studies prior to this time incorrectly assumed that protoporphyrin IX was the substrate of all ferrochelatases.

There are a number of assay procedures that have been used to evaluate ferrochelatase activity and determine kinetic parameters.

Oleh Khalimonchuk (ed.), *Iron Metabolism: Methods and Protocols*, Methods in Molecular Biology, vol. 2839,
https://doi.org/10.1007/978-1-0716-4043-2_14,

Two of these assays are discontinuous and quantitate product either as its oxidized minus reduced pyridine hemochromagen [3, 4] or via radioactive iron incorporation [5]. The radioactive assay, while most sensitive, is seldom utilized due to its arduous nature and issues associated with obtaining ^{59}Fe or ^{55}Fe. The continuous assay for ferrochelatase activity is based on changes in the spectrum of free porphyrin versus metalated porphyrin. The direct spectroscopic assay uses selected wavelength-specific absorbance, zeroed to absorbance at isosbestic points to measure in real time the disappearance of porphyrin substrate or appearance of metalated product [2, 6, 7]. Additionally, one may follow the loss of fluorescence of the free porphyrin as it's converted to the nonfluorescent metalated porphyrin [8]. While the quantitation of product by pyridine hemochromagen is more sensitive than the direct spectroscopic determination of product formation or substrate disappearance, the continuous assay is now the one most commonly employed since it is more rapid and avoids the use of pyridine and the need for a split beam spectrophotometer. Herein, we summarize details of the real-time direct spectroscopic ferrochelatase assay and its use for both protoporphyrin and coproporphyrin ferrochelatases.

2 Materials

Prepare all solutions using ultrapure water at room temperature.

Assay reagents:

1. Ferrochelatase assay buffer: 100 mM Tris–HCl, pH 8.0, and 0.5% Tween 20 (*see* **Note 1**). For 100 mL of assay buffer, 10 mL of 1 M Tris–HCl, pH 8.0, 5 mL of Tween 20 Surfact-Amps Detergent 10% Solution, and 85 mL of ultrapure water are mixed.
2. Reductant: 100 mM β-mercaptoethanol (*see* **Note 2**). A stock solution of 100 mM β-mercaptoethanol is prepared by diluting 70 microliters of β-mercaptoethanol in 10 mL of ultrapure water.
3. Porphyrin stock solution: 1 mM porphyrin. Porphyrin stock solutions should be freshly prepared and stored cold in the dark. Use only dark-colored glass tubes (black or brown) or those covered in aluminum foil to protect from light. Do not use polypropylene or polystyrene, as porphyrins stick to these. To make ~1 mM solution of porphyrins, measure out between 2.5 to 4 mg of porphyrins (i.e. mesoporphyrin IX, protoporphyrin IX, deuteroporphyrin IX, coproporphyrin III, etc.). Add 50 μL of concentrated ammonium hydroxide. Vortex well to make a paste. Add 500 μL of 10% Triton X-100 and

mix until there is a clear solution, i.e., no undissolved porphyrin. Add 5 mL of ultrapure water and vortex to mix (*see* **Note 3**).

4. Iron stock solution: 1 mM ferrous ammonium sulfate. For metals, use a 15 mL conical tube, either polypropylene or polystyrene. To make an ~1 mM solution of iron, weigh out 2 mg of ferrous ammonium sulfate. Add 5 mL of ultrapure water and vortex to mix. Add a small amount (a crystal or pinch) of ascorbic acid. Iron solutions should be freshly prepared (*see* **Note 4**).

3 Methods

Carry out all assays at a temperature appropriate to the source of enzyme. This may be room temperature or in a thermostated cuvette. Assay solutions should be held at the appropriate temperature prior to initiation of the assay to avoid warming during the assay period. Mesoporphyrin IX or deuteroporphyrin IX are used except for kinetic measurements since those two porphyrins are more soluble and stable in solution than protoporphyrin IX. Coproporphyrin III, which is used only for coproporphyrin ferrochelatases, is similar in properties to deuteroporphyrin IX and mesoporphyrin IX with regard to its solubility. Early ferrochelatase assays were frequently run under anaerobic conditions, but this has been shown not to be a necessity.

3.1 Porphyrin and Iron Quantitation

1. To quantitate porphyrins, dilute porphyrin 1:1000 in a glass tube in the appropriate concentration of HCl and scan from 300 to 500 nm (*see* **Note 5**).
2. To quantitate the iron, dilute iron 1:100 in a conical tube. Add 100 μL of 10 mM ferrozine solution. Measure the absorbance of 1 mL of the iron:ferrozine solution at 562 nm and use an extinction coefficient of 0.278 mM^{-1} (*see* **Note 6**).

3.2 Assay

1. Add the following to a cuvette and mix:

    900 μL of assay buffer.

    100 μL of 100 mM β-mercaptoethanol (or 50 mM dithiothreitol).

    100 μL of 1 mM of selected porphyrin.

    100 μL of 1 mM ferrous ammonium sulfate (or other selected metal).

    Insert cuvette into spectrophotometer.

2. Add 10 μL of ~100 μM purified ferrochelatase or an appropriate amount of cell extract to allow one to follow changes in the observed spectrum.

Mix rapidly with a sealed, bent capillary tube and start recording absorbance immediately (*see* **Note 7**). For a purified enzyme, scans covering the appropriate wavelengths should be run every 30 s for approximately 3 min. For less active preparations, longer times may be required. However, one should avoid leaving the assay cuvette in the light beam for excessive lengths of time since porphyrins are light-sensitive and will degrade.

3. For assays in which the disappearance of substrate is being measured, the slope will be negative.
4. For kinetic measurements, change the amount of one substrate while holding the other constant (*see* **Note 8**).

4 Notes

1. The Tween 20 Surfact-Amps Detergent Solution is a highly purified Tween-20 detergent stabilized and supplied as a 10% solution in 10 mL glass ampules. While Tris–HCl is employed herein, any buffer that functions at pH 7.5–8.0 is acceptable.
2. A reductant is also included in all assay buffer to maintain the iron ion in a reduced state. This reductant may be either dithiothreitol or β-mercaptoethanol. For either, solutions are freshly prepared.
3. If measuring kinetic parameters, porphyrin solutions should be used immediately since porphyrins will stack and precipitate out of aqueous solution. For routine assays, stock solutions of mesoporphyrin IX, deuteroporphyrin IX, or coproporphyrin III may be stored overnight in cold. Protoporphyrin should always be freshly prepared.
4. Other metals, such as Zn, Co, and Ni, may be prepared in advance and stored at room temperature.
5. For different porphyrins, the concentration of HCl and the extinction coefficient differs. A full list can be found here [9] and specific porphyrins are described herein. For protoporphyrin IX, dilute in 2.7 N HCl and use the extinction coefficient of 13.5 mM^{-1} at 554 nm. For coproporphyrin III, dilute in 0.1 N HCl and use the extinction coefficient 16.8 mM^{-1} at 548 nm. For mesoporphyrin IX, dilute in 0.1 N HCl and use the extinction coefficient of 15.1 mM^{-1} at 547 nm, and for deuteroporphyrin IX, dilute in 0.1 N HCl and use the extinction coefficient of 13.7 mM^{-1} at 548 nm.
6. For other divalent metals, such as nickel, copper, cobalt, and zinc, make a 10 mM stock and dilute down to 1 mM for assays. For routine assays, we frequently employ Ni and follow product formation since it avoids the problem of metal oxidation

that one has with ferrous iron. Ni-porphyrin yields a product whose absorbance maximum is well shifted from free porphyrin. Zn insertion is employed by some since one can follow the product formation using fluorescence. However, Zn-porphyrin is considerably less stable than other metals. Early continuous assays employed Co [6], but for some ferrochelatases, Co is less active than other metals.

7. The isosbestic point for each metal with each porphyrin differs [2]. Wavelength pairs are deuteroheme IX, 496–484, $\Delta\varepsilon_{mM} = 3.5$; mesoheme IX, 497–486, $\Delta\varepsilon_{mM} = 7.5$; and protoheme, 497–486, $\Delta\varepsilon_{mM} = 10.5$. Alternatively, one may follow protoporphyrin disappearance at 385 nM (58 mM^{-1} cm^{-1}) [10], and coproporphyrin disappearance at 385 (128 mM^{-1} cm^{-1}) [11]. For Ni insertion into mesoporphyrin IX, use nickel chloride and follow an increase in absorbance at 550 nm. Scan the assay mixture every 30 s for 3 min.
8. Appropriate volumes of variable substrate to use for kinetics are 100, 80, 60, 50, 40, 30, 25, 20, 15, and 10 μL. Perform all assays in triplicate.

References

1. Obi CD, Bhuiyan T, Dailey HA, Medlock AE (2022) Ferrochelatase: mapping the intersection of iron and porphyrin metabolism in the mitochondria. Front Cell Dev Biol 10:894591. https://doi.org/10.3389/fcell.2022.894591
2. Dailey HA, Gerdes S, Dailey TA, Burch JS, Phillips JD (2015) Noncanonical coproporphyrin-dependent bacterial heme biosynthesis pathway that does not use protoporphyrin. Proc Natl Acad Sci USA 112(7): 2210–2215. https://doi.org/10.1073/pnas.1416285112
3. Dailey HA, Fleming JE (1983) Bovine ferrochelatase. Kinetic analysis of inhibition by N-methylprotoporphyrin, manganese, and heme. J Biol Chem 258(19):11453–11459
4. Porra RJ, Jones OT (1963) Studies on ferrochelatase. 1. Assay and properties of ferrochelatase from a pig-liver mitochondrial extract. Biochem J 87(1):181–185. https://doi.org/10.1042/bj0870181
5. Jones OT (1968) Ferrochelatase of spinach chloroplasts. Biochem J 107(1):113–119. https://doi.org/10.1042/bj1070113
6. Jones MS, Jones OT (1969) The structural organization of haem synthesis in rat liver mitochondria. Biochem J 113(3):507–514. https://doi.org/10.1042/bj1130507
7. Sellers VM, Wu CK, Dailey TA, Dailey HA (2001) Human ferrochelatase: characterization of substrate-iron binding and proton-abstracting residues. Biochemistry 40(33): 9821–9827. https://doi.org/10.1021/bi010012c
8. Olsson U, Billberg A, Sjovall S, Al-Karadaghi S, Hansson M (2002) In vivo and in vitro studies of Bacillus subtilis ferrochelatase mutants suggest substrate channeling in the heme biosynthesis pathway. J Bacteriol 184(14): 4018–4024. https://doi.org/10.1128/JB.184.14.4018-4024.2002
9. Smith KM (ed) (1975) Porphyrins and metalloporphyrins, 1st edn. Elsevier/North-Holland Biomedical Press, Amsterdam
10. Lobo SA, Scott A, Videira MA, Winpenny D, Gardner M, Palmer MJ, Schroeder S, Lawrence AD, Parkinson T, Warren MJ, Saraiva LM (2015) Staphylococcus aureus haem biosynthesis: characterisation of the enzymes involved in final steps of the pathway. Mol Microbiol 97(3):472–487. https://doi.org/10.1111/mmi.13041
11. Gabler T, Sebastiani F, Helm J, Dali A, Obinger C, Furtmuller PG, Smulevich G, Hofbauer S (2022) Substrate specificity and complex stability of coproporphyrin ferrochelatase is governed by hydrogen-bonding interactions of the four propionate groups. FEBS J 289(6): 1680–1699. https://doi.org/10.1111/febs.16257

Chapter 15

Reverse Thiol Trapping Approach to Assess the Thiol Status of Metal-Binding Mitochondrial Proteins

Hui Zhong, Eva Nyvltova, and Antoni Barrientos

Abstract

Thiol-disulfide interconversions are pivotal in the intricate chemistry of biological systems. They play a vital role in governing cellular redox potential and shielding against oxidative harm. These interconversions can also act as molecular switches within an expanding array of redox-regulated proteins, facilitating dynamic and responsive processes. Furthermore, metal-binding proteins often use thiols for coordination. Reverse thiol trapping is a valuable analytical tool to study the redox state of cysteines in biological systems. By selectively capturing and stabilizing free thiol species with an alkylating agent, reverse thiol trapping allows for their subsequent identification and quantification. Various methods can be employed to analyze the trapped thiol adducts, including electrophoresis-based methods, mass spectrometry, nuclear magnetic resonance spectroscopy, and chromatographic techniques. In this chapter, we will focus on describing a simple and sensitive method to sequentially block thiols in their cellular state with a cell-permeant agent (iodoacetamide), and following reduction and denaturation of the samples, trap the native disulfides with a second blocker that shifts the apparent molecular weight of the protein. The oxidation status of proteins for which suitable antibodies are available can then be analyzed by immunoblotting. We present examples of mitochondrial proteins that use cysteine thiols to coordinate metal factors such as iron-sulfur clusters, zinc, and copper.

Key words Cysteine oxidation, Thiol trapping, Cysteine disulfide, Redox biology, Mitochondria, AMS, IAA

1 Introduction

Proteins frequently utilize cysteine thiols as coordination sites for binding metal prosthetic groups such as iron-sulfur clusters, zinc, or copper. These metals play crucial roles in the structural integrity and catalytic function of these proteins. The thiol group of cysteine provides a strong nucleophile, capable of forming coordinate bonds with metal ions through the donation of its lone pair of electrons. This interaction between the thiol group and metal ions is often facilitated by nearby amino acid residues within the protein's active

Oleh Khalimonchuk (ed.), *Iron Metabolism: Methods and Protocols*, Methods in Molecular Biology, vol. 2839,
https://doi.org/10.1007/978-1-0716-4043-2_15,

site, creating a specific and well-defined binding pocket. Through this coordination, cysteine residues can effectively anchor metal ions, allowing for the formation of metal clusters or the stabilization of metal centers within the protein structure. Metal-binding proteins that employ cysteine thiols as coordination sites display a wide range of functionalities. Iron-sulfur proteins, for instance, are involved in electron transfer reactions and serve as essential components in processes such as respiration and photosynthesis. Zinc-binding proteins participate in diverse biological processes, including enzyme catalysis, DNA binding, and gene regulation. Copper-binding proteins are crucial for oxygen transport, oxidative stress regulation, and electron transfer reactions [1–4]. Metals can also play structural roles in proteins. For example, recent studies have shown that human mitochondrial ribosomes contain proteins that coordinate [2Fe-2S] clusters or zinc that maintain the structural integrity of the ribosome [5–8].

Studies of metal-binding proteins often require assessing the redox status of their cysteines: the presence of free thiols (sulfhydryl groups) versus metal-bound thiols. Methods for assessing the presence of free thiols in proteins in vivo or in the cellular context require careful sample handling to maintain the native redox status. This involves minimizing thiol and disulfide rearrangements and preventing unintended oxidation during sample preparation [9]. Furthermore, protein denaturation may be necessary to fully expose the target species, enhancing their reactivity with the reagents employed [9, 10].

Here, we describe a reverse thiol-trapping approach suitable for assessing the coordination of metals in mitochondrial proteins, by determining the redox status of the cysteine residues involved (*see* Fig. 1). A critical first step is quenching the biological samples to trap the cellular or mitochondrial thiol–disulfide status. For this purpose, isolated highly purified mitochondrial samples are first incubated with iodoacetamide (IAA), a cell- and mitochondrion-permeable sulfhydryl alkylating compound that can interact with any free sulfhydryl group from cysteine residue and form covalent bonds [11]. The cysteine residues engaged in disulfide bonds or metal coordination cannot interact with IAA in this step. Then, the samples are denatured with sodium dodecyl sulfate (SDS) and reduced by a strong reducing agent, tris(2-carboxyethyl)phosphine (TCEP). After this step, the proteins have lost their secondary structures and metal coordination, exposing all cysteines. Those previously free cysteines are now covalently bound to IAA, and the cysteines that natively engaged in disulfide bonds or metal binding now have reduced sulfhydryl groups following treatment with TCEP. The samples are subsequently treated with another thiol-modifying compound, 4-acetamido-4′-maleimidylstilbene-2,2′-disulfonic acid (AMS). AMS reacts with the reduced free

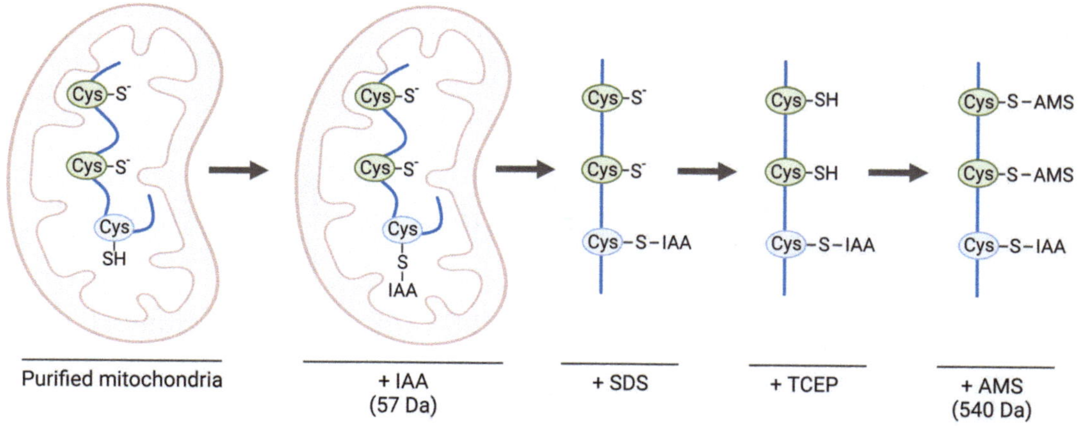

Fig. 1 The reverse thiol trapping approach. The depicted method enables the detection of the native cysteine residues inaccessible to a cell-permeable alkylating compound (2-iodoacetamide, IAA). Upon denaturation with SDS (sodium dodecyl sulfate) and full reduction with TCEP (tris(2-carboxyethyl)phosphine), the IAA-inaccessible cysteine is bound to AMS (4-acetamido-4′-maleimidylstilbene-2,2′-disulfonic acid), which adds ~540 Da per thiol group

cysteines to form thioether bonds, increasing the molecular weight of the proteins by around 540 Da per AMS molecule added [12]. The 540 Da mass difference can be detected by Western blot or mass spectrometry.

We present examples of the successful use of the method to determine the coordination of Fe–S clusters, and zinc in human mitochondrial ribosome proteins (*see* Fig. 2), as well as copper coordination by human mitochondrial cytochrome *c* oxidase copper chaperones (*see* Fig. 3). As native thiol blocking is the most crucial step of the approach, we provide tips on the selection of the optimal reaction conditions required for the efficient quenching of the native thiols via IAA labeling (pH, incubation time, and alkylating agent concentration) and the use of appropriate controls to determine the quenching efficiency. A comprehensive exploration of different alkylating reagents and treatments, as well as potential pitfalls, can be found elsewhere [9, 10].

2 Materials

Prepare all the solutions in ultrapure water (*see* **Note 1**) and analytical grade reagents. Prepare and store all reagents at room temperature unless indicated otherwise. Avoid the use of detergent in washing glassware and plasticware because residual detergent might disrupt the integrity of the mitochondria and native protein structures. Diligently follow all waste disposal regulations when disposing of waste materials.

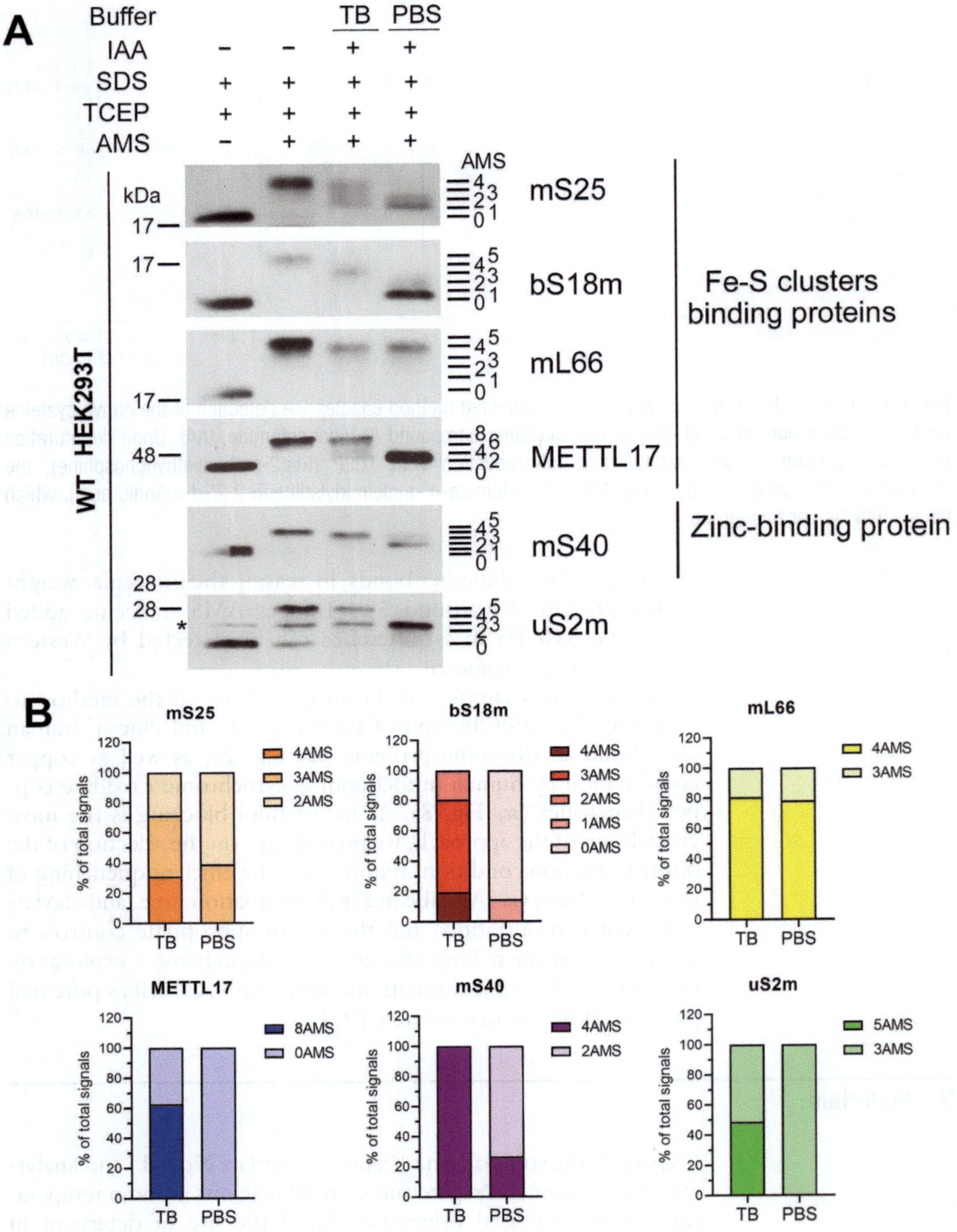

Fig. 2 Application of the reverse thiol approach to study the redox state of mitoribosome metal-binding proteins. (**a**) Immunoblot analysis of reverse thiol trapping of [2Fe-2S] clusters-binding mitoribosomal proteins (mS25, bS18m, and mL66), [4Fe-4S] cluster-binding mitoribosome assembly factor (METTL17), zinc-binding mitoribosomal protein (mS40), and a mitoribosomal protein containing cysteines not involved in metal coordination (uS2m), in mitochondria isolated from WT HEK293T cells. (**b**) The bar graphs represent the number of AMS-bound cysteines (unavailable to IAA binding *in isolated mitochondria*). As a comparison, IAA

2.1 Mitochondria Isolation from Cultured Cells

1. Human embryonic kidney 293 T (HEK293T) cells.
2. T-K-Mg buffer: 10 mM Tris–HCl, pH 7.4, 10 mM KCl, 0.5 mM $MgCl_2$. Store at 4 °C.
3. Sucrose stock solution: 1 M sucrose, 10 mM Tris–HCl, pH 7.4. Store at 4 °C.
4. ST Resuspension buffer: 0.32 M sucrose, 20 mM Tris-HCl, pH 7.4, 1 mM phenylmethylsulfonyl fluoride (PMSF), 1× EDTA-free protease inhibitor cocktail. Store at 4 °C.
5. 10 mL Glass/Teflon homogenizer.
6. PBS: 1× Dulbecco's Phosphate Buffered Saline solution, no calcium, no magnesium, pH 7.2.
7. 50× EDTA-free protease inhibitor cocktail.

2.2 Reverse Thiol Trapping

1. 2× Translation mix: 100 mM mannitol (*see* **Note 2**), 10 mM sodium succinate, 80 mM KCl, 5 mM $MgCl_2$, 1 mM sodium phosphate buffer, pH 7.4, 25 mM HEPES, pH 7.4.
2. 100× amino acid mix: 6 mg/mL each alanine, arginine, aspartic acid, histidine, lysine, glutamic acid, serine, threonine, asparagine, glutamine, glycine, proline, valine, isoleucine, leucine, phenylalanine, and tryptophan. Dissolve in H_2O. Store at −20 °C.
3. Translation buffer: 1× translation mix, 1× amino acid mix, 60 μg/mL cysteine (*see* **Note 3**), 60 μg/mL tyrosine (*see* **Note 4**), 60 μg/mL methionine (*see* **Note 5**), 5 mM ATP, 20 μM GTP, 6 mM phosphocreatine disodium salt hydrate, 60 μg/mL creatine phosphokinase from bovine heart (*see* **Note 6**).
4. Iodoacetamide (IAA) buffer: 80 mM solution. Weight 14.80 mg of iodoacetamide powder and dissolve in 1 mL of translation buffer (*see* **Notes 6** and 7).
5. Sodium dodecyl sulfate (SDS): 10% (w/v) solution dissolved in H_2O.
6. Tris(2-carboxyethyl)phosphine (TCEP) buffer: 75 mM solution. Weight 1.075 mg of TCEP powder and dissolve in 50 μL of 1 M Tris-HCl, pH 8.0. Store at −20 °C (*see* **Note 8**).
7. 4-Acetamido-4′-Maleimidylstilbene-2,2’-Disulfonic Acid, Disodium Salt (AMS) buffer: 25 mg/mL solution in DMSO. Protect from light. Store at −20 °C.

Fig. 2 (continued) was dissolved in either TB (translation buffer) or PBS. Compared to the samples with PBS, the amounts of bound AMS in TB are higher in most of the cases and more consistent with the structure data generated by cryo-EM [6, 13]. The difference may be caused by a lack of divalent cation in the PBS buffer. *, nonspecific band

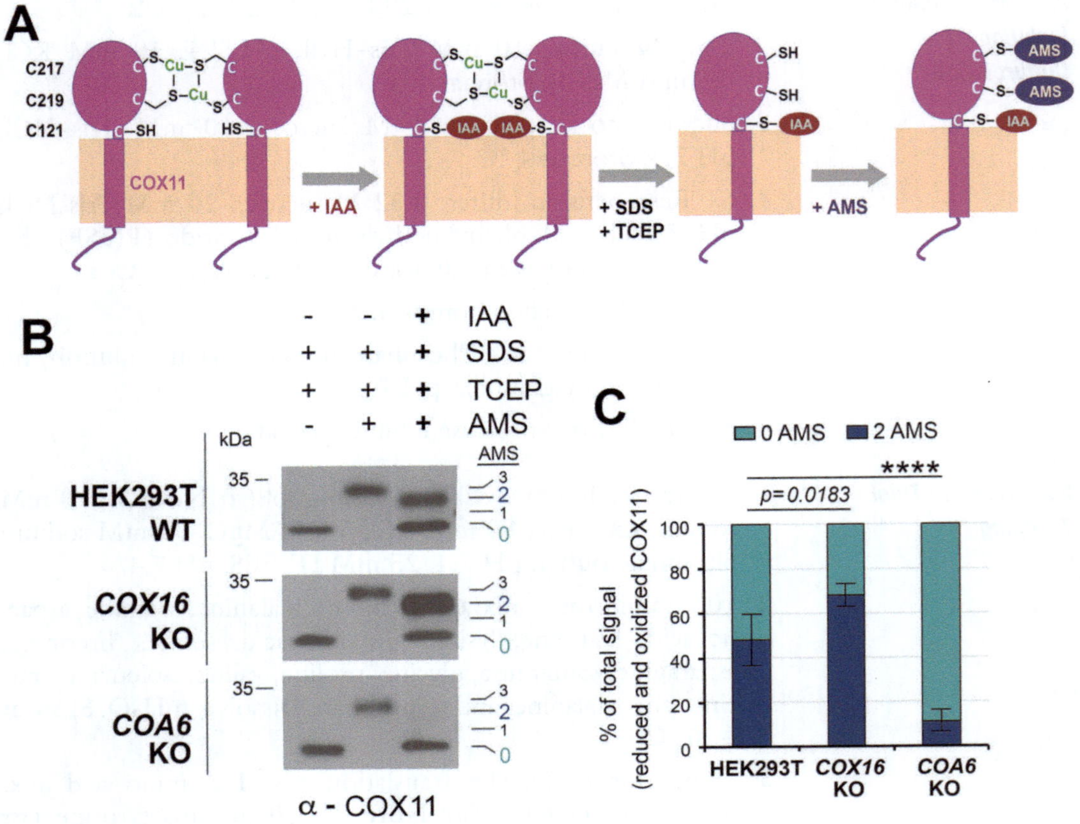

Fig. 3 Reverse thiol trapping of the copper binding-protein COX11. (**a**) Reverse thiol trapping approach used to detect the native cysteine residues in the copper (Cu)-bound COX11 dimer. COX11 is a cytochrome *c* oxidase assembly factor that delivers copper to COX1 [14]. (**b**) Reverse COX11 thiol trapping in wild-type HEK293T cells and the indicated cell lines carrying mutations in other cytochrome *c* oxidase assembly factors. (**C**) The bar graphs represent the average ± SD of four independent experiments. Two-sided unpaired *t*-test, $^{****}p < 0.0001$

8. PBS: 1× Dulbecco's Phosphate Buffered Saline solution, no calcium, no magnesium, pH 7.2.
9. 4× Laemmli buffer: 200 mM Tris-HCl, pH 6.8, 8% (w/v) sodium dodecyl sulfate (SDS), 40% (v/v) glycerol, 4% (v/v) β-mercaptoethanol, 0.4% (w/v) bromophenol blue. Store at −20 °C.

3 Methods

3.1 Isolation of Mitochondria

To obtain a clear signal from the mitochondrial proteins, mitochondria are purified to exclude the noise from the cytoplasm and other organelles. The protocol for the isolation of mitochondria from exponentially growing cells is based on the method reported by Enriquez and Attardi [15] with slight modifications [16].

1. Grow HEK293T cells in complete growth media to obtain ten to fifteen 150 mm plates at 80–90% confluency. Alternatively, HEK293T cells can be grown in suspension in 1 L of liquid culture until reaching 80–90% cell confluency.
2. To collect the cells grown in plates, remove the media, wash once with 10 mL PBS, and add 1.5 mL trypsin to each plate. Incubate at 37 °C for 2–3 min or until the cells start detaching from the plate. Neutralize the trypsin with 6 mL of growth media and transfer the cell suspension to a 50-mL centrifuge tube. Collect the cells by centrifugation at 400× *g* for 5 min. If the cells are grown in suspension, collect them by direct centrifugation at 400 × *g* for 5 min. Discard the supernatant.
3. Wash the cell pellets twice with 10 mL of cold PBS. Discard the supernatant. Before the final centrifugation step, transfer the cells to a pre-weighted tube and centrifuge at 400× *g* for 5 min.
4. Weigh the collecting tubes containing the pelleted cells and calculate the weight of the wet cell pellets.
5. Resuspend the cell pellets with ice-cold T-K-Mg buffer using 1 mL for 0.15 g of wet cells.
6. Incubate on ice for 10 min to allow the swelling of cells.
7. Disrupt the cell membranes with 16 strokes in a Teflon/glass homogenizer until ~70% are broken (*see* **Note 9**).
8. Add sucrose stock solution to the homogenate to reach a final concentration of 0.25 M sucrose (*see* **Note 10**).
9. Centrifuge the homogenate at 1200× *g* for 3 min at 4 °C to pellet unbroken cells, debris, and nuclei.
10. Carefully transfer the supernatant to a new tube (*see* **Note 11**). Centrifuge the supernatant again at 1200× *g* for 3 min at 4 °C to ensure complete elimination of heavy contaminants.
11. Carefully transfer the supernatant to a new tube. Centrifuge the supernatant at 10,000× *g* for 10 min at 4 °C.
12. Discard the supernatant. Gently resuspend the pelleted mitochondria with 1 mL of resuspend buffer and transfer them into a 1.5-mL microtube.
13. Centrifuge at 10,000× *g* for 10 min. at 4 °C.
14. Discard the supernatant. Gently resuspend the pellets with 300 μL of ST resuspension buffer.
15. Quantify the concentration of mitochondria proteins with the standard protein quantification method used in your laboratory (e.g., Lowry of Bradford reagents).
16. Store mitochondria on ice until use. Unused mitochondria can be aliquoted and snap frozen in liquid N_2 and stored at −80 °C for additional uses.

3.2 Reverse Thiol Trapping Approach

In each experimental set, three kinds of samples are included: a negative control, a positive control, and the experimental samples. The experimental samples are first incubated with IAA before being denatured, reduced, and labeled with AMS. However, for the negative and positive control samples, omit the labeling with the IAA step and proceed directly to the washing step. Following denaturation with SDS and full sample reduction with TCEP, the samples are incubated (positive control) or not (negative control) with AMS. In this way, during SDS-PAGE, proteins in the negative control will undergo no shift in mobility, while proteins in the positive control sample will undergo a maximum shift, corresponding to the number of thiols present in the protein. It is important to note that each sample contains the same amount (30 μg) of isolated mitochondria.

3.2.1 Labeling with IAA (See Note 12)

Quenching the samples with an alkylating agent before lysing the cells or the isolated mitochondria is a critical step to trap the native thiol–disulfide status of proteins. It does prevent artificial oxidation during mitochondrial lysis or sample preparation and needs to be rapid to avoid perturbation of the thiol–disulfide equilibria.

1. Aliquot 30 μg of mitochondria into a microtube. Centrifuge at 8000× *g* for 5 min at 4 °C. Discard the supernatant.
2. Resuspend the mitochondria with 300 μL of IAA buffer (*see* **Note 13**), with gentle pipetting to mix homogeneously.
3. Incubate at room temperature for 30 min. Invert the tube from time to time to resuspend the sedimented mitochondria.

3.2.2 Washing

1. Add 1 mL of PBS to the mitochondria. Centrifuge at 8000× *g* for 5 min at 4 °C. Discard the supernatant and keep the pellets.
2. Repeat washing with 1 mL of PBS two more times. Discard the supernatant and keep the pellets.

3.2.3 Denaturation, Reduction, and Labeling with AMS (See Notes 14 and 15)

AMS is light-sensitive; thus, all reactions should be performed in the dark.

1. The mitochondria pellet is resuspended using 30 μL (negative control samples) or 26 μL (positive control and IAA-treated samples) of PBS.
2. Add 6 μL of 10% SDS and 1.5 μL of TCEP buffer to the mitochondria. Gentle pipet to mix well.
3. Incubate at 95 °C for 12 min, and then put on ice briefly to cool down to room temperature. The samples can also be incubated at 4 °C overnight.
4. Add 4 μL of AMS buffer to the positive control sample and the IAA-incubated sample. Do not add AMS to the negative control sample.

5. Incubate at room temperature for 1 h in a dark space.
6. Add 12.5 μL of 4× Laemmli buffer to the samples and pipet to mix well.
7. Load all 50 μL of samples to the SDS-PAGE followed by immunoblot analysis (*see* **Note 16**).

4 Notes

1. Ultrapure water can be prepared by purifying deionized water to attain a sensitivity of 18 MΩ-cm at 25 °C using a water GenPurePro purifier (Thermo).
2. Mannitol is dissolved in water, although its solubility is only 18% (w/v) at 25 °C. Therefore, to facilitate the process, heat the water up to 80 °C and shake the solution vigorously.
3. Pure cysteine powder needs to be dissolved in 1 M HCl. If L-cysteine hydrochloride monohydrate is used, the powder can be dissolved in water directly.
4. The solubility of tyrosine in water at a pH lower than 2 or higher than 9 is much better than at physiological pH (around 2.5 mM) [17]. We recommend adding NaOH to the solution to facilitate tyrosine solubilization.
5. Generally, the solubility of methionine in water is known to be approximately 50–55 g/L at 25 °C. If necessary, use pre-heated water to facilitate methionine solubilization.
6. Prepare a fresh solution each time.
7. Work under the biochemical hood. Do not inhale the substance/mixture.
8. TCEP is soluble in water and odorless. The pH of pure TCEP solution is very acidic, around pH 2.5. To avoid the interference of low pH to AMS interaction, Tris–HCL pH 8.0 buffer is recommended to dissolve TCEP. TCEP in Tris–HCL or HEPES solution has been shown to be stable, with less than 20% being oxidized after 21 days at pH values between 1.5 and 11.1 [18]. TCEP is not very stable in phosphate-buffered saline and other phosphate-containing reagents, particularly at neutral pH values. Fifty to 100% of TCEP is oxidated after 72 h. Therefore, if dissolving TCEP in PBS is needed, prepare fresh buffer every time before use.
9. For optimal homogenization, it is essential to create a vacuum within the dounce using the pestle. This negative pressure will effectively rupture the plasma membrane. Depending on the tightness of the pestle, the number of strokes may require optimization.

10. The volume of the required sucrose stock solution equals one-third of the volume of the T-K-Mg buffer.
11. If the yield of mitochondria is too low, it is possible to resuspend the first low-speed pellet with the same volume of T-K-Mg buffer, perform a second homogenization, and follow the same centrifugation steps to collect extra mitochondria.
12. Choosing the appropriate reaction conditions for alkylation reactions depends on the agent used. We propose to use IAA, but N-ethylmaleimide (NEM) is also used often. The pH of the buffer, alkylating agent concentration, and time of incubation need to be determined empirically for each sample or protein that is being analyzed. Details can be found elsewhere [9].
13. Prewarm the IAA buffer to 37 °C.
14. Thiol-specific reagents that modify protein mobility, such as AMS, are used to estimate the thiol–disulfide redox state. After blocking thiols and reducing disulfides, thiol derivatization with these reagents results in proteins with a significantly higher molecular weight. Thus, proteins with oxidized thiols run more slowly on SDS–PAGE than do proteins with reduced thiols. This technique is compatible with immunoblot analysis, where specific cellular proteins can be detected without the need for purification.
15. In addition to AMS, other thiol-specific reagents that modify protein mobility can be used. They include polyethylene glycol (PEG) derivatives of N-ethylmaleimide (NEM) with molecular masses of 2 and 5 kDa [9]. AMS adds 540 Da per thiol, and then it is useful when working with relatively small proteins. PEG derivatives are more useful when working with larger proteins to facilitate their proper detection by immunoblotting.
16. Recommended running on a 14-well medium size 13% polyacrylamide gel. However, the gel properties need to be chosen depending on the MW of the protein under study and the number of thiols that it contains.

Acknowledgments

This research was supported by the National Institutes of Health grant R35-GM118141 (AB).

References

1. Johnson DC et al (2005) Structure, function, and formation of biological iron-sulfur clusters. Annu Rev Biochem 74:247–281
2. Lee TT, Agarwalla S, Stroud RM (2004) Crystal structure of RumA, an iron-sulfur cluster containing *E. coli* ribosomal RNA 5-methyluridine methyltransferase. Structure 12(3):397–407
3. Outten CE, O'Halloran TV (2001) Femtomolar sensitivity of metalloregulatory proteins

controlling zinc homeostasis. Science 292(5526):2488–2492
4. Waldron KJ et al (2009) Metalloproteins and metal sensing. Nature 460(7257):823–830
5. Zhong H et al (2023) BOLA3 and NFU1 link mitoribosome iron-sulfur cluster assembly to multiple mitochondrial dysfunctions syndrome. bioRxiv, p 2023.05.27.542581
6. Harper NJ, Burnside C, Klinge S (2023) Principles of mitoribosomal small subunit assembly in eukaryotes. Nature 614(7946):175–181
7. Itoh Y et al (2021) Mechanism of membrane-tethered mitochondrial protein synthesis. Science 371(6531):846–849
8. Saurer M et al (2023) Molecular basis of translation termination at noncanonical stop codons in human mitochondria. Science 380(6644): 531–536
9. Hansen RE, Winther JR (2009) An introduction to methods for analyzing thiols and disulfides: reactions, reagents, and practical considerations. Anal Biochem 394(2): 147–158
10. Hawkins CL, Davies MJ (2019) Detection, identification, and quantification of oxidative protein modifications. J Biol Chem 294(51): 19683–19708
11. Gundry RL et al (2010) Preparation of proteins and peptides for mass spectrometry analysis in a bottom-up proteomics workflow. Curr Protoc Mol Biol 90(1):10.25.1–10.25.23
12. Botello-Morte L et al (2016) Cysteine mutational studies provide insight into a thiol-based redox switch mechanism of metal and DNA binding in FurA from anabaena sp. PCC 7120. Antioxid Redox Signal 24(4):173–185
13. Itoh Y et al (2022) Structure of the mitoribosomal small subunit with streptomycin reveals Fe-S clusters and physiological molecules. elife 11:e77460
14. Nývltová E et al (2022) Coordination of metal center biogenesis in human cytochrome c oxidase. Nat Commun 13(1):3615
15. Enríquez J, Attardi G (1996) [17] Analysis of aminoacylation of human mitochondrial tRNAs. In: Methods in enzymology. Elsevier, pp 183–196
16. Maiti P, Fontanesi F (2023) Metabolic labeling of mitochondrial translation products in whole cells and isolated organelles. In: The mitoribosome: methods and protocols. Springer, pp 193–215
17. Hitchcock DI (1924) The solubility of tyrosine in acid and in alkali. J Gen Physiol 6(6): 747–575
18. Han JC, Han GY (1994) A procedure for the quantitative determination of tris (2-carboxyethyl) phosphine, an odorless reducing agent more stable and effective than dithiothreitol. Anal Biochem 220(1):5–10

Chapter 16

A Chemoproteomic Approach to Monitor Native Iron–Sulfur Cluster Binding

Daniel W. Bak and Eranthie Weerapana

Abstract

Iron–sulfur (Fe-S) clusters are essential redox-active metallocofactors participating in electron transfer, radical chemistry, primary metabolism, and gene regulation. Successful trafficking and incorporation of Fe-S clusters into target proteins are critical to proper cellular function. While biophysical studies of isolated Fe-S proteins provide insight into the structure and function of these inorganic cofactors, few strategies currently exist to directly interrogate Fe-S cluster binding within a cellular environment. Here, we describe a chemoproteomic platform to report on Fe-S cluster incorporation and occupancy directly within a native proteome, enabling the characterization of Fe-S biogenesis pathways and the identification of undiscovered Fe-S proteins.

Key words Iron–sulfur clusters, Cysteine reactivity, Protein abundance, *Escherichia coli*, Mass spectrometry, Chemoproteomics, Activity-based protein profiling (ABPP), Reductive dimethylation (ReDiMe)

1 Introduction

Iron–sulfur (Fe-S) proteins harbor one or more Fe-S clusters, which are inorganic metallocofactors of iron and sulfide. Fe-S clusters are widely distributed and are evolutionarily ancient redox-active cofactors [1]. As such, Fe-S proteins have become central components of primary metabolism, where they perform diverse functions including (1) electron transfer reactions in respiratory chain complexes [2], (2) generation of reactive radical species for the synthesis of cofactors and the modification of tRNA/rRNA [3–5], (3) regulation of gene transcription and DNA repair [6–8], and (4) non-redox catalysis of primary metabolites, such as aconitate and fumarate [9, 10]. Due to the high reactivity and potentially damaging toxicity of their constitutive parts, iron and sulfide, organisms have developed complex pathways for the protected biosynthesis and delivery of these cofactors [11–13]. In

Oleh Khalimonchuk (ed.), *Iron Metabolism: Methods and Protocols*, Methods in Molecular Biology, vol. 2839,
https://doi.org/10.1007/978-1-0716-4043-2_16,

E. coli, Fe-S cluster biogenesis is performed by the housekeeping Isc pathway (expressed from the *isc* operon) under standard growth conditions. Briefly, Fe-S clusters are synthesized on the scaffold protein, IscU, with inputs of (1) iron, (2) electrons (from ferredoxin, Fdx), and (3) sulfide (liberated by the cysteine desulfurase, IscS). With the aid of the HscA/HscB chaperone/cochaperone system, clusters on IscU are shuttled to A-type scaffolds, such as IscA, for assembly into [4Fe-4S] clusters and subsequent delivery to client Fe-S proteins.

The field of Fe-S biology has been handicapped by a lack of techniques to efficiently identify novel Fe-S proteins and easily interrogate Fe-S biogenesis pathways directly within a cellular environment. The field has traditionally relied on in vitro characterization of the cluster-bound (holo) state of Fe-S proteins, which can be difficult to isolate due to the highly transient non-covalent nature of cluster binding. While some aspects of Fe-S cluster biogenesis have been reconstituted in vitro [11, 12, 14, 15], it remains challenging to validate these findings in cellular systems, and contradicting data from in vitro and in vivo experiments have been observed. Currently, only a few methodologies exist that can report on Fe-S cluster dynamics in native systems, including the use of radioisotopes [16, 17] and the assaying of downstream Fe-S enzyme activity [18, 19]. Along with being technically challenging and having only modest sensitivity, the major limitations of these studies are the lack of site specificity and dependence on preexisting knowledge of potential Fe-S protein targets and biogenesis components. As Fe-S clusters are primarily ligated by protein cysteine residues, we hypothesized that cysteine-targeted chemoproteomics could provide information on Fe-S cluster binding in native systems. Cysteine-targeted chemoproteomic platforms, such as isoTOP-ABPP, have been employed to great success to identify and quantify cysteine post-translational modifications (PTMs) [20, 21], cysteine oxidation [22, 23], and small-molecule binding [24, 25]. By leveraging the power of isoTOP-ABPP to quantify Fe-S cluster binding across the *E. coli* proteome, we identified unique functional perturbations to *E. coli* Fe-S proteins in response to iron depletion or deletion of components of the Isc pathway and identified two novel *E. coli* Fe-S proteins that had not previously been characterized [26]. Here, we provide a detailed protocol for applying this platform to quantitatively monitor cysteine reactivity changes in Fe-S cluster-disrupted proteomes (e.g., iron depletion or *isc* deletion strains).

Initially, bacterial cell lysates representing an Fe-S cluster-replete proteome are generated from cultures of wild-type (WT) *E. coli* grown in standard MOPS minimal media and lysates representing an Fe-S cluster-disrupted proteome generated from either WT *E. coli* grown in iron-depleted MOPS minimal media or *isc* deletion strains grown in standard MOPS minimal media. These

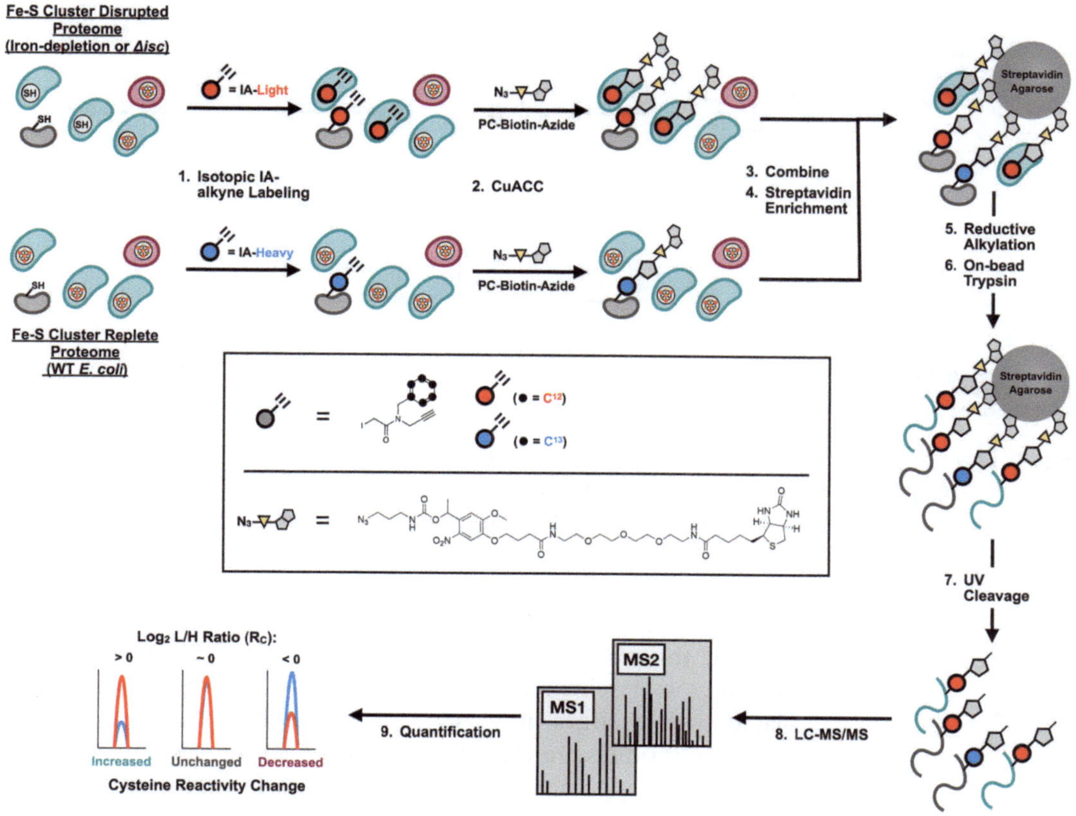

Fig. 1 Chemoproteomic workflow for a typical isoTOP-ABPP experiment comparing an Fe-S cluster-disrupted proteome (light) with a Fe-S cluster-replete proteome (heavy) to quantify changes in cysteine reactivity. Peptide $\log_2$ L/H ratios (R_c) > 1 indicate an increase in cysteine reactivity upon cluster disruption, while R_c values <1 indicate a decrease in reactivity. Inset: Structures of IA-Light (IAL), IA-Heavy (IAH), and photo-cleavable (PC) biotin-azide. (CuAAC—copper(I)-catalyzed azide–alkyne cycloaddition)

proteomes are first labeled with isotopically light (Fe-S cluster-replete proteome) or heavy (Fe-S cluster-disrupted proteome) iodoacetamide alkyne (IA) and subject to isotopic tandem orthogonal proteolysis–activity-based protein profiling (isoTOP-ABPP) LC-MS/MS analysis in order to quantify the changes in cysteine reactivity that occur upon loss of Fe-S cluster binding (Fig. 1) [27, 28]. Growth conditions that result in Fe-S cluster disruption will significantly impact cellular protein abundance, as a result of *E. coli* transcriptional programs geared to both increase intracellular iron levels [29] and optimize Fe-S cluster utilization [30]. Consequently, the extent of cysteine labeling by the IA probes reported by isoTOP-ABPP can reflect both changes in cysteine reactivity and/or protein abundance. Therefore, protein abundance changes are quantified in parallel by reductive dimethylation (ReDiMe) LC-MS/MS analysis, in which bacterial cell lysates are labeled with isotopically light (Fe-S cluster-replete proteome) or heavy

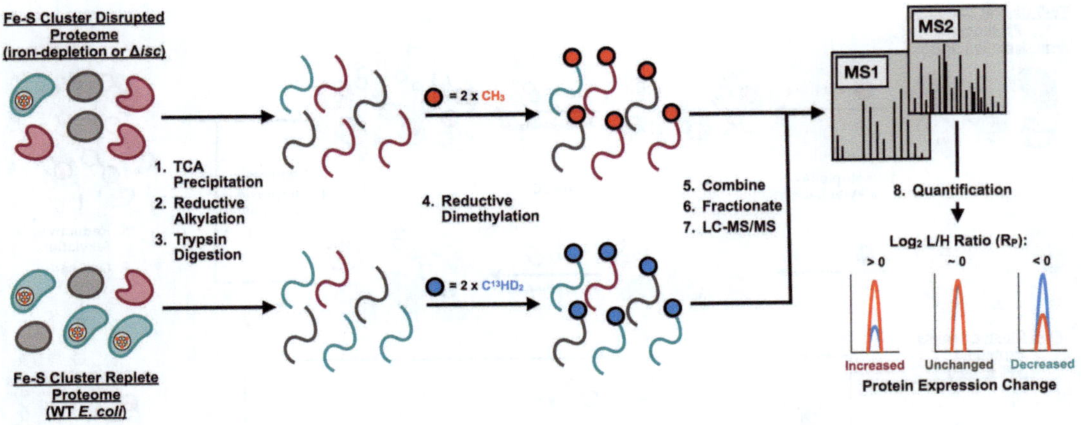

Fig. 2 Chemoproteomic workflow for a typical ReDiMe experiment comparing an Fe-S cluster-disrupted proteome (light) with a Fe-S cluster-replete proteome (heavy) to quantify changes in protein abundance. Protein $\log_2$ L/H ratios (R_p) > 1 indicate an increase in protein abundance upon cluster disruption, while R_p values <1 indicate a decrease in abundance

(Fe-S cluster-disrupted proteome) formaldehyde (Fig. 2) [31]. To enable the quantification of protein abundance-corrected (net) changes in cysteine reactivity, protein expression changes are correlated with changes in cysteine reactivity (Fig. 3a). Fe-S ligating cysteine residues are expected to display net increases in reactivity, while those cysteine residues unaffected by Fe-S cluster disruption should display changes in reactivity that track linearly with protein abundance. A typical experiment can quantify net changes in cysteine reactivity for 30–50 Fe-S ligands, as well as the unbiased quantification of cysteine reactivity changes across the *E. coli* proteome, potentially enabling the identification of novel Fe-S proteins.

2 Materials

All buffers and solutions are made with ultrapure Milli-Q water unless indicated (e.g., LC and LC-MS buffers, which instead use MS-grade solvents).

2.1 *E. coli* Growth Conditions and Generation of Lysates

1. LB Agar Plates: 10 g tryptone, 5 g NaCl, 5 g yeast extract, and 16 g agar autoclaved in 1 L water. Once cool enough to handle, add appropriate antibiotics and pour ~25 mL into each 10-cm petri dish. Cover and store at 4 °C.
2. Minimal Media Recipe [32]: 40 mM MOPS and 4 mM tricine (adjusted to pH 7.4 with KOH), 0.1 M NaCl, 10 mM NH_4Cl, 1.32 mM KH_2PO_4, 0.523 mM $MgCl_2$, 0.276 mM Na_2SO_4, 0.1 mM $FeSO_4$, trace micronutrients (4×10^{-4} mM H_3BO_3, 8×10^{-5} mM $MnCl_2$, 1×10^{-5} mM $ZnSO_4$, 1×10^{-5} mM $CuSO_4$, 3×10^{-6} mM $CoCl_2$, and 3×10^{-6} mM $(NH_4)_6Mo_7O_{24}$), and 0.1% glucose.

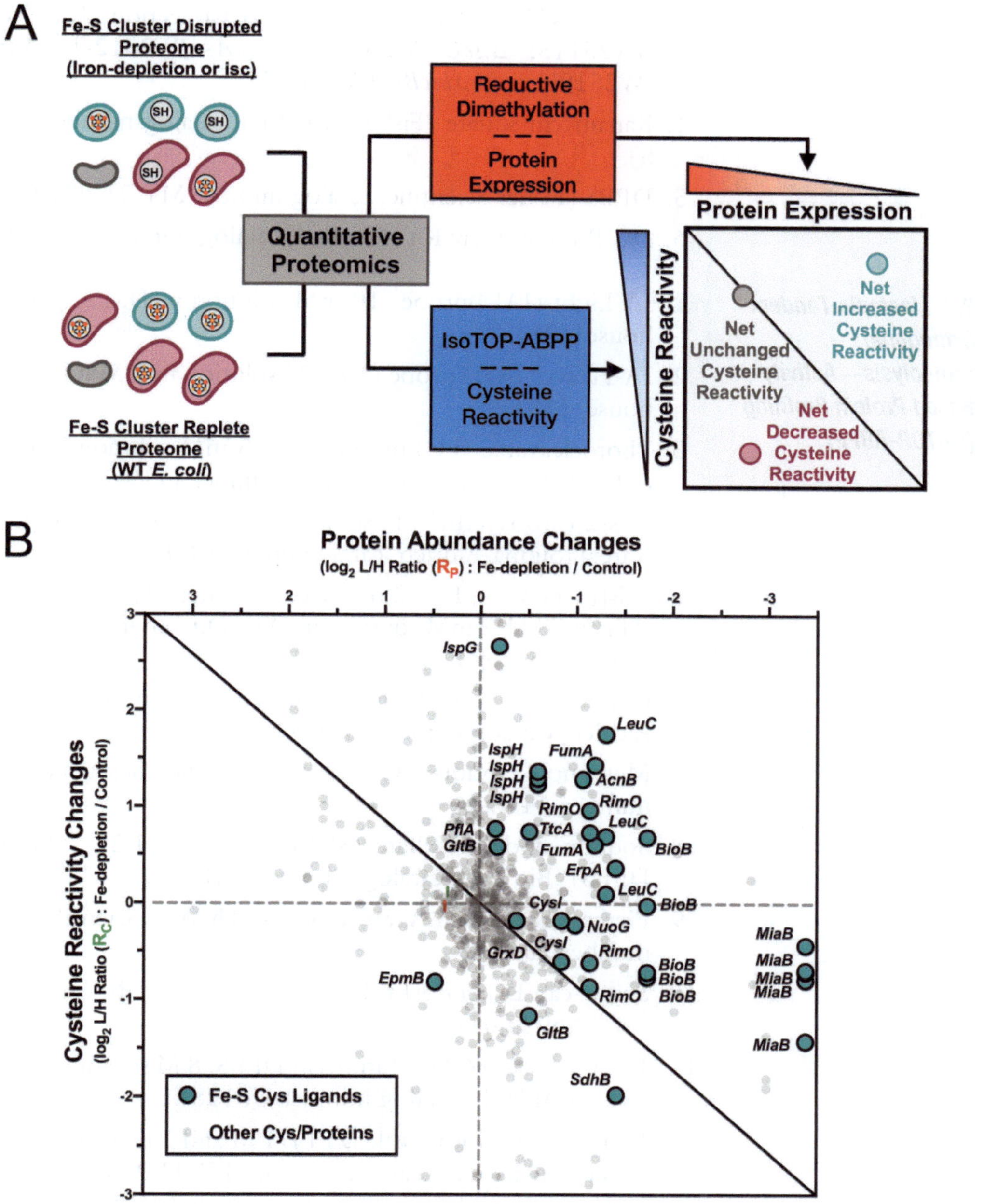

Fig. 3 Two-dimensional chemoproteomic analysis of Fe-S cluster occupancy. (**a**) Correlation of ReDiMe proteomic data (R_p—x-axis) for protein expression changes with isoTOP-ABPP proteomic data (R_c—y-axis) for cysteine reactivity changes. Peptides with isoTOP-ABPP ratios (R_c) which lie above the $R_c = R_p$ correlation line are considered to display a net change in cysteine reactivity that is proportional to the distance above the correlation line. (**b**) Two-dimensional proteomic dataset for the *E. coli* proteome grown in iron-depleted media. All quantified cysteine residues are plotted as light gray circles, while annotated Fe-S cysteine ligands are shaded in teal. Data for the plot obtained from [26]

3. *E. coli* K12 Bacterial Strains (Keio Collection) [33]: WT—BW25113; *ΔiscU*–JW2513-1; *ΔiscA*—JW2512-1; *ΔhscA*—JW2510-1; and *ΔhscB*—JW2511-1.
4. Kanamycin sulfate (Fisher Scientific, catalog number BP906-5).
5. DPBS (Fisher Scientific, catalog number MT-21-031-CM).
6. DC Protein Assay Kit (Bio-Rad, catalog number 5000111).

2.2 Isotopic Tandem Orthogonal Proteolysis—Activity-Based Protein Profiling (isoTOP-ABPP)

1. IA-Light (IAL) probe (10 mM solution in DMSO) (made in house) [28].
2. IA-Heavy (IAH) probe (10 mM solution in DMSO) (made in house) [28].
3. Photocleavable (PC) biotin azide (5 mM solution in DMSO) (Click Chemistry Tools, catalog number 1119-25).
4. Tris-2-carboxyethyl phosphine (TCEP) (50 mM solution in water) (Sigma-Aldrich, catalog number C4706).
5. Tris[(1-benzyl-1H-1,2,3-triazol-4-yl)methyl]amine ("ligand") (1.7 mM solution in DMSO:t-butanol 1:4) (Sigma-Aldrich, catalog number 678937).
6. Copper (II) sulfate (100 mM solution in water) (Sigma-Aldrich, catalog number C1297).
7. Methanol, Optima LC/MS grade (Fisher Scientific, catalog number A456-4).
8. Sodium dodecyl sulfate (SDS) (0.2% and 1.2% solutions in DPBS) (Bio-Rad, catalog number 1610302).
9. Pierce Streptavidin Agarose beads (Thermo Scientific, catalog number 20353).
10. Screw-cap Eppendorf tubes (1.5 mL) (Sarstedt, catalog number 72.692.005).
11. Urea (2 M and 6 M solutions in DPBS, 8 M solution in TEAB) (Sigma-Aldrich, catalog number U5128).
12. Ultrapure dithiothreitol (DTT) (1 M and 200 mM solutions in water) (Invitrogen, catalog number 15508-013).
13. Iodoacetamide (500 mM and 400 mM solutions in water) (Acros Organics, catalog number 122270050).
14. Calcium chloride (100 mM solution in water) (Sigma-Aldrich, catalog number C1016).
15. Trypsin solution: 20 μg of sequencing grade trypsin reconstituted in 40 μL of trypsin buffer (provided in package) or 100 mM TEAB (Promega, catalog number V5111).
16. Stirbar, Flea Micro (Fisher Scientific, catalog number 14-513-64).

17. Black-Ray High-Intensity Inspection 365-nm UV Lamp (140 Watt/1.3 Amps/115 V) (VWR, catalog number 36595-025).
18. Protein LoBind Eppendorf tube (1.5-mL) (Eppendorf, catalog number 022431081).

2.3 Reductive Dimethylation (ReDiMe)—Protein Abundance

1. Trichloroacetic acid (100% w/v in water) (Sigma-Aldrich, catalog number T9159).
2. Acetone, Optima LC/MS grade (Fisher Scientific, catalog number A929SK-4).
3. Triethylammonium bicarbonate buffer (TEAB) (1 M diluted to 100 mM in water) (Sigma-Aldrich, catalog number T7409-100ML).
4. Formaldehyde-^{12}C solution (20% in H_2O) (Sigma-Aldrich, catalog number 606758).
5. Formaldehyde-^{13}C, d_2 solution (20% in D_2O) (Sigma-Aldrich, catalog number 596388).
6. Sodium cyanoborohydride (0.6 M solution in water) (Fisher Scientific, catalog number AC16855-0100).
7. Ammonium hydroxide (28–30% solution in water) (Sigma-Aldrich, catalog number 1054231011).
8. Formic acid, Optima LC/MS grade (Fisher Scientific, catalog number A1171-50).
9. Low-pH Buffer A: 95% water (Optima LC/MS grade), 5% acetonitrile (Optima LC/MS grade), and 0.1% formic acid (Optima LC/MS grade).
10. Low-pH Buffer B: 20% water (Optima LC/MS grade), 80% acetonitrile (Optima LC/MS grade), and 0.1% formic acid (Optima LC/MS grade).
11. Sep-Pak C18 classic cartridges (Waters, catalog number WAT051910).
12. Acetonitrile, Optima LC/MS grade (Fisher Scientific, catalog number A955-4).
13. Water, Optima LC/MS grade (Fisher Scientific, catalog number W6-4).
14. Ammonium bicarbonate (MP Biomedicals, catalog number 150107).
15. High pH Buffer A: 95% water (Optima LC/MS grade), 5% acetonitrile (Optima LC/MS grade), 10 mM ammonium bicarbonate.
16. High pH Buffer B: 10% water (Optima LC/MS grade), 90% acetonitrile (Optima LC/MS grade), 10 mM ammonium bicarbonate.

17. Agilent 1100 Series HPLC System (Agilent).
18. Manual injection valve (Agilent).
19. ZORBAX 25 cm Extend-C18 Column (Agilent, catalog number 770450-902).
20. Gilson FC 203B Fraction Collector (Gilson).
21. 96 deep-well microplates (Fisher Scientific, catalog number 12-566-611).
22. Protein LoBind Eppendorf tube (2,0 mL) (Eppendorf, catalog number 022431102).

2.4 LC-MS/MS Analysis

1. SureSTART Screw Top LC Microvials (2 mL) (Thermo Scientific, catalog numbers 03-452-259) and caps (Thermo Scientific, catalog numbers 03-376-483).
2. Orbitrap Exploris 240 Mass Spectrometer (Thermo Fisher Scientific).
3. Xcalibur v4.4 Software Package (Thermo Fisher Scientific).
4. Dionex UltiMate 3000 RSLCnano System (Thermo Fisher Scientific).
5. MS Buffer A: 100% water (Optima LC/MS grade), 0.1% formic acid (Optima LC/MS grade).
6. MS Buffer B: 20% water (Optima LC/MS grade), 80% acetonitrile (Optima LC/MS grade), 0.1% formic acid (Optima LC/MS grade).
7. Acclaim PepMap 100 Trap Column (Thermo Fisher Scientific, catalog number 164199).
8. Acclaim PepMap RSLC Analytical Column (Thermo Fisher Scientific, catalog number 164534).

2.5 Data Workup, Statistical Filtering, and Graphical Visualization

1. Proteome Discoverer v2.4 (Thermo Fisher Scientific).

3 Methods

This procedure describes a two-dimensional (isoTOP-ABPP and ReDiMe) quantitative chemoproteomics workflow employed to quantify protein abundance-corrected (net) cysteine reactivity changes that occur in *E. coli* upon disruption of Fe-S cluster biosynthesis and delivery. This Fe-S cluster disruption is achieved in one of two ways, either (1) through iron depletion of the growth media or (2) through deletion strains of the iron–sulfur cluster (*isc*) biosynthetic genes; *iscU*, *iscA*, *hscA*, and *hscB*. These Fe-S cluster-disrupted samples are compared to wild-type (WT) *E. coli* grown in iron-replete growth media to quantitatively determine in parallel both the proteomic changes in cysteine reactivity and protein abundance.

3.1 *E. coli* Growth Conditions and Generation of Lysates

1. Make LB agar plates. Prepare standard MOPS minimal media according to the recipe, omitting the addition of $FeSO_4$ for iron-depleted MOPS minimal media (*see* **Note 1**).
2. Streak out bacterial culture (from frozen glycerol stocks) onto LB agar plates with appropriate antibiotic (for wild-type *E. coli*, no antibiotic is necessary, but for *Δisc* strains, the LB agar plates should be kan^+: 50 μg/mL). Grow colonies overnight at 37 ° C.
3. Select single bacterial colonies to inoculate 4 mL overnight cultures of standard MOPS minimal media (do this even for eventual iron-depletion samples). Grow overnight cultures at 37 °C with 200 rpm rotation.
4. Inoculate 100 mL of standard or iron-depleted MOPS minimal media (no antibiotics) with 1 mL of appropriate overnight culture. Grow cultures at 37 °C with 200 rpm rotation to an O.D. of 0.6 (*see* **Note 2**).
5. Harvest cells by centrifugation of the growth culture at 5000 g for 10 min at 4 °C. Discard the supernatant and wash the cell pellet by gentle resuspension in 10 mL of ice-cold DPBS followed by centrifugation at 5000 g for 10 min at 4 °C. After discarding the supernatant, cell pellets can be frozen and stored at −80 °C until ready for sample preparation and analysis.
6. ***Optional:*** For iron-depletion studies, it may be necessary to confirm the successful reduction of cellular iron levels. This can be achieved by measuring the change in intracellular iron concentration by ICP-OES/MS (not described here).
7. To generate bacterial cell lysates appropriate for proteomic analysis, resuspend the thawed cell pellet in 1 mL of DPBS in a 1.5-mL Eppendorf tube and lyse by sonication (3 × 10 one-second pulses at 90% power). Clarify cell lysates by centrifugation at 10,000 g for 10 min and transfer the supernatant to a clean 1.5-mL Eppendorf or 15-mL conical tube. Determine the concentration of the lysate by Bradford assay (DC Protein Assay Kit; Bio-Rad) as described in the manufacturer's instructions, and dilute to 3 mg/mL with DPBS for immediate use in proteomic preps (*see* **Note 3**).

3.2 Isotopic Tandem Orthogonal Proteolysis–Activity-Based Protein Profiling (isoTOP-ABPP)—Cysteine Reactivity (Fig. 1) [27]

1. For each sample, label two 500 μL aliquots of Fe-S cluster-disrupted (either iron-depleted WT strain or *Δisc* strain) *E. coli* cell lysates (3 mg/mL in DPBS) by the addition of 5 μL of 10 mM IA-Light (IAL) probe (100 μM final concentration) [28]. In parallel, label two 500 μL aliquots of Fe-S cluster-replete WT *E. coli* cell lysates (3 mg/mL in DPBS) by the addition of 5 μL of 10 mM IA-Heavy (IAH) probe (100 μM final concentration) [28]. Vortex all four tubes briefly and incubate for 1 h at 25 °C in the dark (*see* **Note 4**).
2. To each of the four IAL or IAH-labeled aliquots of cell lysate perform Cu(I)-catalyzed azide–alkyne cycloaddition (CuACC) by adding, in order, 10 μL of 5 mM photocleavable (PC) biotin azide (100 μM final concentration), 10 μL of freshly prepared 50 mM TCEP (1 mM final concentration), 30 μL of 1.7 mM TBTA "ligand" (100 μM final concentration), and 20 μL of 100 mM copper sulfate (4 mM final concentration). Vortex and incubate click chemistry reaction for 1 h at 25 °C in the dark, vortexing reactions every 15 min (*see* **Note 5**).
3. After the click chemistry reaction is complete, quickly vortex and combine tubes pairwise (one IAL-labeled with one IAH-labeled tube). Centrifuge tubes at 6500 g for 5 min at 25 °C to pellet precipitated protein and discard the supernatant. There should now be two tubes/samples.
4. Resuspend pellets in 500 μL of ice-cold methanol by gentle sonication (5–10 pulses at 50% power) and combine the remaining two tubes. Centrifuge tube at 6500 g for 5 min at 4 °C and discard the supernatant. There should now be only one tube/sample.
5. Repeat ice-cold methanol wash of pellet (500 μL), discarding supernatant after centrifugation. Make sure that methanol is completely removed at this point (*see* **Note 6**).
6. Resuspend protein pellet in 1 mL of 1.2% SDS in DPBS by gentle sonication (5–10 pulses at 50% power) and subsequent heating at ~90 °C for 5 min.
7. Centrifuge sample at 6500 g for 5 min at 25 °C in order to pellet excess copper. Being careful to avoid disturbing the copper pellet, transfer the solubilized protein sample (1 mL) to a 15-mL conical tube with 5 mL of DPBS (0.2% SDS final concentration).
8. Aliquot 200 μL of streptavidin agarose slurry (~100 μL of beads) into a 1.5-mL Eppendorf and add 500 μL of DPBS. Pellet beads by centrifugation at 1400 g for 1 min at 25 °C and remove the supernatant. Wash beads two more times with 500 μL DPBS, discarding supernatant afterward. Add washed streptavidin beads to the 15-mL conical tube containing solubilized protein sample (*see* **Note 7**).

9. Incubate the sample, with rotation, overnight in the dark at 4 ° C, then transfer the sample to 25 °C, and incubate with rotation for an additional 2–3 h in the dark to resolubilize SDS.
10. Centrifuge samples at 1400 g for 5 min at 4 °C and remove the supernatant from beads by careful decanting or aspiration (*see* **Note 8**).
11. Wash beads one time with 5 mL of 0.2% SDS in DPBS, three times with 5 mL of DPBS, and three times with 5 mL of water. Between washes, pellet beads by centrifugation at 1400 g for 5 min at 4 °C and discard the supernatant as before.
12. Transfer beads to 1.5-mL screw-cap Eppendorf tubes using three sequential 166 μL aliquots of 6 M urea in DPBS (total of 500 μL) (*see* **Note 9**).
13. Add 25 μL of freshly prepared 200 mM dithiothreitol (10 mM final concentration) and heat the sample for 20 min in the dark at 65 °C, mixing once by inversion after 10 min.
14. Remove from heat and let the sample cool for a couple of minutes. Add 25 μL of freshly prepared 400 mM iodoacetamide (20 mM final concentration) and incubate the samples with rotation for 30 min in the dark at 37 °C (*see* **Note 10**).
15. Add 950 μL of DPBS (2 M urea final concentration) and centrifuge samples at 1400 g for 5 min at 25 °C. Remove the supernatant completely using a micropipette.
16. Resuspend beads in 200 μL of 2 M Urea in DPBS. Add 2 μL of 100 mM $CaCl_2$ (1 mM final concentration) and 4 μL (2 μg) of sequence grade-modified trypsin solution.
17. Mix by inversion and incubate with rotation overnight in the dark at 37 °C.
18. Centrifuge samples at 1400 g for 1 min at 25 °C and completely remove the supernatant (*see* **Note 11**).
19. Wash beads three times with 500 μL of DPBS and three times with water. Between washes pellet beads by centrifugation at 1400 g for 5 min at 4 °C and remove the supernatant as before.
20. Add 250 μL of water to beads, along with a micro-stir bar (flea), and incubate the sample uncapped directly under a 365 nm UV lamp for 2 h with gentle stirring (50 rpm).
21. Centrifuge samples at 1400 g for 1 min, and transfer supernatant to a clean 1.5-mL low-bind Eppendorf tube (*see* **Note 12**).
22. Add 75 μL of water to the beads, mix by inversion, centrifuge at 1400 g for 1 min, and transfer supernatant to the same 1.5-mL low-bind Eppendorf tube with the previous ~200 μL. Repeat wash with a second 75 μL aliquot of water (*see* **Note 13**).

23. Dry the sample overnight on a SpeedVac and then store the dried peptide pellet at −20 °C until ready for LC-MS/MS analysis.

3.3 Reductive Dimethylation (ReDiMe)—Protein Abundance (Fig. 2) [31]

1. For each sample, dilute Fe-S cluster-disrupted (either iron-depleted WT strain or *Δisc* strain) and Fe-S cluster-replete WT *E. coli* cell lysates threefold to 1 mg/mL (33 μL diluted to 100 μL with DPBS) into two clean 1.5-mL Eppendorf tubes.
2. Precipitate protein by the addition of 5 μL of 100% TCA solution. Vortex vigorously and freeze aliquots at −80 °C for at least 1 h or overnight.
3. Remove samples from the freezer, allow to thaw completely, and pellet protein by centrifugation at 15,000 g for 10 min. A visible white pellet should be present at the bottom of each tube.
4. Completely remove the supernatant without disturbing the protein pellet and resuspend in 500 μL ice-cold acetone by bath sonication.
5. Pellet protein by centrifugation at 5000 g for 10 min. Completely remove acetone without disturbing the protein pellet and allow to air dry (*see* **Note 14**).
6. Resuspend protein pellet in 30 μL of 8 M Urea (stock in 100 mM TEAB) with vortexing and bath sonication.
7. Add 70 μL of 100 mM TEAB followed by 1.5 μL of freshly prepared 1 M dithiothreitol (15 mM final concentration) and incubate the sample at 75 °C for 15 min.
8. Allow samples to cool for a couple of minutes and add 2.5 μL of freshly prepared 500 mM iodoacetamide (12.5 mM final concentration) and incubate the sample at room temperature in the dark for 30 min.
9. Add 120 μL of 100 mM TEAB, 2 μL of 100 mM $CaCl_2$ (1 mM final concentration), and 4 μL (2 μg) of sequence grade-modified trypsin solution (*see* **Note 15**).
10. Mix by inversion and incubate overnight at 37 °C.
11. After tryptic digestion, perform ReDiMe labeling with the sequential addition of 4 μL of 20% light (Fe-S cluster-disrupted tube) or 20% heavy (Fe-S cluster-replete tube) formaldehyde followed by 20 μL of freshly prepared 0.6 M sodium cyanoborohydride. Vortex samples and incubate at 25 °C for 2 h (*see* **Note 16**).
12. Quench the ReDiMe labeling reactions with the addition of 4 μL of 28–30% ammonium hydroxide. Gently mix, spin down, and incubate on ice for 15 min. Add 8 μL of formic acid, and gently mix samples by tapping on the side of the tube. Spin down and allow to sit on ice for 5 min (*see* **Note 17**).

13. Combine light (Fe-S cluster-disrupted) and heavy (Fe-S cluster-replete) ReDiMe-labeled reactions into a clean low-bind Eppendorf tube (*see* **Note 18**).
14. Desalt sample on a C18 Sep-Pak according to the manufacturer's instructions. Briefly, in sequence, (1) condition Sep-Pak with 3 × 1 mL acetonitrile, (2) equilibrate Sep-Pak with 3 × 1 mL low-pH buffer A (95% water, 5% acetonitrile, 0.1% formic acid), (3) pass peptide sample (~ 0.5 mL) through Sep-Pak twice, (4) wash Sep-Pak with 3 × 1 mL low-pH buffer A, and (5) elute from Sep-Pak into a low-bind Eppendorf tube with 3 × 0.5 mL buffer B (20% water, 80% acetonitrile, 0.1% formic acid).
15. Dry the sample overnight on a SpeedVac and then store the dried peptide pellet at −20 °C until ready for off-line fractionation of the sample.
16. Resuspend peptide pellet in 400 μL of high pH Buffer A (95% water, 5% acetonitrile, 10 mM ammonium bicarbonate) by bath sonication. Load the entire peptide sample onto a 500 μL manual injection loop connected to an Agilent 1100 Series HPLC.
17. Separate peptides on a 25 cm Agilent Extend-C18 column using a two-buffer gradient system at a flow rate of 500 μL/min. A 77-min gradient method was employed: 100% high pH buffer A for 2 min, a 60-min linear gradient from 20% to 35% high pH buffer B (10% water, 90% acetonitrile, 10 mM ammonium bicarbonate), 100% high pH buffer B for 5 min, and 100% high pH buffer A for 10 min [34].
18. Collect peptide fractions on a Gilson FC203B fraction collector into a 96 deep-well plate (0.6 min/well) from 11.5 to 69.5 min. A total of 96 fractions will be collected.
19. The 96-well fractions were concatenated into six pooled fractions by transferring every sixth well into a clean 2-mL low--bind Eppendorf tube. The total volume being combined into each pooled fraction is often >2 mL; therefore, it may be necessary to do the liquid transfer in two stages with SpeedVac drying in between (*see* **Note 19**).
20. Dry the six sample fractions overnight on a SpeedVac and then store the dried peptide pellets at −20 °C until ready for LC-MS/MS analysis.

3.4 LC-MS/MS Analysis

1. Resuspend dried peptide samples in low-pH buffer A (20 μL for isoTOP-ABPP samples and 50 μL per fraction for ReDiMe samples) and incubate at 25 °C for 30 min to fully resolubilize peptides. Bath sonicate and centrifuge at 15,000 g for 10 min to pellet any undissolved particulate. Transfer peptide samples to clean, labeled LC vials (1 for isoTOP-ABPP samples and 6 for ReDiMe samples).

Table 1
Ultimate 3000 RSLCnano LC gradient for Orbitrap Exploris 240 LC-MS/MS analysis

Time	Flow [μL/min]	% B	Curve
Equilibration			
−20.00	0.300	5.0	5
Start run			
0.00	0.300	5.0	5
10.00	0.300	5.0	5
130.00	0.300	25.0	5
150.00	0.300	65.0	5
151.00	0.300	99.0	5
160.00	0.300	99.0	5
161.00	0.300	5.0	5
Stop run			

2. Perform LC-MS/MS analysis on an Orbitrap Exploris 240 mass spectrometer running Xcalibur v4.4 (Thermo Scientific) coupled to a Dionex Ultimate 3000 RSLCnano System (*see* **Note 20**).

3. Each sample (5 μL) is injected directly onto an Acclaim PepMap 100 trap column, followed by peptide separation on an Acclaim PepMap RSLC with a 2-h gradient from 5% to 25% of MS Buffer B (20% water, 80% acetonitrile, 0.1% formic acid) in MS Buffer A (100% water, 0.1% formic acid) at a flow rate of 0.3 μL/min (Table 1).

4. With the positive spray voltage set to 2.1 kV, one full MS1 scan (120,000 resolution, 350–1800 m/z, RF lens 65%, AGC target 300%, automatic maximum injection time, profile mode) is obtained every 2 s with dynamic exclusion, isotopic exclusion, and apex detection enabled. A variable number of MS2 scans (15,000 resolution, AGC 75%, maximum injection time 100 ms, centroid mode) are obtained between each MS1 scan based on the highest precursor masses, filtered for monoisotopic peak determination, theoretical precursor isotopic envelope fit, intensity, and charge state. MS2 analysis consisted of precursor ion isolation (isolation window 2 m/z) followed by higher-energy collision dissociation (HCD, collision energy 30%) (Table 2).

5. Each sample/fraction was subjected to a second injection and LC-MS/MS run in order to generate two technical replicates

Table 2
Summary of MS method parameter for Orbitrap Exploris 240 LC-MS/MS analysis

Parameter	Value
Global parameters: ion source	
Ion source type	NSI
Spray voltage	Static
Positive ion (V)	2100
Negative ion (V)	0
Gas mode	Static
Sweep gas (Arb)	0
Ion transfer tube temp. (°C)	250
Global parameters: MS global settings	
Infusion mode	Liquid chromatography
Expected LC peak width (s)	15
Advanced peak determination	True
Default charge state	2
Experiment: master scan: full scan	
Orbitrap resolution	120,000
Scan range (m/z)	350–1800
RF lens (%)	65
AGC target	Custom
Normalized AGC target (%)	300
Maximum injection time mode	Auto
Microscans	1
Data type	Profile
Polarity	Positive
Source fragmentation	Disabled
Experiment: master scan: filters: charge state	
Include charge state(s)	2–6
Include undetermined charge states	False
Experiment: master scan: filters: dynamic exclusion	
Dynamic exclusion mode	Custom
Exclude after n times	1
Exclusion duration (s)	20

(continued)

Table 2
(continued)

Parameter	Value
Mass tolerance	ppm
Low	10
High	10
Exclude isotopes	True
Perform dependent scan on single charge state per precursor only	True
Experiment: master scan: filters: intensity	
Intensity threshold	5.0e4
Experiment: master scan: filters: MIPS	
Monoisotopic peak determination	Peptide
Relax restriction when too few precursors are found	True
Experiment: master scan: filters: apex detection	
Desired apex window (%)	30
Experiment: master scan: filters: isotope exclusion	
Isotope exclusion	Assigned
Experiment: master scan: filters: precursor fit	
Fit threshold (%)	70
Fit window (m/z)	2
Experiment: master scan: filters: data dependent	
Data-dependent mode	Cycle time
Time between master scans (s)	2
Experiment: scan: ddMS^2	
Multiplex ions	False
Isolation window (m/z)	2
Isolation offset	Off
Collision energy mode	Fixed
Collision energy type	Normalized
HCD collision energy (%)	30
Orbitrap resolution	15,000
Scan range mode	Auto
AGC target	Custom
Normalized AGC target (%)	75

(continued)

Table 2 (continued)

Parameter	Value
Maximum injection time mode	Custom
Maximum injection time (ms)	100
Microscans	1
Data type	Centroid

per biological sample. In general, three biological replicates (six injections) were analyzed for isoTOP-ABPP experiments and two biological replicates (six fractions each, two injections/fraction, 24 injections total) for ReDiMe experiments (*see* **Note 21**).

3.5 Data Processing, Statistical Filtering, and Graphical Visualization

1. Using the Thermo Proteome Discoverer V2.4 software package, analyze the isoTOP-ABPP tandem-MS data (six replicate .raw files) using the built-in SequestHT [35] and Percolator [36] algorithms, searching against a reverse-concatenated UniProtKB *Escherichia coli* (strain K12) (K12/MG1655/ATCC47076) database (www.uniprot.org—Entry UP000000625). Set the following parameters for the SequestHT and Percolator peptide identification analysis: enzyme specificity, trypsin; maximum missed cleavages, 2; peptide precursor mass tolerance, 10 ppm; fragment mass tolerance, 0.02 Da; and peptide false discovery rate (FDR), 1%. Allow the following dynamic modifications: oxidation of methionine (+15.995), protein N-terminal acetylation (+42.011), and/or methionine loss (+131.040), cysteine alkylation (+57.021), and either IAL-PC Biotin (+285.159) or IAH-PC Biotin (+291.179) modification as two independent SequestHT searches. Calculate peptide light/heavy (L/H) ratios as the ratio of IAL and IAH-modified peptide precursor ion intensities (*see* **Note 22**).
2. Using the Thermo Proteome Discoverer V2.4 software package, analyze ReDiMe tandem-MS data (four replicate .raw files) using the built-in SequestHT [35] and Percolator [36] algorithms, searching against a reverse-concatenated UniprotKB *Escherichia coli* (strain K12) (K12/MG1655/ATCC47076) database (www.uniprot.org—Entry UP000000625). Set the following parameters for the SequestHT and Percolator peptide identification analysis: enzyme specificity, trypsin; maximum missed cleavages, 2; peptide precursor mass tolerance, 10 ppm; fragment mass tolerance, 0.02 Da; and peptide false discovery rate (FDR), 1%. Allow the following dynamic

modifications: oxidation of methionine (+15.995), protein N-terminal acetylation (+42.011), and/or methionine loss (+131.040); and the following static modifications: cysteine alkylation (+57.021) and either light (+28.031) or heavy (+34.063) dimethylation of lysine residues and peptide N-termini modification as two independent SequestHT searches. Calculate protein light/heavy (L/H) ratios as the median value of all relevant peptide light/heavy ratios, calculated from the ratio of light and heavy dimethylated peptide precursor ion intensities.

3. Export the isoTOP-ABPP and ReDiMe data in Excel spreadsheet format as $\log_2$ peptide and protein L/H ratio, respectively (*see* **Note 23**).

4. For the isoTOP-ABPP dataset, determine the modification position(s) for each quantified cysteine residue entry (e.g., GrxD C30) [37] and then combine redundant entries corresponding to the same site of modification. Next, for each of the six replicate runs, determine an average $\log_2$ L/H ratio for sites with redundant entries (ignore singleton values when non-singleton ratios are available), which will later be used for data correction and calculating L/H ratio averages. This analysis can either be done manually in Excel or with a simple Python script (*see* **Note 24**).

5. For the isoTOP-ABPP data, identify Fe-S cluster ligands as those quantified peptides with modification sites that correspond to the protein position of Fe-S cluster cysteine ligands listed in a database of annotated Fe-S proteins (Table 3).

6. For both isoTOP-ABPP and RediMe data, determine the median $\log_2$ L/H ratio for each replicate. Generate corrected $\log_2$ L/H ratios for each replicate by subtracting the corresponding median $\log_2$ L/H value from all ratios (do not correct singleton values).

7. For both isoTOP-ABPP and ReDiMe data, remove peptide or protein singleton values for entries where a non-singleton value exists in at least one replicate. This represents the strictest singleton removal criteria, for a less strict analysis requires more non-singleton values from additional replicates.

8. For both isoTOP-ABPP and ReDiMe data, average the six or four corrected technical and biological replicates into a single average $\log_2$ L/H ratio. Additionally, calculate confidence intervals for each peptide or protein based on either the coefficient of variance (CV) for average $\log_2$ L/H ratios $> |1.0|$ or standard error of measurement (SEM) for average $\log_2$ L/H ratios $< |1.0|$: high, <0.5; medium, 1.0–0.5; and low, >1.0. Singleton values are assigned a confidence interval of medium if $n > 2$ or low otherwise (*see* **Note 25**).

Table 3
Database of annotated *Escherichia coli* Fe-S proteins

Name	ID	Ligands	Name	ID	Ligands	Name	ID	Ligands
fdx	P0A9R4	C42, C48, C51, C87	*fixX*	P68646	?	*ydiT*	P77714	?
ygcO	Q46905	?	*yjdI*	P0AF59	?	*yfaE*	P0ABW3	C37, C42, C45, C74
yfhL	P52102	C9, C12, C15, C54 C19, C38, C41, C50	*xdhC*	Q46801	C44, C49, C52	*paoA*	P77165	C99, C104, C105, C107 C158, C161, C208, C210
paoB	p77324	C119, C129, C138, C157	*narG*	P09152	C50, C54, C58, C93	*narZ*	P19319	C50, C54, C58, C93
dmsA	P18775	C63, C67, C71, C104	*ynfE*	P77374	C56, C60, C64, C96	*ynfF*	P77783	C59, C63, C67, C99
napA	P33937	C46, C49, C53, C81	*fdoG*	P32176	C50, C53, C57, C92	*fdnG*	P24183	C50, C53, C57, C92
fdhF	P07658	C8, C11, C15, C42	*ydeP*	P77561	C49, C52	*ydhV*	P76192	C307, C310, C314, C558
nuoG	P33602	C34, C45, C48, C67 H99, C103, C106, C112 C151, C154, C157, C201 C228, C231, C235, C263	*hyaA*	P69739	C62, C65, C160, C194 H232, C235, C260, C266 C275, C294, C297	*hybO*	P69741	C59, C62, C157, C191 H229, C232, C257, C263 C272, C292, C295
nuoB	P0AFC7	C63, C64, C129, C158	*hyfI*	P77668	C41, C47, C111, C141	*hycG*	P16433	C45, C51, C115, C145
narH	P11349	C16, C19, C22, C263 C26, C244, C247, C259 C184, C187, C192, C227 C196, C217, C227	*narY*	P19318	C16, C19, C22, C262 C26, C243, C246, C258 C183, C186, C191, C226 C195, C216, C226	*dmsB*	P18776	C14, C17, C20, C145 C24, C126, C129, C141 C67, C70, C75, C109 C79, C99, C102, C105
ynfG	P0AAJ1	C14, C17, C20, C145 C24, C126, C129, C141 C67, C70, C75, C109 C79, C99, C102, C105	*ydhX*	P77375	C46, C49, C52, C171 C56, C151, C154, C167 C92, C95, C100, C134 C104, C124, C127, C130	*ydhY*	P0AAL6	C68, C71, C74, C193 C78, C183, C186, C189 C123, C126, C131, C166 C135, C156, C159, C162
ygfS	Q46819	C12, C15, C18, C133 C22, C117, C120, C129 C55, C58, C63, C96 C67, C86, C89, C92	*fdoH*	P0AAJ5	C39, C42, C45, C179 C49, C160, C163, C175 C100, C103, C108, C143 C112, C133, C136, C139	*fdnH*	P0AAJ3	C39, C42, C45, C179 C49, C160, C163, C175 C100, C103, C108, C143 C112, C133, C136, C139

(continued)

Table 3
(continued)

Name	ID	Ligands	Name	ID	Ligands	Name	ID	Ligands
nrfC	P0AAK7	C46, C49, C52, C172 C56, C152, C155, C168 C92, C95, C100, C135 C104, C125, C128, C131	*hybA*	P0AAJ8	C47, C50, C53, C197 C57, C174, C177, C193 C112, C115, C120, C155 C124, C145, C148, C151	*hyfA*	P23481	C12, C15, C18, C162 C22, C146, C149, C158 C51, C54, C59, C92 C63, C82, C85, C88
hycB	P0AAK1	C12, C15, C18, C159 C22, C143, C146, C155 C51, C54, C59, C92 C63, C82, C85, C88	*hydN*	P0AAK4	C12, C15, C18, C147 C22, C131, C134, C143 C58, C61, C66, C99 C70, C89, C92, C95	*ysaA*	P56256	C12, C15, C18, C134 C22, C118, C121, C130 C58, C61, C66, C99 C70, C89, C92, C95
ygfT	Q46820	C12, C15, C18, C129 C22, C112, C115, C125 C56, C59, C64, C97 C68, C87, C90, C93 C210, C213, C219, C223 C258, C262, C268, C272	*aegA*	P37127	C12, C15, C18, C137 C22, C121, C124, C133 C56, C59, C64, C97 C68, C87, C90, C93	*gltD*	P09832	C47, C50, C55, C59
preT	P76440	?	*ygfK*	Q46811	C938, C941, C944, C948	*fadH*	P42593	C335, C338, C342, C354
napF	P0AAL0	C37, C40, C43, C47 C69, C72, C75, C79 C141, C144, C147, C151	*napG*	P0AAL3	C61, C64, C67, C71 C99, C102, C107, C111 C139, C147, C150, C154 C186, C189, C192, C196	*napH*	P33934	C226, C229, C232, C236 C260, C263, C266, C270
yccM	P52636	C251, C254, C257, C288 C261, C278, C281, C284	*rsxB*	P77223	C49, C52, C57, C74 C117, C120, C123, C157 C127, C147, C150, C153	*rsxC*	P77611	C377, C380, C383, C426 C387, C416, C419, C422
nuoI	P0AFD6	C60, C63, C66, C109 C70, C99, C102, C105	*hyfH*	P77423	C40, C43, 46, C50 C75, C78, C81, C85	*hycF*	P16432	C40, C43, 46, C50 C75, C78, C81, C85
ykgF	P77536	C312, C315, C318, C372 C322, C362, C365, C368	*ydiJ*	P77748	?	*glpC*	P0A996	C9, C12, C15, C66 C19, C56, C59, C62

glcF	P52074	C25, C28, C31, C85 C35, C75, C78, C81	***ydjY***	P76220	?	***preA***	P25889	C344, C347, C350, C388 C354, C378, C381, C384
ydbK	P52647	C689, C692, C695, C755 C699, C745, C748, C751 C819, C822, C847, C1071	***queG***	P39288	C193, C196, C199, C253 C203, C246, C249	***sdhB***	P07014	C55, C60, C75 C159, C206, C212 C149, C152, C155, C216
frdB	P0AC47	C58, C63, C66, C78 C159, C205, C211 C149, C152, C155, C215	***gltB***	P09831	C1102, C1108, C1113	***nuoE***	P0AFD1	C92, C97, C133, C137
nuoF	P31979	C351, C354, C357, C398	***cntA***	P0ABR7	C89, H91, C109, H112	***hcaE***	P0ABR5	C85, H87, C105, H108
cntB	P76254	C270, C275, C278, C308	***paaE***	P76081	C299, C304, C307, C337	***hcr***	P75824	C273, C278, C281, C311
ycbX	P75863	?	***hcaC***	0ABW0	C42, H44, C62, H65	***fdhE***	P13024	?
bioB	P12996	C53, C57, C60 C97, C128, C188, R260	***lipA***	P60716	C94, C98, C101 C68, C73, C79, S308	***moaA***	P30745	C24, C28, C31 C257, C260, C274
thiC	P30136	C581, C584, C589	***thiH***	P30140	C85, C89, C92	***hemN***	P32131	C62, C66, C69
hemW	P52062	C16, C20, C23	***miaB***	P0AEI1	C157, C161, C164 C12, C49, C83	***rimO***	P0AEI4	C150, C154, C157 C17, C53, C82
ygiQ	Q46861	C386, C390, C393	***rlmN***	P36979	C125, C129, C132	***queE***	P64554	C31, C35, C38
epmB	P39280	C120, C124, C127	***pflA***	P0A9N4	C30, C34, C37	***pflC***	P32675	C47, C51, C54
pflE	P75794	C25, C29, C32	***yjjW***	P39409	C31, C35, C38 C47, C50, C53, C86 C57, C76, C79, C82	***nrdG***	P0A9N8	C26, C30, C33
aslB	P25550	C21, C25, C28 C276, C282, C297 C339, C342, C348, C352	***ydeM***	P76134	C12, C16, C19 C257, C263, C278 C28, C315, C318, C324	***yhcC***	P0ADW6	C33, C45, C48
phnJ	P16688	?	***acnA***	P25516	C435, C501, C504	***acnB***	P36683	C710, C769, C772
ybhJ	P75764	C360, C421, C424	***leuC***	P0A6A6	C347, C407, C410	***fumA***	P0AC33	C105, C224, C318
fumB	P14407	C105, C224, C318	***sdaA***	P16095	?	***sdaB***	P30744	?

(continued)

Table 3
(continued)

Name	ID	Ligands	Name	ID	Ligands	Name	ID	Ligands
tdcG	P42630	?	*ilvD*	P05791	C122, C195	*edd*	P0ADF6	C154, C221
yjhG	P39358	?	*yagF*	P77596	?	*ttdA*	P05847	C71, C190, C277
dinG	P27296	C120, C194, C199, C205	*yoaA*	P76257	C108, C168, C173, C179	*mutY*	P17802	C192, C199, C202, C208
nth	P0AB83	C187, C194, C197, C203	*iscR*	P0AGK8	C92, C98, C104	*nsrR*	P0AF63	C91, C96, C102
soxR	P0ACS2	C119, C122, C124, C130	*yeiL*	P0A9E9	C68, C91, C93, C116	*fnr*	P0A9E5	C20, C23, C29, C122
feoC	P64638	C56, C61, C64, C70	*uvrC*	P0A8G0	?	*cysI*	P17846	C434, C440, C479, C483
nirB	P08201	C641, C647, C681, C685	*hcp*	P75825	C3, C6, C18, C25 C317, C405, C433, C458	*rlmC*	P75817	C3, C11, C14, C87
rlmD	P55135	C81, C87, C90, C162	*ttcA*	P76055	C122, C125, C213	*trhP*	P76403	?
rlhA	P76104	?	*nadA*	P11458	?	*ubiU*	P45527	C169, C176, C193, C232
ubiV	P45475	C39, C180, C193, C197	*ispG*	P62620	C270, C273, C305, C312	*ispH*	P62623	C12, C96, C197
fhuF	P39405	C244, C245, C256, C259	*bfd*	P0AE56	C4, C6, C39, C42	*dppD*	P0AAG0	?
dppF	P37313	?	*oppD*	P76027	?	*oppF*	P77737	?
iscU	P0ACD4	?	*iscA*	P0AAC8	C35, C99, C101	*sufB*	P77522	?
sufA	P77667	?	*erpA*	P0ACC3	C42, C106, C108	*nfuA*	P63020	C149, C152

9. In a new Excel sheet, correlate $\log_2$ L/H peptide cysteine reactivity ratios and confidence intervals with corresponding protein ratios and confidence intervals (not every quantified peptide will have corresponding protein quantification data). Plot high-confidence peptide and protein ratios against one another in a 2-dimensional plot with peptide $\log_2$ L/H ratios on the y-axis and protein $\log_2$ L/H ratios on the x-axis. Highlight peptides with cysteine modifications assigned as annotated Fe-S cluster ligands (Fig. 3).
10. Calculate protein abundance-corrected (net) cysteine reactivity ratios by subtracting the corresponding protein $\log_2$ L/H ratios from peptide $\log_2$ L/H ratios. Use these net cysteine reactivity ratios to compare global changes in Fe-S cluster ligand reactivity between study conditions (i.e., iron depletion or *isc* deletions) as a violin plot or average net cysteine reactivity ratios for all ligands of a given Fe-S cluster to compare individual cluster occupancy changes across study conditions in the form of a heat map (Fig. 4).

4 Notes

1. Other commercial or homemade media formulations, such as LB or M9 minimal media, can be used for proteomic studies for Δ*isc* strain studies, but commercial versions of these media will contain iron and cannot be employed for iron depletion studies. Traditionally, MOPS minimal media is made and stored as a 10× stock, omitting the addition of glucose and K_2HPO_4. The 10× stock can then be diluted to 1× along with the addition of 0.1% glucose and the K_2HPO_4. All media should be prepared from filter-sterilized solutions and filter-sterilized just before use. Additionally, for experiments employing iron-depleted MOPS minimal media, only metal-free plastic or acid-washed glassware should be used so as not to introduce any iron contamination.
2. The doubling time of *E. coli* in both standard and iron-free MOPS media is ~1 h, and it should be expected to take ~5–6 h to reach an O.D. 0.6 from a 1:100 dilution of overnight culture grown in standard MOPS minimal media. For some of the Δ*isc* strains, the doubling time can be significantly slower and could take 8–12 h to reach O.D. 0.6.
3. Generation of cell lysates from frozen cell pellets can be performed on the benchtop, as no significant change in Fe-S cluster binding was observed by performing these steps in an anaerobic glovebox or under reducing conditions [26]. This result may not hold true for more oxygen-sensitive bacteria or with *E. coli* samples grown under anaerobic or microaerobic

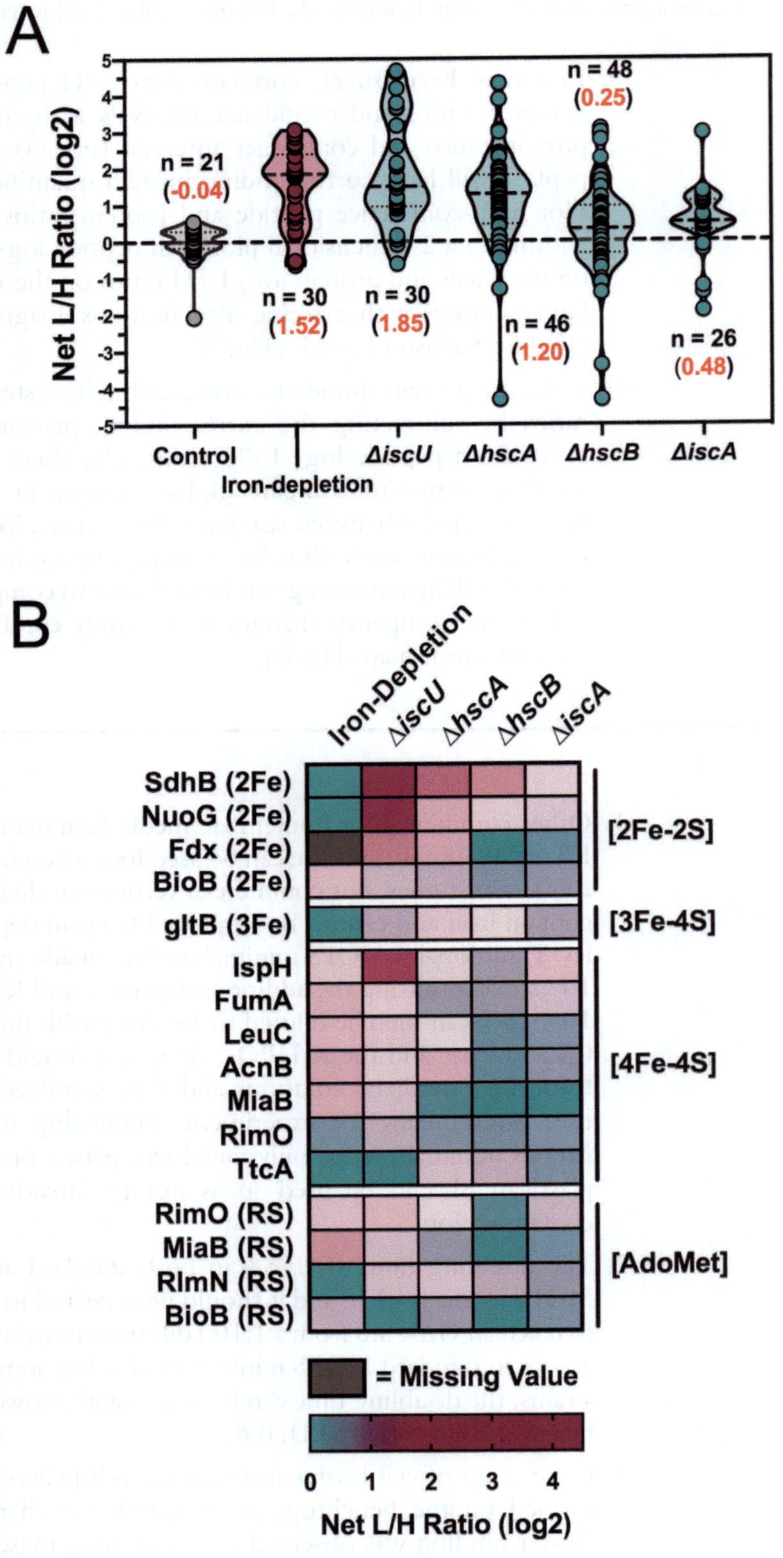

Fig. 4 Net changes in the cysteine reactivity of Fe-S cluster ligands across the *E. coli* proteome upon disruption of proper Fe-S cluster biosynthesis and delivery. (**a**) Violin plot of global net changes in cysteine reactivity for all quantified Fe-S cluster ligands under control (gray), iron depletion (magenta), or *isc* deletion (teal—Δ*iscU*, Δ*hscA*, Δ*hscB*, Δ*iscA*) growth conditions. The median R value is displayed as a dashed line in the violin plot

conditions and should be tested by the user for newer sample types. As Fe-S clusters can degrade over time, it is important to quickly move through the lysis and dilution steps and quickly get to the proteomic labeling steps (Subheadings 3.2 and 3.3) within about ~1 h of thawing the cell pellets.

4. There are many published variations on the traditional isoTOP-ABPP protocol that could be employed to measure Fe-S-dependent changes in cysteine reactivity, including but not limited to the use of a) alternative cysteine reactive probes, b) alternative cleavable biotin azide reagents, and c) isobaric TMT tags. These other strategies will not be expanded on here, but further work adapting this platform to alternative chemoproteomic protocols could take advantage of future advances in the proteomics field.
5. The addition of copper during the CuACC reaction will result in protein precipitation, which necessitates the resuspension of proteins by vortexing to ensure that the click chemistry reaction proceeds efficiently.
6. The methanol washes are especially important for completely removing unreacted protein-free PC biotin azide, which would later compete for streptavidin binding sites with biotin-labeled proteins.
7. Use 100 μL aliquots of the 6 mL protein sample to resuspend the beads and transfer them over to the protein sample in the 15-mL conical tube. Repeat as needed until all beads are moved from 1.5-mL Eppendorf tube to 15-mL conical tube.
8. It is important during these wash steps not to accidently decant or aspirate off any streptavidin–agarose beads. If necessary, leave a small amount (< 100 μL) of supernatant remaining after each wash to ensure complete sample retention.
9. It is necessary to use screw-cap Eppendorf tubes as the high urea concentration causes traditional Snap-Cap Eppendorf tubes to leak during the subsequent rotation steps resulting in loss of sample.

Fig. 4 (continued) and enumerated in red, with the number of Fe-S ligand data points (n) indicated for each condition. Data for the plot obtained from [26]. **(b)** Heat map of net cysteine reactivity changes for a selection of Fe-S clusters under iron depletion or *isc* deletion (*ΔiscU, ΔhscA, ΔhscB, ΔiscA*) growth conditions. Each cell represents the average of all quantified net cysteine reactivity changes (net L/H ratios) for ligands of that particular cluster. Increases in net cysteine reactivity are represented with magenta shading of the cell, no change in net reactivity by teal shading, and missing data by gray shading. Data for the plot obtained from [26]

10. The addition of DTT is necessary to reduce any native or advantageous cysteine disulfide bonds. The reaction with IA is employed to cap all reduced cysteine residues, preventing reoxidation and disulfide bond formation during subsequent steps of the MS preparation, which would hinder MS peptide identification.
11. This supernatant will contain unlabeled tryptic peptides, but any IAL/IAH-labeled peptides will remain bound to the streptavidin agarose beads. It is not usually necessary to save or analyze this fraction, unless troubleshooting the cysteine labeling or enrichment procedure.
12. The volume of the UV-irradiated supernatant will likely be less than the starting 250 μL (closer to 200 μL) due to evaporation that occurs as a result of the heat generated by the UV lamp. This loss of sample volume is not an issue. Additionally, it is easier to leave the stir bar in the tube during centrifugation steps, as trying to remove it may lead to loss of sample.
13. If necessary, the eluted peptide sample and washes can be filtered through an empty micro-bio-spin column into the 1.5-mL low-bind Eppendorf tube, in order to prevent the transfer of any streptavidin–agarose beads.
14. Do not allow the protein pellet to over-dry or it will be difficult to fully solubilize in the next step.
15. For ReDiMe samples, the lyophilized trypsin should be resuspended in 40 μL of 100 mM TEAB rather than the supplied trypsin buffer solution. This will prevent any incompatibility with the subsequent reductive dimethylation reaction. Additionally, if modifying the protocol, all buffers and reagents should be free of any primary amines that could react with formaldehyde, thereby reducing labeling efficiency.
16. Formaldehyde solutions and vapors are toxic; as such, all subsequent steps should be performed in a fume hood until samples have been combined and desalted.
17. Addition of ammonia and formic acid can produce HCN and H_2 gas in an exothermic reaction. It is important to do this quenching reaction in the hood, as well as on ice to prevent frothing of the samples.
18. If the samples are overly frothy and it is difficult to completely transfer the sample, it is possible to wash out both the light and heavy tubes with up to a total of 500 μL of low-pH buffer A. These low-pH buffer A washes should then be transferred to the same low-bind Eppendorf tube as the original reactions, and the whole sample (~ 1 mL) desalted as described.

19. It is possible to skip this off-line fractionation step and proceed directly to analyze the unfractionated sample for changes in protein abundance. This will significantly reduce the overall coverage of the proteome at the protein level and therefore reduce the total number of protein abundance-corrected (net) cysteine reactivity changes observed.
20. While the use of an Orbitrap Exploris 240 mass spectrometer is described in this protocol, any standard LC-MS/MS instrumentation setup should be amenable to analysis of these samples with appropriated parameters and filters applied for standard shotgun proteomic experiments.
21. The combination of three biological replicates for isoTOP-ABPP and two biological replicates for ReDiMe experiments allows for robust statistical analysis of the data while optimizing instrument run time.
22. Alternative proteomics analysis and quantification software can be used (e.g., ProLuCID or MaxQuant), but should employ similar parameters to those described for the Proteome Discoverer SequestHT and Percolator nodes used in this study.
23. $\log_2$ L/H ratios are preferred when datasets are expected to have both biologically relevant increases and decreases in ratios. This allows for the magnitude of increased and decreased peptide or protein ratios to be visually similar. For datasets with expected unidirectional changes (i.e., inhibitor treatment), linear values are often more informative.
24. Singleton ratios occur when a light- or heavy-labeled peptide is identified, but there is no corresponding heavy or light peak intensity with which to calculate a L/H ratio, so a value of 100 ($\log_2$: 6.64) or 0.01 ($\log_2$: −6.64) is automatically assigned. These values are not real and therefore should not be subject to data manipulation or ratio correction. Additionally, singleton values often occur erroneously and should only be considered when other non-singleton data are unavailable for a peptide or protein.
25. Average $\log_2$ L/H ratios with a high-confidence interval have consistent ratios across multiple technical and biological replicates. The use of SEM (the standard deviation/the square root of n, where n is the number of replicates with quantified ratios) weighs peptides or proteins with a larger number of quantified replicates more strongly than those with fewer. The use of CV (SEM / average $\log_2$ L/H ratio) for those peptides and proteins with $\log_2$ L/H ratio $> |1.0|$ allows for wider standard deviation when larger and smaller ratios are considered, where small changes in the intensity of the MS peptide signals would significantly impact the calculated $\log_2$ L/H ratio.

Acknowledgments

This work was supported by NIH R35GM134964 to EW.

References

1. Beinert H, Holm RH, Münck E (1997) Iron-sulfur clusters: nature's modular, multipurpose structures. Science 277(5326):653–659
2. Hirst J (2013) Mitochondrial complex I. Annu Rev Biochem 82:551–575
3. Broderick WE, Broderick JB (2019) Radical SAM enzymes: surprises along the path to understanding mechanism. J Biol Inorg Chem 24(6):769–776
4. Landgraf BJ, McCarthy EL, Booker SJ (2016) Radical S-adenosylmethionine enzymes in human health and disease. Annu Rev Biochem 85:485–514
5. Vey JL, Drennan CL (2011) Structural insights into radical generation by the radical SAM superfamily. Chem Rev 111(4):2487–2506
6. Boal AK, Genereux JC, Sontz PA, Gralnick JA, Newman DK, Barton JK (2009) Redox signaling between DNA repair proteins for efficient lesion detection. Proc Nat Acad Sci USA 106(36):15237–15242
7. Lukianova OA, David SS (2005) A role for iron–sulfur clusters in DNA repair. Curr Opin Chem Biol 9(2):145–151
8. Crack JC, Green J, Thomson AJ, Le Brun NE (2012) Iron–sulfur cluster sensor-regulators. Curr Opin Chem Biol 16(1–2):35–44
9. Beinert H, Kennedy MC, Stout CD (1996) Aconitase as iron–sulfur protein, enzyme, and iron-regulatory protein. Chem Rev 96(7):2335–2374
10. Flint DH, Allen RM (1996) Iron–sulfur proteins with nonredox functions. Chem Rev 96(7):2315–2334
11. Braymer JJ, Freibert SA, Rakwalska-Bange M, Lill R (2020) Mechanistic concepts of iron-sulfur protein biogenesis in biology. Biochim Biophys Acta, Mol Cell Res 1868(1):118863
12. Rouault TA, Maio N (2017) Biogenesis and functions of mammalian iron-sulfur proteins in the regulation of iron homeostasis and pivotal metabolic pathways. J Biol Chem 292(31):12744–12753
13. Pérard J, de Choudens SO (2018) Iron–sulfur clusters biogenesis by the SUF machinery: close to the molecular mechanism understanding. J Biol Inorg Chem 23(4):581–596
14. Lill R, Freibert SA (2020) Mechanisms of mitochondrial iron-sulfur protein biogenesis. Annu Rev Biochem 89:471–499
15. Maio N, Jain A, Rouault TA (2020) Mammalian iron–sulfur cluster biogenesis: recent insights into the roles of frataxin, acyl carrier protein and ATPase-mediated transfer to recipient proteins. Curr Opin Chem Biol 55:34–44
16. Upadhyay AS, Stehling O, Panayiotou C, Rösser R, Lill R, Överby AK (2017) Cellular requirements for iron–sulfur cluster insertion into the antiviral radical SAM protein viperin. J Biol Chem 292(33):13879–13889
17. Pallesen LJ, Solodovnikova N, Sharma AK, Walden WE (2013) Interaction with Cfd1 increases the kinetic lability of FeS on the Nbp35 scaffold. J Biol Chem 288(32):23358–23367
18. Py B, Gerez C, Huguenot A, Vidaud C, Fontecave M, de Choudens SO, Barras F (2018) The ErpA/NfuA complex builds an oxidation-resistant Fe-S cluster delivery pathway. J Biol Chem 293(20):7689–7702
19. Waller JC, Ellens KW, Hasnain G, Alvarez S, Rocca JR, Hanson AD (2012) Evidence that the folate-dependent proteins YgfZ and MnmEG have opposing effects on growth and on activity of the iron-sulfur enzyme MiaB. J Bacteriol 194(2):362–367
20. Kulkarni RA, Bak DW, Wei D, Bergholtz SE, Briney CA, Shrimp JH, Alpsoy A, Thorpe AL, Bavari AE, Crooks DR, Levy M, Florens L, Washburn MP, Frizzell N, Dykhuizen EC, Weerapana E, Linehan WM, Meier JL (2019) A chemoproteomic portrait of the oncometabolite fumarate. Nat Chem Biol 15(4):391–400
21. Martin BR, Wang C, Adibekian A, Tully SE, Cravatt BF (2012) Global profiling of dynamic protein palmitoylation. Nat Methods 9(1):84–89
22. Deng X, Weerapana E, Ulanovskaya O, Sun F, Liang H, Ji Q, Ye Y, Fu Y, Zhou L, Li J, Zhang H (2013) Proteome-wide quantification and characterization of oxidation-sensitive cysteines in pathogenic bacteria. Cell Host Microbe 13(3):358–370
23. Bechtel TJ, Li C, Kisty EA, Maurais AJ, Weerapana E (2020) Profiling cysteine reactivity and oxidation in the endoplasmic reticulum. ACS Chem Biol 15(2):543–553
24. Maurais AJ, Weerapana E (2019) Reactive-cysteine profiling for drug discovery. Curr Opin Chem Biol 50:29–36

25. Backus KM, Correia BE, Lum KM, Forli S, Horning BD, González-Páez GE, Chatterjee S, Lanning BR, Teijaro JR, Olson AJ, Wolan DW (2016) Proteome-wide covalent ligand discovery in native biological systems. Nature 534(760):570–574

26. Bak DW, Weerapana E (2023) Monitoring Fe–S cluster occupancy across the *E. coli* proteome using chemoproteomics. Nat Chem Biol 19: 356–366

27. Weerapana E, Wang C, Simon GM, Richter F, Khare S, Dillon MB, Bachovchin DA, Mowen K, Baker D, Cravatt BF (2010) Quantitative reactivity profiling predicts functional cysteines in proteomes. Nature 468:790–795

28. Abo M, Li C, Weerapana E (2017) Isotopically-labeled iodoacetamide-alkyne probes for quantitative cysteine-reactivity profiling. Mol Pharm 15:743–749

29. Seo SW, Kim D, Latif H, O'Brien EJ, Szubin R, Palsson BO (2014) Deciphering Fur transcriptional regulatory network highlights its complex role beyond iron metabolism in *Escherichia coli*. Nat Commun 5:1–10

30. Yeo WS, Lee JH, Lee KC, Roe JH (2006) IscR acts as an activator in response to oxidative stress for the suf operon encoding Fe-S assembly proteins. Mol Microbiol 61:206–218

31. Boersema PJ, Raijmakers R, Lemeer S, Mohammed S, Heck AJ (2009) Multiplex peptide stable isotope dimethyl labeling for quantitative proteomics. Nat Protoc 4:484–494

32. Neidhardt FC, Bloch PL, Smith DF (1974) Culture medium for enterobacteria. J Bacteriol 119(3):736–747

33. Baba T, Ara T, Hasegawa M, Takai Y, Okumura Y, Baba M, Datsenko KA, Tomita M, Wanner BL, Mori H (2006) Construction of *Escherichia coli* K-12 in-frame, single-gene knockout mutants: the Keio collection. Mol Syst Biol 2:1–11

34. Edwards A, Haas W (2016) Multiplexed quantitative proteomics for high-throughput comprehensive proteome comparisons of human cell lines. In: Proteomics in systems biology. Humana Press, New York

35. Eng JK, McCormack AL, Yates JR (1994) An approach to correlate tandem mass spectral data of peptides with amino acid sequences in a protein database. J Am Soc Mass Spec 5:976–989

36. Käll L, Canterbury JD, Weston J, Noble WS, MacCoss MJ (2007) Semi-supervised learning for peptide identification from shotgun proteomics datasets. Nat Methods 4:923–925

37. Bak DW, Pizzagalli MD, Weerapana E (2017) Identifying functional cysteine residues in the mitochondria. ACS Chem Biol 12:947–957

Index

Oleh Khalimonchuk (ed.), *Iron Metabolism: Methods and Protocols*, Methods in Molecular Biology, vol. 2839,
https://doi.org/10.1007/978-1-0716-4043-2,

MIX
Papier aus verantwortungsvollen Quellen
Paper from responsible sources
FSC® C105338
FSC
www.fsc.org

If you have any concerns about our products,
you can contact us on
ProductSafety@springernature.com

In case Publisher is established outside the EU,
the EU authorized representative is:
Springer Nature Customer Service Center GmbH
Europaplatz 3, 69115 Heidelberg, Germany

Printed by Libri Plureos GmbH
in Hamburg, Germany